Parasitism and host behaviour

Parasitism and host behaviour

Edited by

C. J. Barnard and J. M. Behnke
Department of Zoology
University of Nottingham, UK

Taylor & Francis
London • New York • Philadelphia
1990

UK Taylor & Francis Ltd, 4 John St., London WC1N 2ET

USA Taylor & Francis Inc., 1900 Frost Rd., Suite 101, Bristol, PA 19007

British Library Cataloguing in Publication Data
Parasitism and host behaviour.
1. Animals. Parasites. Infections with hosts behavioural aspects
I. Barnard, C.J. (Christopher J) II. Behnke, J.M.
591.5249

ISBN 0-85066-498-5

Library of Congress Cataloging-in- Publication Data
Parasitism and host behaviour / edited by C.J. Barnard and J.M. Behnke.
p. cm.
ISBN 0–85066–498–5
1. Host–parasite relationships. 2. Parasites. 3. Parasites––Ecology. I. Barnard, C.J. (Christopher J.) II. Behnke, J.M.,

QL757.P284 1990
591.52′49--dc20 90–10783
CIP

Typeset in 10/12 Garamond by Chapterhouse, The Cloisters, Formby
Printed in Great Britain by Burgess Science Press, Basingstoke on paper which has a specified pH value on final paper manufacture of not less than 7.5 and is therefore 'acid free'.

Contents

Preface

Most organisms depend for their existence on limited natural resources like nutrients, water and suitable breeding sites. These resources usually require the investment of time, energy and risk of predation for their exploitation. The relative contributions of these costs and the benefit of the resource to the reproductive potential of the organism determines the strength and direction of natural selection on the resource harvesting strategy of that organism. While selection for increased efficiency (benefit:cost ratio) may result in a honing of the perceptual, sampling and resource handling abilities of the organism, the fact that it does so means that the searching (or other) investment of one organism can be exploited by another. The investment of one organism (*producer*) thus provides an adaptive shortcut for selection in another (*scrounger*). The commonest and most apparent forms of scrounging are what have been referred to as strategies of usurpation, in which producers invest time and effort in procuring a resource which is then stolen or usurped in other ways by scroungers. Strategies of usurpation can thus be regarded as parasitic. While the term 'parasite' is normally restricted to those organisms like tapeworms and fleas which have evolved dependence on a host organism for survival, such 'conventional' parasites are only the extreme end of a spectrum of scrounger strategies for which 'parasite' is an appropriate description. Klepto-parasitism (food piracy), brood parasitism and satellite mating strategies, for instance, are other forms of parasitic relationship which occur both between and within species.

Whatever form parasitic relationships take, they are costly to the host and selection is likely to favour counteradaptive measures on the part of the host to reduce the impact of exploitation. The result is a counteradaptive arms race between host and parasite with far-reaching consequences for the physiology, behaviour and ecology of both. While for a long time interest in 'conventional' parasites has focused on the physiological aspects of host–parasite relationships, particularly immunological responses to infection by endoparasites, it is clear that parasites of all kinds have important consequences for the behaviour of their hosts, both in terms of pathological or other changes induced by the relationship and in selecting for decision rules and behavioural counter-responses to exploitation.

Similarly, host behaviour imposes a selection pressure on the life history characteristics of parasites with profound effects on population and community ecology and epidemiology.

This volume brings together work across the disciplines of parasitology and behavioural ecology which reveals a fundamental role of parasitism in the evolution of behaviour and *vice versa*. In doing so, it traverses the phylogenetic scale from enteric Protozoa and nematodes to the colouration and courtship of birds and human cultural traditions, providing the first interdisciplinary synthesis of an exciting and rapidly-developing area of research.

In the first chapter, Barnard reviews some of the ways in which host behaviour is exploited in different parasitic relationships and the reproductive and mortality costs imposed on hosts which are likely to select for host counter-responses. This is followed by a consideration of factors influencing the outcome of arms races between hosts and parasites and why races may sometimes be biased in favour of one or other lineage.

The cost of parasitism is examined in more detail in the next chapter by Holmes and Zohar who discuss some of the pathological consequences of parasitism and their influence on host behaviour. To what extent are behavioural changes associated with parasitism host adaptations, to what extent are they parasite adaptations and to what extent are they simply non-adaptive side-effects of the way parasites ensure their survival? To help answer this, Holmes and Zohar consider three main categories of pathology: organ malfunction, reduction in nutrition and changes in neuroendocrine control. The physiological bases of some of these, and other, changes are discussed by Thompson (Chapter 3) who also considers their implications for our understanding of the host–parasite relationship in general. Maintenance of a parasite in a living host is often accomplished by alterations in the physiology of the host at various levels, resulting in the formation of an integrated host–parasite complex. The metabolism, behaviour and ecology of the host may be changed permanently as a result. Most of the physiological alterations reflect the normal functioning of the regulatory mechanisms of the host, but parasites themselves may affect the regulatory process. Failure of the host–parasite complex to maintain a homeostatic relationship leads to disease, but much remains to be understood about the mechanisms of pathogenicity and the way in which selection shapes the subsequent integration of host and parasite.

Metabolic changes in the host as a result of parasitism can influence its behaviour directly through such effects on physiological control systems or indirectly by influencing host decision-making. An animal weakened by disease is likely to be less competitive than a healthy individual and its reduced competitive ability may affect the decisions it makes in a number of different contexts. In Chapter 4, Milinski reviews some of the experimental evidence for changes in foraging decisions by predators as a result of infection with different parasites.

Read (Chapter 5) is also concerned with decision-making in relation to parasitic infection but in the context of host sexual behaviour. Several lines of argument suggest that parasites have played an important part in the evolution of sexual reproduction and sexual behaviour. One of the most recent hinges on the assumption that outward signs of infection may be a basis for choosing between potential mates and passing on resistance to disease to offspring. Read provides a penetrating critique both of the hypothesis itself and its assumptions and the evidence that has accrued for and against it.

The following two chapters consider the relationship between parasitism and host behaviour from a life-cycle perspective. Tinsley (Chapter 6) looks at how selection has produced a remarkable tuning of parasite transmission strategies to take advantage of sometimes astonishingly narrow windows of opportunity in the behaviour and ecology of host organisms. While carefully-timed opportunism is one way of enhancing transmission, modifying host behaviour to *create* opportunities is another. The idea that some altered host behaviours reflect manipulation by parasites to their own ends is one that has generated a great deal of discussion. Although there is evidence in favour of the manipulation hypothesis, there are good reasons for caution in interpreting altered host behaviour in this way. One important problem is the phylogenetic history of host and parasite strategies. Do observed characteristics reflect current adaptation or are they simply inherited ancestral characters which may have been adaptive in the past but are not longer? In Chapter 7, Moore and Gotelli review some of the evidence for altered behaviour patterns as manipulation and point to the importance of phylogenetic analyses in drawing conclusions about adaptive significance.

Chapters 8 and 9 focus on the role of host behaviour in the epidemiology of helminth infections in humans. Nelson considers human behaviour in relation to transmission at the individual host and local community levels, discussing some of the cultural practices which have been responsible for intense foci of infection with trichinosis, hydatid disease and schistosomiasis in particular parts of the world. Bundy and Blumenthal, on the other hand, analyze the problems inherent in quantifying selected types of human behaviour at the population level and relating these to exposure to infection, age intensity/prevalence profiles and the development of resistance.

The volume concludes with a discussion by Murray of the effectiveness of host maintenance and social behaviours in regulating ectoparasite infestations among birds and mammals.

C. J. Barnard
J. M. Behnke

Contributors

C. J. Barnard Department of Zoology, University of Nottingham, University Park, Nottingham NG7 2RD, UK.

U. J. Blumenthal Department of Epidemiology and Population Sciences, London School of Hygiene and Tropical Medicine, Keppel Street, London WC1E 7HT, UK.

D. A. P. Bundy Parasite Epidemiology Research Group, Department of Pure and Applied Biology, Imperial College, Prince Consort Road, London SW7 2BB, UK.

N. J. Gotelli Department of Zoology, University of Oklahoma, Norman, Oklahoma 73019, USA.

J. C. Holmes Department of Zoology, University of Alberta, Edmonton, Alberta, Canada T6G 2E9.

M. Milinski Abteilung Verhaltensökologie, Zoologisches Institut, Universität Bern, Wohlenstrasse 50 a, CH-3038 Hinterkappelen, Switzerland.

J. Moore Department of Biology, Colorado State University, Fort Collins, Colorado 80523, USA.

M. D. Murray (CSIRO) 17 Ashmore Avenue, Pymble, New South Wales 2073, Australia.

G. S. Nelson Liverpool School of Tropical Medicine, Liverpool L3 5QA, UK.

A. F. Read Department of Zoology, South Parks Road, Oxford OX1 3PS, UK.

S. N. Thompson Department of Entomology, University of California, Riverside, California 62521, USA.

R. C. Tinsley School of Biological Sciences, Queen Mary and Westfield College, University of London, Mile End Road, London E1 4NS, UK.

S. Zohar Department of Zoology, University of Alberta, Edmonton, Alberta, Canada T6G 2E9.

1. Parasitic relationships

C. J. Barnard

INTRODUCTION

To most people a parasite is an organism like a tapeworm or a flea which has evolved to become dependent on a host organism for survival and/or reproduction. Parasites are exploitative, taking from their host nutrients and energy made available through the latter's foraging effort, as well as perhaps benefitting from transport, protection and a thermally-regulated environment provided by the host's body. Tapeworms, fleas, flukes and other parasites, however, represent only the extreme end of a range of exploitative relationships between organisms which can be called parasitic (Barnard, 1984a). Organisms can capitalize on the resource procurement of others without becoming physiologically dependent on them. Opportunistic kleptoparasitism (food stealing or piracy), intraspecific egg dumping and kleptogamy ('sneaky' breeding) are obvious examples (see Barnard, 1984a and chapters therein for detailed discussions).

Barnard and Sibly (1981) coined the general terms 'producer' and 'scrounger' to describe parties (individuals or species) involved in exploitative relationships. Scroungers in some way use the behavioural or physiological investment of producers, for instance in foraging, building, incubating, displaying, avoiding predators or defending territories to reduce the cost of obtaining limited resources. This exploitation is likely to impose a cost on the producer which, if severe enough, will be reflected in some form of counteradaptation. Counteradaptation by producers sets up a selection pressure for counter-counteradaptation by scroungers and the result is likely to be an escalating arms race (see Dawkins and Krebs, 1979; Barnard, 1984b). Later we shall discuss some of the reasons why one or other party might win such a race. First, however, it is worth outlining the range of relationships which fall within the producer/scrounger definition and looking at the nature of the costs that may be inflicted by scroungers and the selection pressures which may lead to behavioural and other counteradaptations by producers.

PRODUCER/SCROUNGER RELATIONSHIPS

As Parker (1984) notes, the central notion of the producer/scrounger model is that alleles code for the build-up of resources such as energy reserves, tissues, bodies (or phenotypic extensions of bodies such as nests, territories, food stores etc.) which act as aids to allele replication (through increased individual reproductive success). Such producer strategies are vulnerable to parasitic scrounger strategies which can be favoured if parasitism of the producer's resources is more profitable (leads to a greater rate of replication) than gathering the equivalent resources independently.

A fundamental property of producer/scrounger systems is that the fitness of the two strategies is likely to be frequency dependent (see Dawkins, 1980; Sibly, 1984). Scrounger strategies do best when they are rare and worst when common relative to producer strategies. As scroungers increase in frequency, competition for the limited resources provided by producers becomes more intense and an evolutionarily stable producer : scrounger ratio, where the fitness of the two strategies is equal, is a likely outcome. However, there may be conditions where the fitnesses of the two strategies are not the same, as for instance where scrounging reflects a phenotype limited (Parker, 1983) 'best of a bad job' strategy adopted by individuals of low efficiency as producers. Barnard (1984c) distinguishes between two broad classes of scrounger strategy: scrounging by usurpation and scrounging by deception.

Scrounging by usurpation

Usurpation occurs where producers invest time and effort in procuring some resource which is then appropriated by a scrounger to the latter's advantage. Any investment of time and energy in resource procurement is in principle vulnerable to exploitation by non- or low-investing usurpers. Unless there is strong retaliation by the producer or some other risk or cost to the scrounger, usurpation is likely to have an immediate and large selective advantage because it side-steps the major costs of acquiring resources. The adaptive short cut provided by producers means that, once they are available and as long as the appropriate mutations arise, usurpers may be almost inevitable even though ultimately many may not be maintained in the population/community.

Because usurpation involves capitalizing on another organism's investment, usurpation strategies are parasitic and organisms conventionally recognized as parasites provide perhaps the most extreme and familiar examples. However, parasites are part of a wide range of strategies of usurpation which involve not only food but space, mates, parental care and other resources including, in the case of viruses, the cellular machinery of the host itself.

Scrounging food

Since organisms are thermodynamically open systems which are energetically costly to maintain, they depend for their existence and reproduction on taking in energy and nutrients as food from the environment. Food is rarely limitless and its restricted availability is for most organisms one of the most important factors limiting survival, reproductive output and thus population growth and stability. The fact that food is usually limited and, when in the form of other organisms, often takes steps to avoid being eaten, means that it can be expensive to obtain. Searching for, catching and ingesting food items are likely to cost time and energy, costs which have to be offset against the nutritional benefits derived from the items. Much can be understood about the foraging decisions of predators by assuming that such decisions have evolved to maximize the ratio of benefits to costs in these terms (for recent reviews see e.g. Stephens and Krebs, 1986; Kamil *et al.*,1987).

REDUCING THE COSTS OF FOOD PROCESSING

One obvious way an organism can increase the benefit : cost ratio is to try to circumvent the cost element altogether and live in an environment in which it is bathed in easily ingested or absorbed food. This is an adequate description of the lifestyle of many gut endoparasites some of which have access to such readily assimilable food that some multicellular forms have even dispensed with an alimentary canal (cestodes and acanthocephalans). Because they are able to minimize their energy expenditure on activities that most other predators have to perform, gut parasites like tapeworms can channel a large proportion of the energy they assimilate into growth and reproduction. For example, Bailey (1975) has shown that of the energy assimilated by the cestode *Hymenolepis diminuta* in rats (*Rattus norvegicus*), 45% is used in production (growth and reproduction) and 93% of this goes into reproduction. However, endoparasitism is far from cost-free. High reproductive output and limited habitat availability within the host can result in intense inter- and intraspecific competition among parasites with crowding leading to a dramatic reduction in individual growth and gamete output (Kennedy, 1984). Competition is also reflected interspecifically by resource partitioning with niche segregation by active site discrimination in several species (e.g. Schad, 1963; Crompton, 1973). Some authors have argued that competitive interspecific interactions have little to do with site segregation and that hosts are in fact under-utilized resources (Price, 1980; Rohde, 1979). Rohde (1979) suggests that, at least in dioecious species, apparent interspecific site segregation may actually be a reflection of species-specific aggregation to maximize mating opportunities. However, species-specific aggregation may also be a strategy for evading host immune responses. Where evasion is effected by the secretion of soluble immunomodulatory factors, collective secretion in an aggregation might produce a local area of immunodepression and thus a safe haven from immune attack (see Behnke, 1987). Whether or not there is competition at high densities, high infestations may increase the risk of immune attack via a threshold effect which is not triggered at lower infestations, or the damage and pathology resulting from high infestations may lead to the death of the host. Both effects are likely to curtail reproductive output from the parasite.

Often, of course, host immune responses limit the establishment and survival of gut (and other) endoparasites and may succeed in eradicating them from the host body altogether. The risk of failure in transmission from one host to another is a further problem (Chapter 6). Gut endoparasites thus reduce their food searching and handling costs but the price is increased costs of habitat acquisition (dependent on transmission), competition (low potential for migration between feeding sites) and elimination (by host immune responses).

Gut endoparasites reduce their foraging costs by taking advantage of the host's digestive processes. Sometimes the digestion of food is begun externally and in these cases free-living species may gain similar feeding advantages to gut parasites. Robinson and Robinson (1977), for example, report small neotropical michiliid flies (*Phyllomyza* spp.) landing on the cephalothorax of the large orb-web spider *Nephila clavipes*. When the spiders catch prey and envelope it with digestive fluid, the flies move onto the prey and imbibe the juices. These or related flies parasitize other spiders (e.g. McMillan, 1965; Sivinski and Stowe, 1980) and also predatory insects such as assassin bugs (Reduviidae; Robinson and Robinson, 1977) and robber flies (Asilidae; Vollrath, 1984) in a similar way.

The webs of spiders provide opportunities for many parasitic species by providing both a trap for otherwise highly mobile and unobtainable prey and/or food in a predigested, more assimilable form. The polistine wasp, *Mischocyttarus drewseni*, pounces on insect prey entangled in webs and chews it into a ball for transportation to the wasp's nest. Some individuals of *M. drewseni* appear to specialize in obtaining food in this way. Although the wasps appear to be opportunistic parasites, they show some evidence of being adapted morphologically to stealing from spiders' webs; their mid and hind legs show an asymmetry of the tarsal lobes which may enable them to walk on the sticky silk strands of webs (Jeanne, 1972). In a similar way to the wasps, scorpion flies (*Panorpa* spp.) invade webs from the surrounding vegetation or land directly on trapped prey (Thornhill, 1975). Sometimes they remove the prey packet, but more often they consume it *in situ*. If scorpion flies become entangled in the web themselves, a fluid is regurgitated at the host spider should it attack the fly. Damselflies, dragonflies and hummingbirds are also regular raiders of spiders' webs (see Vollrath, 1984).

The wasps, scorpion flies and other examples discussed above are visitors to webs, taking trapped prey items if they happen to be there. In addition to these opportunists, however, webs can also house some residential parasites. Vollrath (1984) describes some small hemipteran bugs (Miridae, Phylinae) which live in the webs of the social spider *Anelosimus eximius*. They are tiny relative to their host and able to move about freely within the tangled strands of the web. The bug waits until prey has been overpowered by a spider then approaches a spot not occupied by feeding spiders, inserts its proboscis and sucks. Dead spiders are also consumed. The bug freezes, if it is discovered by a spider, reducing the risk of being located by vibrations on the strands of the web, or flees. All developmental stages of the bug live in the webs of *Anelosimus* and have not been recorded in the webs of any other spiders (Vollrath, 1984), suggesting that the relationship is finely tuned and involves a high degree of specificity.

Webs may also harbour other species of spider. The tiny symphytognatid spider *Curimagua bayano* lives in the webs of the large mygalomorph spider *Diplura*

(Vollrath, 1978). *Curimagua* rides on the cephalothorax of its host, climbing down over its chelicerae onto its prey. *Curimagua* appears to be restricted to the webs of *Diplura* and its characteristic mouthparts are adapted to feeding on the predigested, liquid food provided by the host. Spiders of the family Mysmenidae also inhabit the webs of *Diplura* (Platnick and Shadab, 1978) and mount their host's prey while it is feeding, though they take mainly small insects which are overlooked or neglected by the host and are thus more commensal than parasitic. Species of the theridiid spider *Argyrodes* are regularly found in the webs of other spiders of a wide range of genera (see Vollrath, 1984), some specializing on particular hosts, others generalizing to several different hosts. Although they feed on small insects entangled in the web, they also steal prey which have already been predigested by the host spider. Females of the specialist *A. elevatus*, for example, pillage the prey stores of their host *Nephila clavipes* with a very high rate of success. Their success is due in part to fine signal line threads constructed by the parasite to connect its resting place towards the outside of the host web with the hub and several radii of the web.

Like the webs of spiders, social aggregations can also provide concentrations of readily assimilable food which are vulnerable to exploitation by parasites. Many examples can be found among ant societies in the form of various invertebrate symbionts. The strategies adopted by these symbionts range from conventional ectoparasitism to subtle and intricate commensal relationships with host species (figure 1.1; for a detailed discussion, see Wilson, 1971). Even among the ectoparasitic species there can be some bizarre adaptations for coexisting with hosts. The mite *Macrocheles rettenmeyeri*, for example, feeds on blood taken from the terminal membranous lobe of the hind tarsus of a soldier of its exclusive army ant host species *Eciton dulcius*. In the process it effectively allows its body to be used as a substitute for the terminal section of the foot (figure 1.1(d)). Army ants characteristically form clusters by hooking their tarsal claws over the legs or other body parts of companion workers and bivouacs may have clusters several inches in diameter hanging in cavities in the ground. When a worker hooks the leg with the mite onto another worker or the nest, the entire hind legs of the mite act as the ant's tarsi. To this end the hind legs of the mite always appear curved rather than straight thus mimicking the morphology of the tarsi (Rettenmeyer, 1962; Wilson, 1971). There do not appear to be any observable differences in the behaviour of unparasitized ants and those whose hind tarsi are replaced by the mite.

Other mites feed on the regurgitated food of their ant hosts. Several species of the mite genus *Antennophorus* ride on the bodies of *Lasius* ant hosts and imbibe the ant's food either as it is regurgitated and passed to other workers or by soliciting its regurgitation directly by stroking the heads of worker ants with long, antenna-like forelegs, thus imitating the tactile signals used by the ants themselves. A number of staphylinid beetles and some mosquitoes and flies have developed a similar ability to persuade ants to regurgitate for them. The reason for the apparent ease with which the ants are duped seems to be the simplicity of the tactile signal used in the ants' normal social context. Hölldobler (e.g. 1970) has shown that the most susceptible worker ant is one that has just finished a meal and is searching for a nestmate with which to share its crop contents. To gain its attention a nestmate or a symbiont has only to tap the ant's body lightly with its antenna or forelegs to cause it to turn and face the signaller. Repeated light taps on its labium will cause the ant to regurgitate. Other ants use their fore-tarsi for this purpose while symbiotic species may use either their tarsi or their antennae. The larvae of the staphylinids

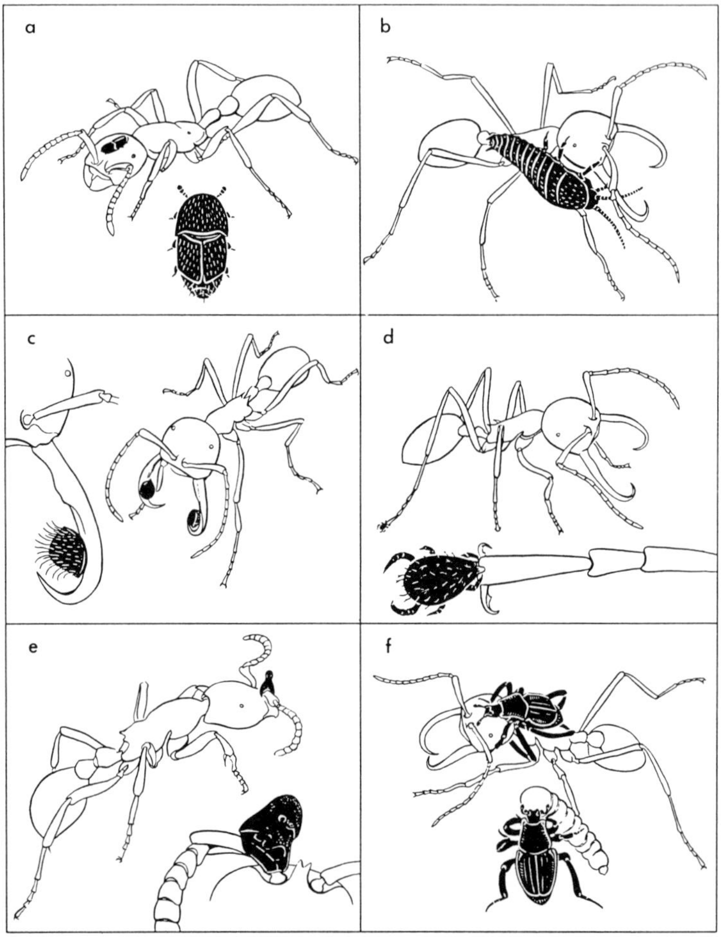

Figure 1.1. Examples of arthropod symbionts of army ants (a) Limuloid beetle, *Paralimuloides wasmanni,* spends most of its time riding on the bodies of its host (*Neivamyrmex nigrescens*). (b) Nicoletiid silverfish, *Trichatelura manni,* scrapes and licks the body secretions of its host, *Eciton* and shares its prey. (c) *Circocylliba* mite specialized for riding on the inner surface of the mandibles of major workers of *Eciton*. (d) The mite *Macrocheles rettenmeyeri* serves as a 'foot' for workers of *Eciton dulcius*. (e) The mite genus *Antennequesoma* is highly specialized for attachment to the first antennal segment of army ants. (f) Hysterid beetle *Euxenister caroli* grooms adults of *Eciton burchelli* and feeds on their larvae. Reproduced from Wilson (1971) with the kind permission of the author and the Belknap Press, Harvard.

Atemeles and *Lomechusa* simply curve the front part of their bodies upward and push their labia against those of the host worker, this clumsy imitation apparently being sufficient to elicit regurgitation if the host's crop is heavily laden.

The species exploiting spiders' webs and social insects are making use of pre-digested or otherwise readily assimilable food sources and thus, at a qualitative level, gaining similar reductions in the food processing costs of foraging to endoparasites. As we shall see later, the parallel with endoparasitism may, in some cases, even extend to the nature of the counteradaptations adopted in the arms race between exploiters and their hosts. Other food-scrounging parasitic strategies, however, focus on the reduction of different foraging costs.

REDUCING THE COSTS OF SEARCHING AND PROCUREMENT

Kleptoparasitism (piracy) among bird species appears to occur most often when the cost of obtaining food directly is likely to be high. Thus the majority of kleptoparasitic relationships between birds documented by Brockmann and Barnard (1979) involved host species which were diving for food at sea or other water bodies or searching for otherwise concealed prey (e.g. soil and turf layer invertebrates in grassland). While some kleptoparasitism between birds involves the theft of partially digested food (see Brockmann and Barnard, 1979; Furness, 1987), such food usually still requires handling and considerable further digestion and is thus not comparable in assimilability with the predigested juices usurped by some of the arthropod scroungers discussed above. Indeed, the theft of thoroughly predigested food in these kinds of associations is unlikely to be feasible. Furness (1987) suggests that one reason kittiwakes (*Rissa tridactyla*) on the island of Foula are attacked only infrequently by kleptoparasitic skuas is that they often regurgitate a well-digested mush of sandeels (*Ammodytes*) and zooplankton which it is impossible for the skuas to handle. The skuas thus avoid wasting time chasing incoming kittiwakes which are likely to prove unprofitable.

The theft of partially digested food regurgitated by hosts occurs in less than 5% of the kleptoparasitic associations in birds reviewed by Brockmann and Barnard (1979); most such associations involving the theft of procured items prior to swallowing. However, within certain associations the theft of regurgitated food may occur more often. In a study of the kleptoparasitic great skuas (*Catharcta skua*), Furness (1978), showed that 45% of chases were directed at birds which had already swallowed their prey.

Even though stolen prey has in most cases still to be handled and digested, the savings in searching and procurement costs may be considerable. Barnard and Thompson (1985) in their study of black-headed gulls (*Larus ridibundus*) stealing earthworms from lapwings (*Vanellus vanellus*) and golden plovers (*Pluvialis apricaria*) on agricultural pasture, showed that the amount of time and effort that plovers spent extracting worms from the ground and handling them increased disproportionately with worm size. The largest worms took on average between 20 and 40 seconds to extract and handle. This was mainly because large worms tended to be deeper down and became more firmly anchored during extraction. However, large worms were the most likely to be stolen by gulls with 70–80% of them being lost by plovers. The high risk of attack following the procurement of a large worm may be one reason why plovers sometimes discard large worms once they have been pulled from the ground (Barnard and Thompson, 1985). While in principle all black-headed gulls have to do is stand around and wait for a plover to discover a worm and make it available, in practice theft is not as easy as this. To steal worms, gulls often have to attack target plovers from a considerable distance which gives the target plenty of time to detect the attack and take evasive action. As a result, gulls can become involved in lengthy aerial chases with average success rates of less than 20% (though success rate depends on, among other things, worm size and host species). In a broader review of kleptoparasitism among seabirds, Furness (1987) recorded average success rates of between 19% and 23% in skuas and frigatebirds and 26% in gulls and terns. When the energetic costs of the various behaviours involved in kleptoparasitism (attack flights, vigilant postures, moving between host flocks etc.) are estimated and taken into account for black-headed gulls stealing from lapwings, gulls just meet or slightly exceed their estimated daily energy requirement through kleptoparasitism (Källander, 1977; Barnard and Thompson, 1985).

Reducing the costs of searching

In some food-scrounging relationships it is only searching costs which appear to be reduced. Two studies of intraspecific food scrounging in Harris (*Zonotrichia querula*) and house (*Passer domesticus*) sparrows suggest that certain individuals within feeding flocks capitalize on the searching effort of others to locate patches of food (Rohwer and Ewald, 1981; Barnard and Sibly, 1981). When Harris sparrows feed on concealed, patchy food supplies, dominant birds tend not to search for food themselves but instead allow subordinates to do the searching and then displace them from the food patches they discover (Rohwer and Ewald, 1981). Indeed, in Rohwer and Ewald's study, dominants defended groups of subordinates against other dominants as if shepherding them for exploitation.

In house sparrows, producer and scrounger strategies are not obviously based on dominance (Barnard and Sibly, 1981). Feeding interactions within flocks are an important way in which individual house sparrows increase their feeding efficiency (Barnard, 1978). Several different forms of interaction are used (see Barnard and Sibly, 1981) but while most birds use all types of interaction at some stage, analysis of interaction frequencies across a number of colour-marked individuals showed that some birds specialized in interacting. Specialization was most pronounced when food supplies were clumped and apparent specialists were referred to as 'copiers' by Barnard and Sibly because 'area copying' (searching the same immediate area as a successful forager) was the commonest form of interaction. Birds which predominantly searched for their own food were referred to as 'searchers'. Barnard and Sibly were able to predict frequency-dependent changes in the pay-off to each strategy on the assumption that individuals were pure (searcher or copier) rather than mixed (searcher and copier) strategists.

The adoption of pure producer/scrounger roles within species such as those suggested by Barnard and Sibly's sparrows seems unlikely to be evolutionarily stable under most conditions of social foraging because changing producer : scrounger ratios within groups as individuals come and go will result in large fluctuations in the pay-off accruing to each strategy (Barnard and Sibly, 1981). A more plausible solution is for producer/scrounger roles to be exchangeable within individuals so that roles can be switched to those that are most profitable at the time. Such exchangeable roles have been reported (Giraldeau and Lefebvre, 1986) in flocks of feral pigeons (*Columba livia*). Giraldeau (1984) suggested that foraging with single species groups may provide individuals with special foraging benefits through a skill pool effect. Individuals within the group may differ in their foraging skills and thus be able to specialize on different sources of food. Each skill, however, is potentially exploitable by those individuals which do not possess it. By foraging in a group, and mixing with other skills, individuals may be able to generalize across food sources but feed on each source at the specialist rate by parasitizing those made available by skills other than their own (Giraldeau and Lefebvre, 1986). In a captive flock of pigeons individuals switched producer/scrounger roles in response to changes in both type of food source and flock composition.

Scrounging sex

Food is one vital limited resource which creates selection pressures for parasitic scrounging strategies. For sexually reproducing species, mates are another. Just as kleptoparasites capitalize on the foraging and/or food assimilation investment of hosts, various scrounger strategies exploit the investment of other individuals in gaining access to mates and mating success. As in the case of scrounging food, sexual scroungers exploit the investment of producers at a number of different levels.

Reducing the costs of mate attraction

In many species, successful mating among members of the competing sex (males in most cases) is achieved through the maintenance and defence of a territory or a harem. The cost of defence may be high in terms of both time and energy and risk of injury (see e.g. Leuthold, 1966). Territories may increase access to mates either because they encompass vital resources such as food or water to which individuals of the limiting sex are obliged to come or because individuals of the limiting sex base their decision to mate on the quality of the territory (from the point of view of raising offspring) or the quality of the defender as evinced by its ability to hold on to its territory. This investment can be exploited by scroungers to gain low cost fertilizations.

A good example is given by Gross (1982) in a study of the bluegill sunfish *(Lepomis macrochirus)*, a common North American freshwater fish which breeds colonially in ponds and lakes. Gross discovered what appear to be two distinct breeding strategies among males which he designated *parental* and *cuckolder* (figure 1.2). *Parental* males are large and construct breeding nests on the pond/lake floor to which they then attract females for mating. When successfully-mated females have spawned the males then provide all subsequent parental care (figure

1.2). *Parental* males become reproductively active late in the life cycle, reaching maturity at about 7 years old. *Cuckolder* males, on the other hand, mature when they are still small (about 2 years old). *Cuckolders* adopt one of two age-dependent scrounger strategies. When they are still small they behave as *sneakers,* lurking in vegetation close to a *parental* male until the *parental* male attracts a female and the moment of spawning and fertilization approaches. At this point the sneaker male rushes out from cover and comes between the *parental* male and the female shedding its own sperm over the eggs. These sneaky fertilizations are achieved at high speed and the *parental* male has little opportunity to prevent them. When they are larger and reach the size of adult females, *sneaker* males become *satellites* and adopt female-like striped markings. Now when *parental* males are shedding sperm on the eggs of a spawning female, the *satellite* male moves in slowly and

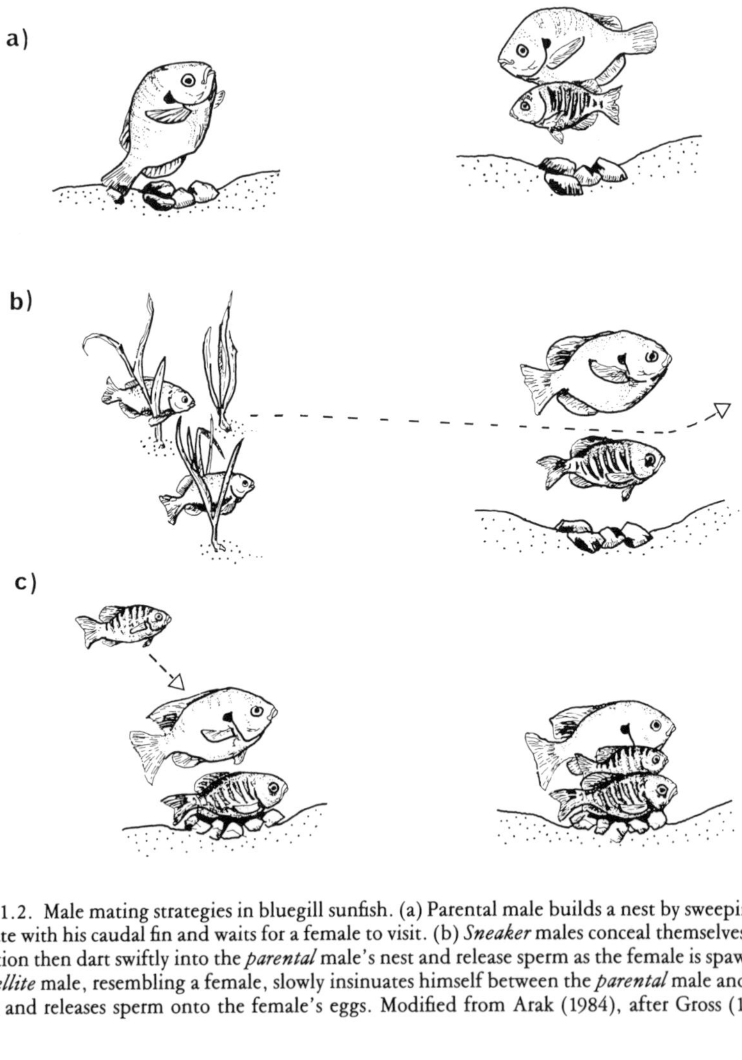

Figure 1.2. Male mating strategies in bluegill sunfish. (a) Parental male builds a nest by sweeping the substrate with his caudal fin and waits for a female to visit. (b) *Sneaker* males conceal themselves in vegetation then dart swiftly into the *parental* male's nest and release sperm as the female is spawning. (c) *Satellite* male, resembling a female, slowly insinuates himself between the *parental* male and the female and releases sperm onto the female's eggs. Modified from Arak (1984), after Gross (1982).

insinuates itself between the mating male and female and sheds its sperm. Presumably because the *satellite* closely mimics a female in size and appearance, the *parental* male appears not to notice it is being cuckolded and allows the *satellite* to proceed. *Parental* and *cuckolder* males represent distinct developmental pathways with no opportunity for switching between the two (Gross, 1982). *Cuckolders* made up about a fifth of the male bluegill population and, in Gross's study, appeared to fertilize about the same proportion of eggs, thus suggesting that *parental* and *cuckolder* males were maintained in an evolutionarily stable ratio with equal average fertilization success rates.

In some other sexual producer/scrounger relationships, the pay-off to scroungers is much lower than that of producers and scrounging appears to be a function of the phenotypic limitations of certain individuals (a 'best of a bad job' strategy). In bullfrogs (*Rana catesbeiana*), for example, males fight for ownership of the best breeding territories to which the victors then attract females by calling (Howard, 1978). Large males are at an advantage in fights and so monopolize the good territories. However, by sitting on their territories and calling and attracting females to them over a distance, territory holders are open to exploitation by the smaller males they defeated for ownership in the first place. Some of these small males (*satellites*) position themselves near the territory of a larger male and wait until his calling attracts a female. As the female approaches the calling male, the *satellite* intercepts her and attempts to mate. While the *satellites* avoid the costs of territory defence and calling, and to that extent obtain free copulations, their success rate is low and they achieve only about 3% of fertilizations in bullfrog populations. However, because they are small, satelliting offers the only opportunity for mating. As they increase in size with age, males which start off as *satellites* become more competitive in contests for territory ownership and may eventually become owners themselves. Unlike the situation in bluegill sunfish, therefore, producing and scrounging are age-dependent strategies which individuals can switch between as their phenotype changes.

REDUCING THE COSTS OF MATE AROUSAL

In the sunfish and bullfrog, scrounger strategists exploit the investment of producers in territory defence and in nesting and mate attraction respectively. Attracting potential mates, however, is only part of the total investment an individual may need to make in reproducing successfully. Time and energy may have to be spent stimulating receptivity before copulation can take place. One of the suggested functions of the sometimes elaborate and costly courtship displays is mate arousal, the motivation to mate increasing with exposure to displays. If courtship displays do have such an effect there is an obvious point of vulnerability to exploitation as the potential mate reaches a state where it is prepared to copulate. If another stepped in, it might secure a mating without having to invest anything in mate arousal. Arak (1984) suggests this as one possible benefit to satelliting among male ruff (*Philomachus pugnax*).

During spring, male ruff aggregate at traditional arenas (leks) where they display vigorously to attract females (Hogan-Warburg, 1966). There are two male mating strategies which are thought to be maintained by a genetic polymorphism: *resident* males defend small territories on the lek and have dark head tufts and ruffs while *satellite* males generally behave non-aggressively, associate with *residents* on their territories and have white tufts and ruffs. While there have been several (not necessarily mutually exclusive) explanations for lekking behaviour and the association between *residents* and *satellites* among ruff (see Arak, 1984), an intriguing possibility is that the very high levels of display afforded by aggregations of males facilitates the sexual arousal of females when they visit the lek and makes them more willing to mate. The tendency for females to mate with males at the centre of the lek may be a result of such a facilitatory effect (rather than, say, of assessment of central males as being of superior quality) because the female is exposed to displays as she approaches the centre and, once there, experiences them on all sides (Davison, 1982) thus perhaps reaching a peak of arousal at the centre. By associating with *resident* males on their territories, *satellites* may be able to capitalize on greater female arousal and sneak copulations when the *resident* is engaged in territorial defence. There is also evidence that tolerating a *satellite* on its territory increases the attractiveness of a *resident* to females (van Rhijn, 1973). This, coupled with the additional aggressive defence of the territory sometimes performed by the *satellite*, may more than offset the cost to the *resident* of occasional cuckoldry.

Reducing investment in copulation

The ultimate in sexual scrounging must be to let someone else do the copulating for you but pass on your gametes. For obvious reasons, given the sexual anatomy of most species, this seems unlikely, but a few insects appear to be able to achieve it.

The males of some species of the insect superfamily Cimicoidea (Hemiptera, Heteroptera) inject sperm directly into the female's body cavity by stabbing through her integument (for a review see Hinton, 1964). Sometimes areas of the abdominal integument (ectospermaleges) have become specialized to receive the copulatory organs of the male and absorption of the sperm is accomplished by another set of specialized cells, the meso-spermaleges; in other cases there do not appear to be such specializations. The latter is the situation in the lyctorinid bug *Xylocoris flavipes* where sperm are injected through the body wall into the haemocoel without either ecto- or meso- spermaleges. An intriguing observation is that, in some species, males regularly inseminate each other. This occurs in *X. maculipennis* where males inject their sperm into the body of other males and the sperm swim to the victim's testes from where they can be passed on the next time the bug mates (Carayon, 1974). Such homosexual 'rape' thus provides an additional outlet for a male's gametes which are transmitted through the copulatory efforts of the victim. Homosexual haemocoelic insemination appears to be particularly prevalent among males of the genus *Afrocimex*. Here, however, males possess ectospermaleges to receive penetrations. Male ectospermaleges thus appear to be an example of an adaptation to facilitate homosexual copulation.

Scrounging parental care

Some of the most familiar producer/scrounger relationships involve parasitism of parental care, cuckoos (Aves: Cuculidae) and cowbirds (Icteridae) perhaps being the obvious examples. However, the usurpation of parental care can sometimes be far more subtle and far-reaching.

REDUCING THE COST OF GAMETIC PARENTAL CARE: ANISOGAMY AS A PRODUCER/SCROUNGER RELATIONSHIP

As Parker (1984) points out, the differences in gamete (and thus zygote) provisioning between males and females in anisogamous species can be viewed as the basis for a producer/scrounger relationship. Females are producers in the sense that they typically invest heavily in provisioning their gametes and may also expend further resources on care of the zygote at later stages of development. Males can be regarded as parasitizing this investment because their contribution to zygote survival is typically zero, consisting simply of a haploid set of chromosomes (Parker, 1984). This is not to say that males do not invest in reproductive effort at all, but that they invest in ways other than parental care, for instance, in intermale competition (Parker, 1984). The parasite analogy rests on the fact that the lack of paternal contribution to the welfare of the zygote places an additional burden on females (see Parker *et al.*, 1972); in other words, females would benefit if males invested. Males also benefit as much as females from the investment of the female although they enjoy the benefit cost-free.

The selection pressures leading to the evolution of anisogamous sex have been modelled by Parker *et al.* (1972) and analyzed further by Bell (1978), Hoekstra (1980), Maynard Smith (1982) and Parker (1978). The argument of Parker *et al.* envisages disruptive selection for gamete size where gametes fuse at random in an external medium (e.g. sea water). The disruptive selection theory proposes that parents producing intermediate-sized gametes are selected against relative to those producing very small or very large gametes because of the advantage of producing, on the one hand, gametes with a high degree of provisioning for the zygote and, on the other, vast numbers of highly motile gametes which increase the number of syngametic fusions. Since producers of small, motile gametes have an advantage over large gamete producers in the race to fuse, selection favours increased provisioning by the producers of the larger, less motile, gametes to make up for the lack of provisioning for the subsequent zygote by the small gamete producers. Selection on gamete size thus rapidly polarizes towards extremes of size and typical anisogamy.

Even in anisogamous species, however, there may be a form of secondary paternal investment in gamete provisioning. In some birds and insects, courtship involves the offering of 'nuptial gifts' by the male (see e.g. Parker and Simmons, 1989). In many cases these gifts are, or contain, food items which are eaten by the female to provision her gametes. The investment by some males in courtship feeding provides the opportunity for parasitism by others. A good example comes from the hanging fly *Bittacus apicalis*. Male hanging flies offer an insect to the female during courtship (Thornhill, 1976). The size of the insect prey determines the duration of copulation and the success of sperm transfer. However, insects have to be caught which means investing time and energy in foraging. It turns out that males sometimes try to avert this cost by robbing other males, or even copulating pairs, of their prey and using the stolen prey themselves to attract and copulate with females. The hanging flies thus appear to provide an example of intrasexual parasitism of gametic parental investment.

REDUCING THE COST OF POSTGAMETIC PARENTAL CARE

While sperm in anisogamous species can be viewed as parasitizing parental investment at the gametic level, postgametic investment is also exploitable. In many sexual species where males continue to provide little or no parental care after fertilization, the notion of male parasitism is still an appropriate one, though with the repeated caveat that this does not imply a lack of reproductive investment elsewhere. Furthermore, investment in parental care may be indirect, for instance through the defence of a food-rich territory which sustains offspring during early postnatal development. In species where prenatal zygotic development takes place outside the body of the female but is dependent on continuing care by the mother or other adults (as in the incubation required by the eggs of most birds), there is scope for both intra- and interspecific parasitism of zygote care.

Among communally-breeding birds, intraspecific parasitism of incubatory effort may take the form of subtle exploitation of one class of individuals by another. A good example comes from a neotropical member of the cuckoo family, the groove-billed ani (*Crotophaga sulcirostris*) (Vehrencamp, 1977, 1978). Anis live in small groups in which each individual helps to defend a group territory. Groups usually comprise between one and four monogamous breeding pairs with occasional non-breeding helpers. Members of the group contribute to the building of a single, communal nest in which all breeding females lay their eggs. However, there is a limit to the number of eggs that can be incubated efficiently; beyond a certain clutch size eggs tend to become buried, fail to be turned regularly and are unable to receive sufficient heat for normal growth and development of the embryo. The percentage of eggs which hatch therefore tends to decrease as the communal clutch size increases.

Vehrencamp discovered that females within ani groups had a linear dominance hierarchy and that the timing of egg laying correlated inversely with rank so that dominant females laid later than subordinates. She also discovered that females visited the nest for several days before laying and that, during visits, they would sometimes remove eggs which had already been laid. Once a female had laid eggs of her own, however, she stopped removing eggs from the nest. Females laying later thus tended to have a greater proportion of eggs in the communal nest because their own were least likely to have been removed. Despite a number of counter-responses by subordinates, dominant females achieved a substantial bias in the final egg ownership of the nest. However, manipulation of egg ownership by dominants was also associated with a reduction in incubation effort and investment in feeding nestlings (though both appeared to be offset to some extent by increased investment of males paired with dominant females). Dominant females were thus parasitic on the incubatory and other parental investment of lower ranking females and what appears on the surface to be a highly cooperative breeding system in fact turns out to involve rank-dependent exploitation. Vehrencamp (1983, 1984) discusses other examples of fitness-biasing within social groups in this and other contexts and proposes a number of models which explore the effects of variation in group composition and degrees of relatedness among group members.

In the anis, parasitism of parental care among females still involves some investment by scrounger individuals, and particularly by their mates, in their own offspring. Parasitism of parental care is, in this sense, only partial. Similar partial parasitism can be inferred from take-overs of breeding burrows among some digger wasps of the family Sphecidae. Female golden digger wasps (*Sphex ichneumoneus*) dig a tube-like burrow in the ground in which they store paralyzed katydids as food for their larvae. As well as digging their own burrows, however, females sometimes enter and provision burrows dug by other females. When unwitting co-owners eventually met they fight for possession and the winner tends to be the female who has provided most food (Brockmann and Dawkins, 1979). Victorious usurpers may thus gain a ready-dug burrow with provisioning thrown in. While the parasitized provisions may benefit the usurper's offspring, the usurper still provisions the burrow herself, perhaps due to limits in her ability to register the presence of prey other than those she has caught.

In other cases, parasitism of investment in burrow-digging and provisioning is complete. An interspecific example comes from the wasps *Stizoides unicinctus* and *Prionyx atratus* (Iwata, 1976). *Stizoides* females seek out burrowing *Prionyx* and wait near the excavation until it is complete and the burrow provisioned with a paralyzed spider on which *Prionyx* has laid her egg. When *Prionyx* leaves, *Stizoides* digs up her nest, eats the egg and lays her own in its place. *Stizoides* in this example is thus completely parasitic on the investment of her *Prionyx* host. The same is true of brood parasitism in some bees. Females of the genus *Nomeda* invade the brood cells of another genus, *Andrena* and oviposit (Iwata, 1976). When the *Nomeda* larvae hatch they kill those of their *Andrena* hosts and feed on their provisions. The destruction of the host brood in the bees is reminiscent of that among some inter- and intraspecific brood parasitic birds (see below) which also exploit the nest construction and offspring provisioning investment of their hosts. Here, however, there are often subtle and complex strategies of adaption and counteradaptation by parasite and host in a long-term parasitic parent/offspring relationship.

In some cases, the exploitation of parental care involves the physical usurping of individual carers. 'Slave-maker' ant species initiate raids on heterospecific colonies, tackling defending workers and guards to gain entry and steal larvae and pupae. These are carried back to the colony of the slavers' where they mature and accept the slavers as conspecifics, cooperating as workers in the rearing of slaver larvae. Slaves may even participate in further slaving raids on other colonies. A different form of slaving involves a reproductive of the slaver species gaining entrance to the colony of a host species and seeking out and killing the queen. The workers of the host species then accept the invader as their queen and care for her eggs and offspring. In this case the parasite usurps both parental care and the physical structure of the colony (for detailed discussions of slaving and brood parasitism in insects see Wilson, 1971; Sudd and Franks, 1987).

Parasite life cycles and usurpation of host behaviour

In the examples of scrounger strategies mentioned above, the behaviour of producers is parasitized at the level of the resources it makes available or the investment it makes in some costly activity. Successful scrounging depends on the normal performance of producer behaviour even though the behaviour may be performed for the benefit of an inappropriate recipient. Even in the case of the slave ant workers, the unwitting producers are simply carrying out their usual (brood care) behaviours though in a foreign colony. In some cases, however, producer behaviour may be subverted by the scrounger and altered to serve a function entirely alien to the producer. Good examples come from some parasite infections in which parasites appear to manipulate the behaviour of their hosts to facilitate the successful completion of their life cycle (see Holmes and Bethel, 1972; Dawkins, 1982; Moore, 1984; Keymer and Read, in press; Chapter 7).

Some of the clearest examples involve parasites which rely on predation of an intermediate host for transmission to their definitive host. Changes in the behaviour of intermediate hosts as a result of infection can take many forms, though referring to them as manipulation is more convincing in some cases than in others. Several infections result in reduced activity among intermediate hosts which can be interpreted as increasing their vulnerability to predation. For instance, fish carrying the plerocercoid of *Ligula intestinalis* tend to be sluggish and float in areas where they are more likely to be spotted by avian predators (Dence, 1958; Holmes and Bethel, 1972). They may also fail to keep up with shoals and become conspicuous by isolation (Holmes and Bethel, 1972). Rau (1983a) found that exploratory and ambulatory activity by mice in an open arena was reduced when animals were infected with *Trichinella spiralis* and that the onset of behavioural changes coincided with the expected time of larval release by adult worms and thus infectivity to potential carnivore hosts. The reduction in activity in infected mice is probably due to muscle damage by migrating and encysting larvae (Campbell, 1983). Recent work by Edwards (1988) suggests that the vulnerability of mice infected with *Trichinella* might also be increased through social discrimination and aggression by uninfected conspecifics which may result ultimately in the eviction of infected individuals from colonies.

While behavioural changes resulting from pathological damage (Arme and Owen, 1968; Holmes and Bethel, 1972) can be interpreted as adaptive to the parasite, some apparently manipulative effects are the result of more subtle interference by the parasite. For example, the intermediate hosts of many trematode and acanthocephalan parasites show marked changes in their responses to physical environmental stimuli such as light, gravity and mechanical disturbance which alter their behaviour in ways that appear tailored to increasing their vulnerability to the parasites' definitive hosts (see Holmes and Bethel, 1972;

Moore, 1984). Batra and Batra (1985) document alterations to host blueberry plant leaves by the fungus *Monilia vaccinii* which involve the display of ultraviolet reflective 'nectar guides' and the secretion of sugar solution to attract the plant's insects pollinators. Exploitation of the insects' feeding behaviour thus enhances transmission of the fungus from the leaves to the flowers and ultimately the berries where it has serious economic consequences. Parasites may also interfere with host reproduction, perhaps to channel the host's metabolic resources into growth and

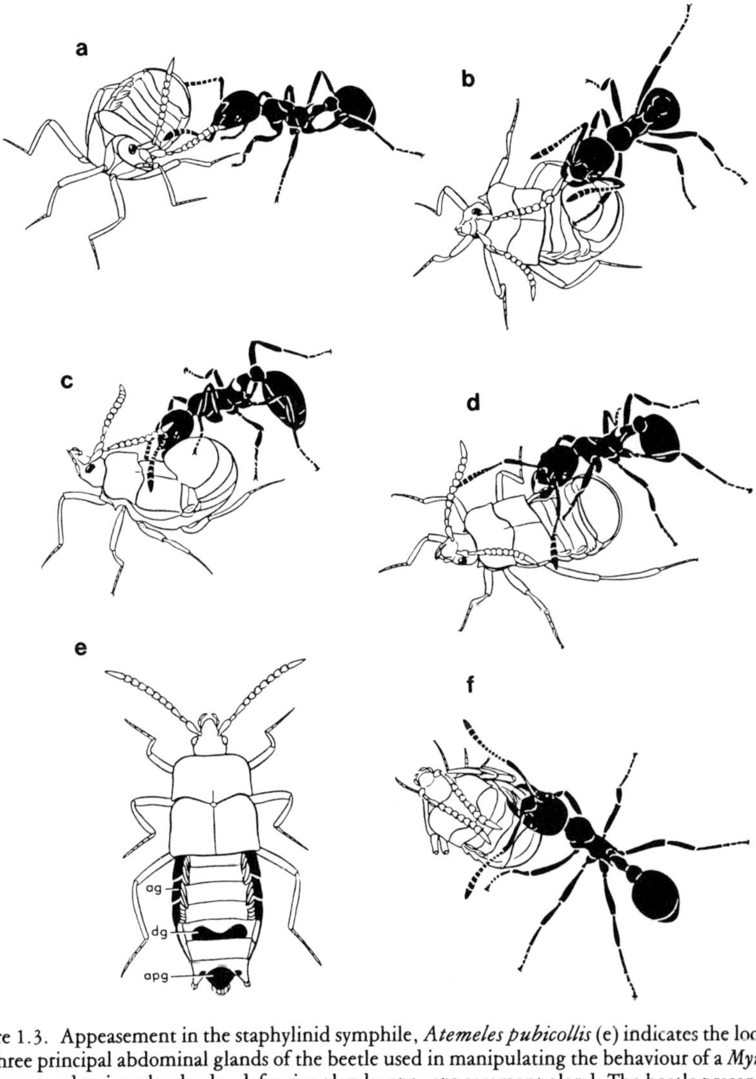

Figure 1.3. Appeasement in the staphylinid symphile, *Atemeles pubicollis* (e) indicates the location of the three principal abdominal glands of the beetle used in manipulating the behaviour of a *Myrmica* worker: ag, adoption glands; dg, defensive glands; apg, appeasement gland. The beetle presents its appeasement gland to a worker that has just approached it (a); after licking the gland opening (b), the worker moves round to lick the adoption glands (c,d) after which it carries the beetle into its nest (f). Reproduced from Wilson (1971) with the kind permission of the author and the Belknap Press, Harvard.

potential nutrition for the parasite. Reproductive interference may involve the destruction of gonadal tissue, as in parasitic castration, physiological/development effects (e.g. the synthesis of a host juvenile hormone mimic by the protozoan *Nosema* to prevent adult moult in its *Tribolium* host) or behavioural changes (such as reduced receptivity to mating attempts) (see Baudoin, 1975; Dawkins, 1982; Edwards and Barnard, 1987; and below). Some other possible physiological bases for the alteration of host behaviour by parasites are discussed by Holmes and Zohar (Chapter 2) and Thompson (Chapter 3).

In contrast to the often complex and elusive physiological machinery behind parasitic manipulation, the physiological mechanisms underlying apparent manipulation of host behaviour in certain other kinds of exploitative relationship are sometimes more obvious. One of the best known concerns so-called appeasement substances (Wilson, 1971) in symphiles of social insects.

Many of the better-integrated symphiles in social insect colonies secrete substances from epidermal glands which appear to be attractive to the hosts. These substances are generally licked up by the host and sometimes even by other symbionts. Wilson (1971) termed them 'appeasement substances' because they appear to be a means of gaining acceptance from the host. The manipulative effect of appeasement substances is illustrated by the relationship between the myrmecophilous staphylinid *Atemeles pubicollis* and its ant host *Myrmica* (figure 1.3). When the beetle approaches the nest of a host colony to seek adoption, it wanders round until it meets a worker whereupon it turns to present its appeasement gland at the tip of its abdomen (figure 1.3(a)). The ant feeds on the material secreted from the gland (figure 1.3(b)) and appears to become less frenetic as it does so. Following this, the ant shifts its attention to the beetle's 'adoption glands' which are situated further forward on the abdomen (figure 1.3(c–e)) and from which the ant also feeds. After feeding from the adoption gland, the ant picks up the beetle and carries it into the nest (figure 1.3(f)). Several symphilic beetles and at least one, presumed symphilid, wasp possess characteristic tufts of red or yellow hairs called trichomes or trichodes. The tufts appear to aid in the dissemination of appeasement substances.

It is not clear whether appeasement substances are simply nutrients or whether they are secondary phagostimulants or specialized compounds that mimic the attraction pheromones of the host. Whatever they are, they are powerful modifiers of host behaviour in favour of the symphiles.

Scrounging by deception

Barnard (1984b) distinguished a second category of scrounger strategy which he referred to as scrounging by deception. While strategies of usurpation involve the direct parasitism of producer resources or behavioural investment, scrounging by deception depends on the existence of what Barnard (1984b) termed cost-increasing models within detector/detectee relationships.

We can envisage organisms as *detectees* which are impinged upon by a number of *detectors* trying to distinguish them from features of the environment (*background*) that are irrelevant or deleterious to the detector's goal. Successful discrimination by a detector enhances the reproductive success of the detector but decreases that of the detectee. The terms *detector, detectee* and *background* apply to any discriminatory relationship which influences the reproductive success of the organisms concerned. A detector may be, for instance, a predator, a host immune system or a member of a mate-choosing sex and the corresponding detectee a potential prey individual, an invading parasite and a member of the candidate sex. Barnard (1984b) discusses this further and develops a general model of the evolution of detector/detectee relationships and the cues used in discrimination.

While detector/detectee relationships can be viewed as the driving force behind the evolution of broad phenotypic characteristics such as crypsis and elaborate secondary sexual characters, they may also be a major selection pressure favouring inter- and intraspecific scrounging. In this case, however, scroungers arise as mimics of inter- or intraspecific cost-increasing models. Such models increase the cost to the detector of making a mistake in its discrimination, good examples being the toxic or distasteful models in Batesian mimetic systems. In the case of Batesian mimicry, taking prey (detectee) which is being selected to resemble a toxic species (model) increases a predator's (detector's) chances of accidentally ingesting a toxic item. Similarly, a female (detector) attempting to discriminate between a good quality, preferred male phenotype (model) and a low quality mimic of that phenotype (detectee) risks rejecting a good quality mate if she rejects an apparent mimic. In both these examples, detectees are capitalizing on the investment of models in characteristics (e.g. toxins, status cues) that influence the preference of detectors and exploiting detector preference without investing themselves. If mimics spread through the population they may begin to impose a cost on the model if sampling by the detector undermines the reliability of the model in the detector's perception (see Barnard and Burk, 1979; Barnard, 1984b). Scrounging by deception is thus a special case of parasitism where the exploitation of investment is effected through a third party. Barnard (1984b) discusses the role of scrounging by deception in the evolution of mimicry generally and intraspecific assessment cues and sexual transvestism in particular.

THE COST OF SCROUNGERS

Exploitation is costly to the victim and the nature and severity of the cost determines the nature and extent of the victim's counter-response. The costs to producers of the kinds of exploitative strategies we have just been discussing can be grouped under a series of broad headings.

Increased metabolic investment

Where resources or behavioural and other metabolic investment are usurped, producer individuals may be forced to invest further to compensate for losses and/or meet the demands of scroungers. Such additional investment may increase the chances of mortality or have negative consequences for other components of the life history such as fecundity.

Parasitic infestations which impose a serious drain on the host's nutritional resources may result in the host increasing its food intake to compensate. One example comes from three-spined sticklebacks (*Gasterosteus aculeatus*) infected with plerocercoids of the cestode *Schistocephalus solidus* which show elevated levels of daily food intake compared with uninfected individuals (see Chapter 4). Barnard and Thomspon (1985) found that a greater proportion of lapwings in winter flocks were feeding at any given time if kleptoparasitic gulls were present, suggesting that birds had to spend more time feeding to compensate for losses of food to the gulls. There was also a positive correlation between the number of hours in a day lapwings spent feeding and the amount of time gulls were present in the flock. If gulls were present late into the afternoon, towards the end of the feeding day, lapwings sometimes continued to feed into the night. The impact of kleptoparasitism by gulls thus appeared to result in increased time and energy investment in foraging by lapwings. When gulls were actually present in a flock they had a negative effect on the rate of energy intake by lapwings and appear to be one reason why lapwings fail to meet their daily energy requirements by daytime feeding alone (Barnard and Thompson, 1985). Since there is little lapwings can do to avoid the presence of gulls, it is not surprising that they have adopted a number of compensatory tactics to reduce the impact of gulls. These include increasing the gulls' attack costs by moving away from their vantage points in the flock, reducing the amount of time spent in tell-tale prey targeting postures prior to pecking and facing away from gulls while handling prey (Barnard and Thompson, 1985).

Inter- and intraspecific brood parasitism may also force greater foraging investment from the host. Data from cuckoo species suggest that the persistent begging of cuckoo fledglings elecits prolonged feeding from the host parent and that, at least among passerine hosts, the period of parental care may sometimes be longer than that for the bird's own young (Payne, 1977). Intraspecific brood parasitism in starlings (*Sturnus vulgaris*) tends to increase brood size (despite at the same time reducing the host parent's own contribution to the brood (see below)) with the result that parental feeding rates may also increase (Evans, 1980; Tinbergen, 1981). However, increased feeding rates may to some extent be offset by changes in diet with hard-pressed parents foraging closer to the nest (Tinbergen, 1981). Along with the direct reproductive costs of brood parasitism, foraging costs may be a selection pressure favouring the various anti-parasite responses shown by host species (see below).

In the above cases, exploitation appears to force increased behavioural investment by the host. In other cases, scroungers may force metabolic investment more directly. The growth-stimulating effects of some parasitic infections are a good example. The larvae of the pseudophyllid cestode *Spirometra mansonoides,* for instance, secrete a substance (sparganum growth factor) which mimics the activities of mammalian growth hormone and stimulates weight gain in laboratory mice and rats acting as model hosts. Similar weight gain occurs when mice and rats respectively are infected with the haemoflagellates *Trypansoma duttoni* and

T. lewisi (Cheng, 1973). Workers of the ant *Pheidole commutata* infected with the nematode *Mermis* become hypertrophied to the extent that the abdomen can sometimes enlarge to eight times the normal size (Wheeler, 1910). Enlargement appears to be due to parasitized ants engorging themselves excessively. Larvae of the beetle genus *Tribolium*, when infected with the sporozoan *Nosema*, often fail to metamorphose into adults. Instead they persist in larval form continuing to grow through six extra stages and ending up as giant larvae weighing over twice as much as uninfected individuals.

In some apparently manipulative relationships between host and parasite, infected hosts show pronounced increases in general activity (e.g. Moore, 1983; Wilson and Edwards, 1986) which may facilitate transmission of the parasite between hosts. These examples of increased or prolonged growth and apparently manipulative increases in activity represent both an increase in and a diversion of host metabolic resources which are likely to impose at least some cost, and sometimes the ultimate one of death, on the host.

Reduced returns

As well as imposing additional costs on hosts, scroungers may also reduce the producer's returns from decisions. In many parasitic infections this may be due to parasites taking nutrients from within the host's body thus reducing directly the benefit to the host of its foraging effort. In other cases, reduction may be the result of interference with intake itself. Male laboratory mice (*Mus musculus*) infected with *Trichinella spiralis*, for example, show some evidence of decreased dominance status (Rau, 1983b), perhaps as a result of muscle damage inflicted by the larvae of the parasite (but see Edwards, 1988). If reduced rank means increased interference and/or reduced access to limited resources then infected individuals may gain less from decisions to feed, mate etc. than those that are uninfected. Milinski (1984), for example, found that three-spined sticklebacks (*Gasterosteus aculeatus*) infected with *Schistocephalus solidus* and/or *Glugea anomala* were more likely to take unprofitable prey than uninfected fish. Milinski interpreted this as a response to the reduced competitive ability of infected fish which would thus fare badly in contests for profitable prey. As Milinski points out in Chapter 4, however, infection does not necessarily result in reduced competitive ability and may indeed have a positive effect on resource acquisition in some circumstances.

Returns to the host may be reduced for other reasons. One of the ways in which lapwings respond to the presence of kleptoparasitic gulls in the flock is by reducing the amount of time spent targeting concealed prey before pecking. This reduces the risk of attack but means that they are restricted to taking smaller, surface-dwelling items which yield less of a return. As a result, lapwings suffer a reduced rate of food intake and undershoot their daily energy requirement to a greater extent when gulls are present (Barnard and Thompson, 1985).

Reduced reproductive output

Scroungers may have both direct and indirect negative effects on their host's reproductive output. Indirect effects can arise through the impact of scroungers on resource acquisition and the diversion of the host's metabolic resources into counteradaptations, both of which may reduce the host's potential to invest in reproduction. As well as losing out through a negative trade-off between life-history components, however, host reproduction may be a direct target of scrounger strategists.

Depending on the nature of transmission between one host and another and the status of the host in its life cycle, it may be adaptive for a parasite to reduce the reproductive investment of its host (see e.g. Baudoin, 1975; Dawkins, 1982; Edwards and Barnard, 1987). There are several ways in which this can be achieved.

Parasitic castration

An effective way of reducing host reproduction is to bring about so-called parasitic castration. The destruction or alteration of the gonads can have several effects, including changes to internal organs and secretions, secondary sexual characters, viability and growth and, in some cases, may cause sex reversal. Castration can be effected in a number of ways. In some cases it is the result of direct ingestion or mechanical interference with the gonads, often in conjunction with damage to other organs (see also Chapter 2). Sometimes the gonads are attacked first, sometimes only later in the sequence of organ damage. In other cases, castration may result from the withdrawal of nutrients or the secretion of 'castration' hormones (see Baudoin, 1975). Rothschild (1941) suggested that an increase in growth of the host was the selective advantage underlying castration of intermediate hosts (the snail *Peringia ulvae*) by trematodes. She demonstrated a positive correlation between frequency of parasitism and host size implying that infected individuals grew faster and to a greater size than uninfected individuals. The adaptive significance to the parasite lay in the fact that the number of cercariae produced within a snail is proportional to its size: larger snails yield more cercariae.

The benefit to the parasite may sometimes arise simply from usurping the body space normally occupied by the gonads or their products. In the parasitic barnacle *Sacculina,* Hartnoll (1967) suggests that sacculinization, resulting in castrated hosts (crabs), provides protection and ventilation to the parasite which occupies the external location usually taken up by the host's eggs. Infected hosts, regardless of sex, behave like ovigerous females. While the trematodes studied by Rothschild benefit from the diversion of host resources from reproduction to growth and increased nourishment, *Sacculina* benefits from the diversion of host parental care.

Parasitic castration, however, may affect host reproduction through more subtle effects on the sexual and parental behaviour of hosts. Some nematode infections in chironomids cause a reduction in the structural complexity of the antennae and tarsi of males (Rempel, 1940) which may impair their mate-searching behaviour (Baudoin, 1975). Ants infected with *Mermis* remain in the nest and neither forage like uninfected individuals nor tend offspring (Wheeler, 1910). Similarly, worker honey bees (*Apis mellifera*) infected with *Nosema apis* do not tend the brood (Wang and Moeller, 1970).

Reduced attractiveness or inclination to mate

While parasitic castration may affect sexual behaviour as well as rendering the host physiologically incapable of reproducing, other infections may influence sexual behaviour and mating success without resorting to destruction of the gonads. Edwards and Barnard (1987) suggested that, where impairing reproduction by the host benefits the parasite, reducing the attractiveness of the host to potential mates or its proceptivity/receptivity when approached sexually may be a means of achieving it. In an experiment with laboratory mice, Edwards and Barnard found that females infected with *Trichinella spiralis* were less likely to be mounted by (uninfected) males than uninfected females. However, when the responses of infected and uninfected females towards approaching males were analyzed, it turned out that infected females were more aggressive to males or showed more escape/avoidance responses when approached than uninfected females. The bias in the sexual attention of males towards uninfected females thus appeared to be due to differential proceptivity/receptivity of females rather than preference by males. Edwards and Barnard (1987) interpreted this in terms of parasitic manipulation (see Chapter 7) in that breeding females become less wide-ranging in their general mobility and may be less vulnerable to predation thus failing to pass on *T. spiralis* to carnivore hosts.

In the case of Edwards and Barnard, apparent manipulation of host reproduction was effected through reduced inclination to mate. An alternative possibility is to reduce attractiveness. Several lines of evidence suggest a relationship between infection and general body condition and appearance and recent theoretical and empirical investigations implicate this in mate choice on the assumption that resistance to infection is heritable (Hamilton and Zuk, 1982; Read, 1987; Chapter 5). While mating preferences based on infection-related appearance can be interpreted as advantageous to the individual exercizing the choice, it may also be advantageous to the parasite if it stands to benefit from reduced involvement of its host in costly sexual activity. Reduced visual attractiveness and/or vigour of infected hosts may thus be interpretable in terms of parasitic manipulation, though distinguishing between the predictions of the 'choose resistant mates' and manipulation hypotheses in particular cases may not be easy.

Brood loss

In the examples of castration and disinclination to mate, parasites prevent reproductive investment by the host to increase resource availability to themselves or their progeny or to facilitate transmission to new hosts. Resources are thus diverted away from hosts towards parasites. Brood parasitism is another class of parasitic relationship which incurs a reproductive cost for the host. The loss to the host can be viewed as a diversion of host resources into parasites though here diversion is of normal parental behaviour to the care of unrelated parasitic offspring (but see Andersson, 1984).

Among birds, brood parasitism usually depresses the breeding success of the host though the means by which it is depressed varies with the strategy of the parasite (Payne, 1977). Females of some cuckoo species may remove a host egg when they lay (Payne, 1977) and cuckoo chicks may themselves oust the host clutch from the nest when they hatch (see Davies and Brooke, 1989). Cuckoos may also cause the host to desert and nest again by eating its eggs if they find the nest after incubation has begun. The non-removal of eggs from many fresh clutches, the eating of eggs nearly ready to hatch and the killing of nestlings without eating them suggests the induction of desertion is a deliberate strategy to increase host availability (Payne, 1977). Cuckoos sometimes lay from a perch above the nest and a thick-shelled cuckoo egg may crack a host egg on impact (Gaston, 1976; Mountfort and Ferguson-Lee, 1966). If parasitic chicks do not evict host eggs they may outcompete them for food and cause their starvation and after fledging young parasites may affect the survival of host young by monopolizing parental care (Payne, 1977). If a parasite is the only fledgling, it may delay another nesting attempt by the host through prolonged demand (see earlier) or exhaust it so that it does not breed again (Gaston, 1976; Payne, 1973). Hosts of the cowbird *Molothrus ater* in North America and *M.bonariensis* in South America generally fledge fewer young than unparasitized pairs for a similar range of reasons (see Payne, 1977).

Facultatively brood parasitic ducks may also depress the breeding success of their host duck species. Redheads (*Aythya americana*) and ruddy ducks (*Oxyura jamaicensis*), although usually incubating their own eggs, sometimes lay in the nests of other ducks. Cinnamon teal (*Anas cyanoptera*) parasitized by redheads and ruddy ducks have smaller clutches than unparasitized conspecifics because host eggs are displaced from the nest (Payne, 1977). Redheads, wood ducks (*Aix sponsa*) and several other duck species also lay fewer eggs if parasitized (Weller, 1959; Clawson *et al.*, 1979; Heusmann *et al.*, 1980). Andersson and Erikson (1982) found that goldeneye (*Bucephala clanga*) responded to the addition of extra eggs to their nests by laying as many fewer as had been added. Ducks in general are indeterminate layers which start incubation when a normal clutch size is present in the nest regardless of the origin of the eggs.

Intraspecific brood parasitism may also reduce the host individual's reproductive output. Brood parasitism among starlings (*Sturnus vulgaris*) is a good example. Electrophoretic evidence from populations in the UK suggests that at least 12–32% of starling nests may be parasitized. Parasitized nests usually contain one parasitic egg but up to four have been recorded (Andersson, 1984). In an extreme case, Evans (1980) recorded an average of 2–3 parasitic eggs per parasitized nest. Intraspecific brood parasitism appears to inflict a reproductive cost on the host. Although total clutch size tends to be increased by the addition of parasitic eggs, the number of eggs produced by the host appears to be reduced. The most likely reason seems to be removal of host eggs by parasitic females. Starling eggs are not infrequently encountered on the ground. Until recently such eggs were attributed to unmated females without a nest (Yom-Tov *et al.*, 1974). However, studies by Feare (1983) involving the marking of all eggs in study nests on the day of discovery showed that eggs on the ground had originally been laid in a nest and must thus have been removed. Karlsson (1983), working in Sweden, also found that eggs were removed and dropped on the ground. Feare (1983) and Karlsson (1983) suggest that parasitic starlings behave like some interspecific brood parasites and remove an egg from the host's nest before laying (see above). From Karlsson's experiments at least it appears that birds do not remove extra eggs that turn up in their nests.

As well as reducing the host female's reproductive output by removing eggs, parasitic starlings may also reduce it by increasing the total clutch size in the host nest. The survival of starling chicks over the first 3 months is inversely related to clutch size (Lack, 1948) and both the amount of food delivered per chick and fledgling success decrease with increasing brood size. When brood size increases, fledglings weigh less and parents lose more weight during the nesting period (Tinbergen, 1981; Westerterp *et al.*, 1982). Large broods may thus have a detrimental effect on parental survival as well. However, while average breeding success is much lower in parasitized nests (see Andersson, 1984), the host's offspring often fare better than those of the parasite. If the parasite's eggs are added after the host begins incubating success rate is low, but even when they are added during the host's laying sequence hatching success tends to be lower and chicks lighter. Parasitic eggs usually seem to be added towards the end of the host's laying sequence. Since embryonic development in starlings starts before the clutch is complete, parasitic eggs may be among the last to hatch thus partly explaining their poor success rate. While there may be disadvantages arising from large broods, it is by no means clear that increased brood size is necessarily detrimental to the host. For example, if there is a dilution effect on the risk of predation or earlier warning of approaching danger, increased brood size may be an advantage and it may pay to accept parasitic eggs (Andersson, 1984).

The advantages of intraspecific brood parasitism *versus* raising its own chicks in female starlings are not clear since the balance between saved incubation effort and chick survival has not been measured quantitiatively. In some other species the advantages are more apparent. In wood duck, for example, the incubation period and breeding season as a whole are extensive (clutches are started over periods of 60 days in the north of the range and 140 days in the south) with the hen incubating alone. Some hens successfully breed twice in a year, the only duck species known to do so (Bellrose, 1976). The extended involvement in nesting and incubation appears to incur a substantial risk to females. Many wood duck nests are destroyed by predators with frequently less than 50% of clutches hatching. Predation of nesting females can be high: Andersson (1984) refers to a study in Illinois in which 103 of 1579 wood duck hens were taken by raccoons (*Procyon lotor*) with other predators inflicting further mortality. Annual mortality among wood duck, at 50–60% is higher than for any other North American duck (Andersson, 1984). The hazards of breeding probably mean it is highest for females and the wood duck, like many duck species, has a male-biased sex ratio (Bellrose *et al.*, 1964). Parasitic hens may thus enjoy enormous benefits in terms of survival by foisting their broods onto other females. As usual, however, there are costs. Hatching is often reduced in parasitized nests with more nests being deserted and a greater proportion of eggs in retained nests failing to hatch, particularly if parasitic eggs are added after the host starts incubating (e.g. Morse and Wight, 1969; but see Clawson *et al.*, 1979).

These examples of brood parasitism involve parasitic females exploiting, inter- or intraspecifically, the parental care of other females. Males too, of course, can be victims of parasitism where they provide exploitable parental care. The *sneaker* male bluegill sunfish discussed earlier effectively exploit the brood care offered by *parental* males when they sneak fertilizations at the latters' nests. As a result of sneaky fertilizations, *parental* males presumably forfeit a number of zygotes which would otherwise have been their own. Copulations arising from other exploitative sneak/satellite relationships between males (see above) may impose a similar reproductive cost on the victim.

Reduced longevity

Reproductive output may also be curtailed through reduced survival. Several of the effects of exploitation by scroungers that have been discussed – increased expenditure of resources, reduced returns, pathology and so on – can have a negative effect on producer longevity. Walkey and Meakins (1970), for example, found that three-spined sticklebacks (*Gasterosteus aculeatus*) infected with *Schistocephalus solidus* died much sooner if starved than uninfected fish. Milinski looks at the consequences of this for foraging decisions by infected and uninfected fish in Chapter 6.

While parasites may kill their hosts through depletion of resources, pathology or mechanical interference, they may also kill them in less direct ways, for instance through the manipulation of host behaviour to facilitate transmission to new hosts. Many instances of apparent parasitic manipulation involve engineering the ingestion of one host, or part of it, by another. Premature host death may thus be an integral part of the exploitative strategy of the parasite. Premature death may also be a form of host retaliation against infection. Smith Trail (1980) suggests that where prevention of infection fails or is not impossible, suicidal responses on the part of infected hosts may be adaptive if they protect the host's close kin from infection. Distinguishing between parasitic manipulation and host suicide as causes of premature death, however, is likely to be difficult and dependent on a detailed understanding of the fitness consequences for manipulation and suicide alleles.

The cost of counteradaptation

While hosts may counter the costly effects of exploitation, counter-responses may themselves be costly and divert resources away from other metabolic demands.

Parasites may stimulate sophisticated host immune responses which can involve a mixture of specific (molecular) and non-specific (largely cellular) responses to invasion. Investment in an effective immune system may detract from investment in other cellular or physiological processes and there is accumulating evidence that mounting an immune response can be costly (see Keymer and Read, in press). Counter-responses to parasites may extend beyond the molecular/cellular mechanisms of immunity. As Read discusses in Chapter 5, arms races between host and parasite may have far-reaching consequences for the evolution of physical appearance and sexual behaviour among hosts. Elaboration of plumage or other body covering characteristics as a sexually-selected strategy for advertising freedom from, and thus apparent resistance to, infection involves the host in the production of structures, pigmentation and associated display behaviours which may impose a considerable drain on metabolic resources. In this case, however, it could be argued that parasites are merely adding to an existing cost of sexually-selected investment driven by the fundamental producer/scrounger relationship inherent in sex itself (see above).

Behavioural responses by hosts in other exploitative relationships may also be costly. The behavioural counter-responses of avian hosts to kleptoparasitic gulls, such as evasion flights, increased flight speed and aerial manoeuvres (see Grant, 1971; Arnason and Grant, 1978, Barnard, 1984b; Barnard and Thompson, 1985), are likely to be energetically demanding as is the tendency for leopards (*Panthera pardus*) to drag their kills up into trees to avoid loss to thieving hyaenas (*Crocuta crocuta*). Counter-responses to brood parasitism, such as the tendency among subordinate female anis to lay more eggs as breeding group size increases and retain late eggs for laying after egg-removing dominants (Vehrencamp, 1977), may also be costly. The additional time and energy drain of counter-responses may have a profound effect on the decision-making of hosts in a variety of contexts (see Chapter 4).

ARMS RACES AND PRODUCER/SCROUNGER RELATIONSHIPS

The impact of scroungers on the survivorship and reproductive success of producers is likely to impose a selection pressure on producers to respond and reduce the impact. This in turn will select for countermeasures by the scrounger. The result is that producer and scrounger lineages become involved in an evolutionary arms race (see Dawkins and Krebs, 1979; Dawkins, 1982), a series of escalating mutual counteradaptations by the two lineages to exploit or prevent/counter exploitation by the other. While it seems clear that selection will result in a process of response and counter-response, however, the speed with which each lineage should respond to the other on an evolutionary time-scale and the extent to which it should respond is less obvious.

Producer/scrounger arms races are asymmetrical in the sense that the currency of adaptation in the two lineages is different (Dawkins and Krebs, 1979; Barnard, 1983). Since different adaptive currencies are used, asymmetrical races are likely to have built-in differences in the pace and magnitude of change by racing lineages simply because of differences in the metabolic pathways involved. However, even if differences in adaptive currency did not themselves give an advantage to one lineage, there are other reasons why a lineage may have an advantage over the others with which it is racing. Following Dawkins and Krebs (1979), we can identify at least three sources of imbalance.

Differences in evolutionary rates

Lineages may differ in the rate at which they evolve (e.g. Stanley, 1979). Races between producers and scroungers of different species may thus be biased if one lineage has an intrinsically greater rate of evolution. This is particularly likely when lineages are phylogenetically widely separated or have very different ecologies. In a race between lineages with differing rates of evolution, the faster evolving lineage may be at an advantage since it would enjoy a longer period before an adaptive step is countered.

Several factors may introduce an asymmetry in the relative rates of evolution of producer and scrounger lineages: differences in population size, the amount of genetic variation within populations, the degree to which populations are fragmented and generation time being obvious ones (Barnard, 1984b). Where scroungers compete for access to producers, producer populations may become partitioned between scrounger individuals. Such partitioning may be maintained by competitive spacing resulting in small populations of scroungers and scrounger life-history strategies that tend to the *K* extreme of the *r*/*K* spectrum relative to producers (as e.g. in some vertebrate klepto- and brood-parasitic relationships). Resultant differences in generation times may create the potential for higher rates of evolution among producers.

In parasitic infections, the generation time advantage is likely to be in the opposite direction and may be considerable. For example, the rate of evolution of viral genes, which is proportional to generation time, is, in at least one documented case, about a million times faster than the rate for the same genes in the host genome (O'Brien and Evermann, 1988). Furthermore, viruses have the opportunity to recombine with each other as well as with cellular genetic information (e.g. the recombinational 'capture' and viral-promoted transcription of oncogenes). Such rapid evolution and genetic plasticity is evident in a number of viruses. The influenza and HIV viruses, for instance, display extreme genetic variation between epidemics and individual viral isolates (O'Brien and Evermann, 1988). Although parasites may enjoy a generation time advantage, the overdispersion which is characteristic of many parasite populations may mean that they adapt to only a small proportion of host genotypes in each generation and are thus limited in their chances of transmission to new hosts. Asexual modes of reproduction within successfully colonized hosts may compound the problem in populations of sexual hosts.

As we have seen, one of the consequences of exploitation for producers may be reduced reproductive output. This can arise through adaptive (to the scrounger) interference with the reproductive anatomy, physiology or behaviour of the producer, as a negative consequence of scrounger-driven investment in the other components of life history or increased offspring mortality. Reduced adult longevity among producers may also limit reproductive output. Reduced rates of reproduction within producer populations will limit the potential for counter-response and increase the period of grace experienced at each stage of the arms race by the scrounger lineage.

Differences in adaptation budgets

Adaptations are costly; they use up metabolic resources. It is thus reasonable to assume that resources invested in one adaptation are not available, or are available in reduced quantities, for investment in other adaptations. In this sense, organisms can be thought of as having a limited adaptation 'budget' (Dawkins and Krebs, 1979) which is dispensed according to the various selection pressures operating on it. The extent to which a lineage can invest in any one adaptation is therefore likely to be compromized by the strength of selection for other adaptations (Dawkins and Krebs, 1979; Barnard, 1984b). This has important consequences for the degree to which each lineage can adapt to the other. In the last section, some counteradaptations by producers were discussed as one of a number of costs inflicted by or arising as a result of the impact of scroungers. Scroungers thus create a drain on producer adaptation budgets which channel resources away from investment in other directions. By the same token, counteradaptation by producers increases the cost of exploitation to scroungers which therefore have to invest in a further round of the arms race, leaving fewer resources available for other activities such as reproducing. Differences between lineages in the resources available for counteradaptation may give an advantage to one side in any particular race. Which lineage benefits, however, is likely to vary across producer/scrounger relationships.

A generalist scrounger involved in arms races with several different producer species is unlikely to show finely-tuned adaptation to any one of them because selection will be exerting conflicting demands on its resources and limiting its response in any particular direction. The conflict in demand may be reduced if producers are phylogenetically closely related or perform similar resource-procuring activities so that similar adaptations are selected for in each case. This may be an important constraint on the range of relationships in which scroungers can become involved.

In many host–parasite relationships, hosts are infected with several species of parasite. Within hosts, therefore, parasites are adapting to a single competing lineage and, within the constraints imposed by adapting to any other hosts they exploit, can specialize. The opposite is true for the host. The range of species with which they are infected means that hosts are racing against diverse adaptations which are likely to compromize their response to any one of them. Host resources are thus spread thinly across parasites while parasite resources are concentrated against hosts. Such asymmetries in resource allocation in a race are likely to put parasites at an advantage. Similar generalization/specialization asymmetries may exist in other interspecific exploitative relationships.

Selection pressures and the commitment of resources may differ between lineages for other reasons. If a producer lineage is exploited by a scrounger only rarely, for example, selection for counter-responses may be weak. Such a 'rare enemy effect' may explain, among other things, why the victims of slavery among ant species appear to have little in the way of effective defences against exploitation and why some hosts of interspecific brood parasites among birds are poor egg discriminators (Dawkins, 1982; but see Davies and Brooke, 1989).

Differences in the cost of failure

Another cause of asymmetry in selection pressure is the cost of failure at a counteradaptive stage. In host–parasite arms races, for instance, parasites which fail to establish in or on a host are likely to die and their reproductive potential is reduced to zero. Hosts which fail to avoid or reject a parasite, however, may suffer only mild discomfort or temporary illness; their reproductive potential may be reduced but not by much. 'Failed' hosts may still contribute offspring to the next generation whereas 'failed' parasites almost certainly will not. As a result selection is likely to act more strongly on parasites to resist rejection by hosts than on hosts to resist attack by parasites. To that extent the parasite lineage has an advantage. By analogy with similar asymmetries in predator–prey relationships, this has been dubbed the 'Life-Dinner Principle' (see Dawkins and Krebs, 1979).

While the Life-Dinner Principle may determine the long-term outcome of producer/scrounger relationships in terms of winners and losers, it may also help to explain some of the characteristics of relationships. One of the striking features of some interspecific brood parasitic relationships among birds is the gross mismatch in size and appearance between the chicks of the parasite and those of the host. Why cannot the host recognize the parasite straight away and reject or abandon it, especially since the same hosts are often good enough at the (presumably more difficult) task of discriminating against parasitic eggs, which forces parasites into detailed egg mimicry? As Dawkins (1982) points out, the answer may lie in the cost of parasitism to the host at different stages of the breeding cycle. Early in the cycle, at the egg stage, the benefit of detecting a parasite is the potential gaining of an entire breeding cycle in the future. The benefit of detecting a nearly fledged parasite is the saving of only a few days and at a time when it may be too late to breed again that year anyway. Selection may thus operate more strongly on host discrimination at the egg stage when the potential cost of parasitism is greatest. The astonishingly convincing egg mimicry among some brood parasites is apparent testimony to such discrimination and its counteradaptive consequences. While the differential cost argument may explain egg mimicry, it is still not clear why hosts show no discrimination against an offspring as obviously alien as is, say, a cuckoo chick to a small passerine parent. After all, time and resources could be saved even if the breeding opportunity that year is lost. One answer may be that the parasite exploits the potentially high cost to the host of withholding parental care from a begging chick on the off-chance that it might be a parasite. If the host gets it wrong it loses its offspring. Parasitic chicks may have taken advantage of this risk factor by exaggerating, via a huge gape and loud begging call, the stimuli which elicit parental feeding among passerine hosts. Dawkins (1982) uses the metaphor of manipulation to characterize the relationship, viewing the effect of the begging display of the cuckoo chick as analogous to that of an addictive drug. That the cuckoo's begging display is highly effective is evidenced by the fact that fledgling European cuckoos (*Cuculus canorus*) sometimes fool adult birds which are not their foster parents into feeding them (Dawkins, 1982).

ACKNOWLEDGEMENTS

I thank Jerzy Behnke and Peter Aldhous for helpful comments on the manuscript.

REFERENCES

Andersson, M., 1984, In *Producers and Scroungers: Strategies of Exploitation and Parasitism* edited by C.J. Barnard (London: Chapman and Hall).
Anderson, M. and Eriksson, M. O. G., 1982, *The American Naturalist*,**120**, 1–16.
Arak, A., 1984, In *Producers and Scroungers: Strategies of Exploitation and Parasitism* edited by C.J. Barnard (London: Chapman and Hall), pp. 154–194.
Arme, C. and Owen, R. W., 1968, *Journal of Parasitology*, **54**, 272–280.
Arnason, E. and Grant, P. R., 1978, *Ibis*, **120**, 38–54.
Bailey, G. N. A., 1975, *International Journal of Parasitology*, **5**, 609–613.
Barnard, C.J., 1978, *Aspects of Winter Flocking and Food-fighting in the House Sparrow (Passer domesticus domesticus)*. D Phil Thesis, Oxford University.
Barnard, C.J., 1983, *Animal Behaviour: Ecology and Evolution* (London: Croom Helm).
Barnard, C.J., 1984a. *Producers and Scroungers: Strategies of Exploitation and Parasitism* (London: Chapman and Hall).
Barnard, C.J. 1984b, In *Producers and Scroungers: Strategies of Exploitation and Parasitism*, edited by C.J. Barnard (London: Chapman and Hall), pp. 95–126.
Barnard, C.J., 1984c, In *Producers and Scroungers: Strategies of Exploitation and Parasitism*, edited by C.J. Barnard (London: Chapman and Hall), pp. 6–33.
Barnard, C.J. and Burk, T.E., 1979, *Journal of Theoretical Biology*, **81**, 65–74.
Barnard, C.J. and Sibly, R.M., *Animal Behaviour*, **29**, 543–550.
Barnard, C.J. and Thompson, D. B. A., 1985, *Gulls and Plovers: the Ecology and Behaviour of Mixed-species Feeding Groups*. (New York: Columbia University Press).
Batra, L. R. and Batra, S. W. T., 1985, *Science*, **228**, 1011–1013.
Baudoin, M., 1975, *Evolution*, **29**, 335–352.
Behnke, J. M., 1987, *Advances in Parasitology*, **26**, 1–71.
Bell, G., 1978, *Journal of Theoretical Biology*, **73**, 247–270.
Bellrose, F. C., 1976, *Ducks, Geese and Swans of North America* (Harrisburg, Pennsylvania: Stackpole Books).
Bellrose, F. C., Johnson, K. L. and Meyers, T. U., 1964, *Journal of Wildlife Management*, **28**, 661–676.
Brockmann, H.J. and Barnard, C.J., 1979, *Animal Behaviour*, **27**, 487–514.
Brockmann, H.J. and Dawkins, R., 1979, *Behaviour*, **71**, 203–245.
Campbell, W. C., 1983, in *Trichinella and Trichinosis* edited by W. C. Campbell (New York: Plenum Press).
Carayon, J., 1974, *Comptes rendus des Séances de L'Academie des Sciences Paris D*, **278**, 2803–2806.
Cheng, T. C., 1973, *General Parasitology* (New York: Academic Press).
Clawson, E. L., Hartman, G. W. and Frederickson, L. H., 1979, *Journal of Wildlife Management*, **43**, 347–348.
Crompton, D. W.T., 1973, *Biological Reviews*, **48**, 27–83.
Davies, N. B. and Brooke, M. de L., 1989, *Journal of Animal Ecology*, **58**, 225–236.
Davison, G. W.H., 1982, *Zeitschrift fur Tierpsychologie*, **58**, 185–202.
Dawkins, R., 1980, In *Sociobiology: Beyond Nature/Nurture?* edited by G. W. Barlow and J. Silverberg (Boulder: Westview Press), pp. 331–367.
Dawkins, R., 1982, *The Extended Phenotype: the Gene as the Unit of Selection* (London: Freeman).

Dawkins, R. and Krebs, J. R., 1979, *Proceedings of the Royal Society of London B*, **205**, 489–511.
Dence, W. A., 1958, *Journal of Parasitology*, **44**, 334–338.
Edwards, J. C., 1988, *Animal Behaviour*, **36**, 529–540.
Edwards, J. C. and Barnard, C. J., 1987, *Animal Behaviour*, **35**, 533–540.
Evans, P. G. H., 1980, *Population Genetics of the European Starling Sturnus vulgaris*. D Phil Thesis, University of Oxford.
Feare, C. J., 1983, *The Starling* (Oxford: Oxford University Press).
Furness R. W., 1978, *Animal Behaviour*, **26**, 1167–1177.
Furness, R. W., 1987, In *Seabirds: Feeding Ecology and Role in Marine Ecosystems*, edited by J. P. Croxall (Cambridge: Cambridge University Press).
Gaston, A. J., 1976, *Journal of Animal Ecology*, **45**, 331–348.
Giraldeau, L-A., 1984, *American Naturalist*, **124**, 72–79.
Giraldeau, L-A. and Lefebvre, L., 1986, *Animal Behaviour*, **34**, 797–803.
Grant, P. R., 1971, *Behaviour*, **40**, 263–281.
Gross, M. R., 1982, *Zeitschrift fur Tierpsychologie*, **60**, 1–26.
Hamilton, W. D. and Zuk, M., 1982, *Science*, **218**, 384–387.
Hartnoll, R. G., 1967, *Journal of the Linnean Society of London* (*Zoology*), **46**, 275–296.
Heusmann, H. W., Belville, R. and Burrell, R. G., 1980, *Journal of Wildlife Management*, **44**, 908–915.
Hinton, H. E., 1964, *Symposium of the Royal Entomological Society of London*, **2**, 95–107.
Hoekstra, R., 1980, *Journal of Theoretical Biology*, **87**, 785–793.
Hogan-Warburg, A. J., 1966, *Ardea*, **54**, 109–229.
Hölldobler, B., 1970, *Zeitschrift fur Vergleichende Physiologie*, **66**, 215–250.
Holmes, J. C. and Bethel, W., 1972, In *Behavioural Aspects of Parasite Transmission* edited by E. U. Canning and C. A. Wright (London: Academic Press), pp. 123–149.
Howard, R. D., 1978 *Evolution*, **32**, 850–871.
Iwata, K., 1976, *The Evolution of Instinct: Comparative Ethology of Hymenoptera* (New Delhi: Amerind)
Jeanne, R. L., 1972, *Bulletin of the Museum of Comparative Zoology*, **144**, 63–150.
Källander, H., 1977, *Fauna och Flora*, **75**, 200–207.
Kamil, A. C., Krebs, J. R. and Pulliam, H. R., 1987, *Foraging Behavior* (New York: Plenum Press).
Karlsson, J., 1983, *Breeding of the Starling (Sturnus vulgaris L.)*. PhD Thesis, Sweden: University of Lund.
Kennedy, C. R., 1984, In *Producers and Scroungers: Strategies of Exploitation and Parasitism* edited by C. J. Barnard (London: Chapman and Hall), pp. 34–60.
Keymer, A. F. and Read, A. F., In *Parasitism: Coexistence or Conflict?* edited by C. A. Toft and A. Aeschlimann (Oxford: Oxford University Press), in press.
Lack, D., 1948, *Evolution*, **2**, 95–110.
Leuthold, W., 1966, *Behaviour*, **27**, 215–258.
McMillan, R. P., 1965, *West Australian Naturalist*, **13**, 96.
Maynard Smith, J., 1982, *Evolution and the Theory of Games* (Cambridge: Cambridge University Press).
Milinski, M., 1984, *Behavioural Ecology and Sociobiology*, **15**, 35–37.
Moore, J., 1983, *Parasitology*, **69**, 1174–1176.
Moore, J., 1984, *The American Naturalist*, **123**, 572–577.
Morse, T. E. and Wight, H. M., 1969, *Journal of Wildlife Management*, **33**, 284–293.
Mountfort, G. and Ferguson-Lees, I. J., 1961, *Ibis*, **103a**, 86–109.
O'Brien, S. J. and Evermann, J. F., 1988, *Trends in Ecology and Evolution*, **3**, 254–259.
Parker, G. A., 1978, In *Behavioural Ecology: an Evolutionary Approach* edited by J. R. Krebs and N. B. Davies (Oxford: Blackwell).
Parker, G. A., 1983, In *Current problems in sociobiology* edited by King's College Sociobiology Group, (Cambridge: Cambridge University Press).
Parker G. A., 1984, In *Producers and Scroungers: Strategies of Exploitation and Parasitism*, edited by C. J. Barnard (London: Chapman and Hall), pp. 127–153.
Parker, G. A., Baker, R. R. and Smith, V. C. F., 1982, *Journal of Theoretical Biology*, **36**, 529–553.
Parker, G. A. and Simmons, L. W., 1989, *Ethology*, **82**, 3–26.
Payne, R. B., 1973, *Ornithological Monographs*, **11**, 1–333.
Payne, R. B., 1977, *Annual Review of Ecology and Systematics*, **8**, 1–28.
Platnick, N. I. and Shadab, M. U., 1978, *American Museum Novitates*, **2661**, 1–22.
Price, P. W., 1980, *Evolutionary Biology of Parasites* (Princeton: Princeton University Press).

Rau, M. E., 1983a, *Parasitology*, **86**, 311–318.
Rau, M. E., 1983b, *Parasitology*, **86**, 319–322.
Read, A. F., 1987, *Nature*, **328**, 68–70.
Rempel, J. G., 1940, *Journal of Experimental Zoology*, **84**, 261–289.
Rettenmeyer, C. W., 1962, *Journal of the Kansas Entomological Society*, **35**, 377–384.
Robinson, M. and Robinson, B., 1977, *Psyche*, **87**, 337–348.
Rohde, K., 1979, *American Naturalist*, **114**, 648–671.
Rohwer, S. and Ewald, P. W., 1981, *Evolution*, **35**, 441–454.
Rothschild, M., 1941, *Parasitology*, **33**, 406–415.
Schad, G. A., 1963, *Nature*, **198**, 404–406.
Sibly, R. M., 1984, In *Producers and Scroungers: Strategies of Exploitation and Parasitism*, edited by C. J. Barnard (London: Chapman and Hall), pp. 267–287.
Sivinski, J. and Stowe, M., 1980, *Psyche*, **87**, 337–348.
Smith Trail, D. R., 1980, *The American Naturalist*, **116**, 77–91.
Stanley, S., 1979, *Macroevolution: Pattern and Process* (San Francisco: Freeman).
Stephens, D. W. and Krebs, J. R., 1986, *Foraging Theory* (Princeton: Princeton University Press).
Sudd, J. H. and Franks, N. R., 1987, *The Behavioural Ecology of Ants* (New York: Blackie/Chapman and Hall).
Thornhill, R., 1975, *Nature*, **258**, 709–711.
Thornhill, R., 1976, *American Naturalist*, **110**, 529–548.
Tinbergen, J. M., 1981, Foraging decisions in starlings (*Sturnus vulgaris L*), *Ardea*, **69**, 1–67.
Van Rhijn, J. G., 1973, *Behaviour*, **47**, 153–299.
Vehrencamp, S., 1977, *Science*, **197**, 403–405.
Vehrencamp, S., 1978, *Behavioural Ecology and Sociobiology*, **4**, 1–33.
Vehrencamp, S., 1983, *Animal Behaviour*, **31**, 667–682.
Vehrencamp, S., 1984, In *Producers and Scroungers: Strategies of Exploitation and Parasitism*, edited by C. J. Barnard (London: Chapman and Hall), pp. 229–266.
Vollrath, F., 1978, *Psyche*, **85**, 347–353.
Vollrath, F., 1984, In *Producers and Scroungers: Strategies of Exploitation and Parasitism*, edited by C. J. Barnard (London: Chapman and Hall), pp. 61–94.
Walkey, M. and Meakins, R. H., 1970, *Journal of Fish Biology*, **2**, 361–372.
Wang, Der I. and Moeller, F. E., 1970, *Journal of Economic Entomology*, **63**, 1539–1541.
Weller, M. W., 1959, *Ecological Monographs*, **29**, 333–365.
Westerterp, K., Gortmarker, W. and Wijngaarden, H., 1982, *Ardea*, **70**, 153–162.
Wheeler, W. M., 1910, *Journal of Experimental Zoology*, **8**, 337–438.
Wilson, E.O., 1971, *The Insect Societies* (Cambridge, Massachusetts: Harvard University Press).
Wilson, K. and Edwards, J. C., 1986, *Animal Behaviour*, **34**, 942–944.
Yom-Tov, Y., Dunnet, G. M. and Anderson, A., 1974, *Ibis*, **116**, 87–90.

2. Pathology and host behaviour

J. C. Holmes and S. Zohar

INTRODUCTION

Many parasites damage their hosts at least some of the time; consequently, there is a vast literature on pathology, most of which emphasizes the physical, biochemical or cellular mechanisms by which organisms damage their hosts (Mackenzie *et al.*, 1987). There is also a rapidly increasing literature on the influences parasites have on the behaviour of their hosts. However, to our knowledge, there has been no attempt to look for patterns relating the kind of damage done to the behavioural consequences of that damage.

Minchella (1985) discussed changes in growth, physiology and the behaviour of snails infected with sporocysts and rediae of trematodes and distinguished three categories of changes: those that benefit the parasites (and therefore may have arisen as parasite adaptations), those that benefit the host (and therefore may have arisen as host adaptations), and those that are side-effects of the infection, with no direct adaptive significance. All three categories have been recognized in other systems (Anderson and May, 1982; Ewald, 1983; Moore, 1984; Hart, 1988), and the distinctions appear useful, even if difficult to make in particular cases (Rowland and Boersma, 1988).

In this chapter, our approach will be to distinguish what we think are the principal general modes of pathology that influence host behaviour: organ malfunction due to direct tissue damage, impaired nutrition and modulation of the neuroendocrine control systems. Thompson discusses the relationship between behavioural changes and physiology in more detail in Chapter 3. Because individual species of parasites can have many effects on their hosts these modes are not independent but they do differ in their principal behavioural consequences. The literature is far too voluminous to attempt a comprehensive review. Instead, we will focus on a few examples selected to illustrate principles and drawn largely (but not exclusively) from the parasitic helminths, the parasites with which we are most familiar; we will also refer to reviews that cover other parasites. Finally, we will discuss briefly the adaptive significance of each type of pathology and the behavioural changes that result.

ORGAN MALFUNCTION

The most direct mode of pathology occurs when a parasite localizes in an organ and interferes with the function of that organ mechanically or through destruction of its tissue. A parasite may occupy space, blocking a tube or exerting pressure on surrounding tissues, or destroy tissues by penetrating through them, by feeding on them, by secreting lytic or toxic substances, or by stimulating the immune or repair responses of the host. Almost all parasites cause some organ malfunction especially when they are present in large numbers, and malfunctions of most organs affect host behaviour directly by affecting receptors, central processing units or effectors or indirectly by affecting nutrition or other features of general body condition. In this section, we focus on those that have direct effects on behaviour through malfunctions of the sensory organs, other parts of the nervous system or the musculature.

Sensory organs

A variety of parasites are located solely or primarily in sensory structures. The most familiar examples are diplostomatid metacercaria (eye flukes) which inhabit the lens, vitreous humor or retina of freshwater fishes, and can cause severe histopathological changes, including cataract formation, lens dislocation, rupture of the lens capsule or detachment of the vitreous humor or retina (Shariff *et al.*, 1980). Light infections with metacercariae of *Diplostomum spathaceum*, which inhabit the lens, have little apparent effect, but severe infections can blind the fish (Erasmus, 1958) and even moderate infections can reduce visual acuity, reducing feeding efficiency and escape responses, and increasing time spent feeding in surface waters, at least in the laboratory (Crowden and Broom, 1980). Other diplostomid metacercariae, such as *D. adamsi*, are found within the retina and cause displacement and detachment of the photoreceptor cells, and must also reduce visual acuity (Lester and Huizinga, 1977).

Nematodes of the genus *Stenurus* localize in the head sinuses and inner ear bullae of cetaceans (Dailey and Stroud, 1978); those in the bullae may interfere with echolocation and orientation (Howard *et al.*, 1983), and this has been suggested as one possible cause of the stranding of parasitized whales and dolphins (Delyamure, 1955; Dailey and Walker, 1978; but see Geraci and St. Aubin, 1987).

Trypanosoma congolense epimastigotes attach by hemidesmosomes to the labral mechanoreceptors of the fly *Glossina*, impairing the sensory function and feeding behaviour of infected flies (Thevenaz and Hecker, 1980). The impairment of normal receptor function, plus the accumulation of trypanosomes in the labrum, also decreases the flow within the food canal (Livesey *et al.*, 1980). Infected flies probe more frequently and take longer to engorge than uninfected flies (for an extensive review see Molyneux and Jefferies, 1986). The mathematical models of Rossignol and Rossignol (1988) indicate that these features increase the transmission rates for the parasites.

A wider variety of parasites usually inhabit other organs, but a small proportion of individuals may occasionally (or even frequently) invade sensory structures, especially the eyes (for partial reviews see O'Connor, 1976; Vercruysse *et al.*, 1988). The most familiar examples are larval ascarids, especially *Toxocara canis* (which can destroy the optic nerves), the microfilariae of *Onchocerca volvulus* (which cause 'river blindness' in humans in Africa) and the 'eyeworm' *Loa loa* (which causes discomfort but no permanent damage). As the examples indicate, some but not all, of these occasional invaders markedly impair sensory function.

These examples, and others, suggest that slight impairment of the sensory abilities of an organism may have little effect but moderate to severe impairment (by larger numbers of parasites) can compromise the host's ability to assess or exploit its environment and may lead to an array of behavioural alterations ranging from impaired feeding efficiency and altered escape behaviour to disorientation.

Central nervous system

Relatively few parasites regularly inhabit the central nervous system (CNS); for a review see Vercruysse *et al.* (1988). Many of the parasites involved are microparasites (small organisms like viruses or protozoans, which multiply within an individual host). Some familiar examples are the rabies virus, which markedly affects the behaviour of infected hosts, and the coccidian *Toxoplasma gondii*, which invades and destroys cells in a wide variety of organs but shows a predilection for the CNS (Hutchison *et al.*, 1980a). Behavioural consequences of infection with *T. gondii* include impaired learning performance (Witting, 1979; but see Hay *et al.*, 1984b), impaired motor performance or loss of coordination (Hay *et al.*, 1983a), reduced responsiveness to environmental stimuli (Hay and Hutchinson, 1983; Hay *et al.*, 1984a) and altered activity levels (Hutchinson *et al.*, 1980b; Hay *et al.*, 1983c).

Adult helminths rarely inhabit nervous tissue. The nematode *Parelaphostrongylus tenuis* and the related *Elaphostrongylus rangiferi* (see Kummeneje, 1974) are well-known exceptions. Larvae migrate through the spinal cord and emerge and mature in the subdural spaces of the brain (Anderson, 1964). In the normal host, white-tailed deer (*Odocoileus virginianus*), larvae migrate rapidly, and infections with few *P. tenuis* cause little damage but infection with many worms can cause extensive damage (W. M. Samuel, personal communication). In other cervids, such as moose (*Alces alces*), migration is slower and damage to the spinal cord and brain can be extensive resulting in visual impairment, circling, incoordination, ataxia, paraplegia and death (Anderson, 1964; Anderson and Prestwood, 1979).

Adult trematodes (and more frequently their ova) of the genus *Nasitrema* (normally present in the air sinuses) have been found in lesions in the cerebrum or cerebellum of stranded dolphins (Ridgway and Dailey, 1972; Dailey and Walker, 1978). Impaired echolocation due to the lesions has been implicated in causing those strandings; Geraci and St. Aubin (1987) regard this as one of the few examples of parasitism clearly linked to strandings.

Larval helminths are commoner inhabitants of the CNS. Metacercariae of several species of digeneans are found normally in the ventricles of the brain or in surrounding tissues of freshwater and marine fishes. Fish infected with some species, such as *Diplostomum phoxini* (Rees, 1955), or with low numbers of metacercariae, may show no effects; those infected with larger numbers or with other species, such as *Ornithodiplostomum ptychocheilus*, may be slower to react and the schools of fish become less cohesive (Radabaugh, 1980a, b), and those infected with still larger numbers or other species, such as *Austrodiplostomum mordax*, may be almost completely incapacitated and uncoordinated, tumbling about on the surface of the water (Szidat, 1969; see Holmes and Bethel, 1972 for other examples).

Metacercariae (such as those of *Dicrocoelium* spp. or *Microphallus papillorobustus*) also occur in the CNS of invertebrate hosts (see review in Helluy, 1983). In most cases, only one or two metacercariae are found in the brain where they are usually unencysted and (at least for *D. dendriticum*: Hohorst, 1964), uninfective to the definitive host. The majority of the metacercariae in each invertebrate are encysted in the haemocoel and are infective. The metacercariae in the brain are associated with v aried, rather specific, altered responses to environmental stimuli, which increase the probability of transfer of the other, infective, metacercariae, to the definitive host: a possible example of self-sacrificial altruism. Because metacercariae in different locations elicit the same behaviour patterns (Romig *et al.*, 1980), the altered behaviour is probably produced by the modulation of control systems (covered below) rather than direct tissue damage.

Many other larval helminths are found in nervous tissue only occasionally or under special circumstances (for partial reviews see Sprent, 1955; O'Connor, 1976; Vercruysse *et al.*, 1988). The best known examples are the larvae of ascarid (or other) nematodes that produce visceral larva migrans in facultative intermediate hosts. Larvae of *Toxocara canis* in mice have been particularly well-studied. These larvae migrate through a wide variety of tissues but accumulate in various parts of the brain, especially myelinated sites (Summers *et al.*, 1983), presumably because the brain is an 'immunologically privileged site' where the parasite is protected against immunological attack (Dunsmore *et al.*, 1983). Infected mice show learning and motor deficiencies (Olson and Rose, 1966; Dolinsky *et al.*, 1981), are less reactive to changes in palatability of diets (Donovick *et al.*, 1981), are hyperactive (Hay *et al.*, 1986), and show preferences for novel areas in maze tests (Hay *et al.*, 1983b; Hay and Aitken, 1984).

Vercruysse *et al.* (1988) noted that most parasites, especially the facultative invaders, have no specific selectivity for any part of the brain and can therefore produce a variety of specific clinical signs. Possibly for the same reason, the severity of the effects of damage to the CNS does not always correlate well with the amount of damage (Hay and Aitken, 1984; Hay *et al.*, 1983a, 1984b). However, damage done to the CNS should be expected to produce sensory, motor or associational deficiencies. More subtle behavioural changes, such as altered but coordinated patterns of behaviour, may be due to modulation of the neuroendocrine system, discussed below.

Musculature

Many parasites inhabit the musculature of both invertebrates and vertebrates (for a comprehensive review of protozoan and helminthic infection of muscles see Garnham, 1972). However, few parasites establish in muscle tissue exclusively (Despommier, 1976). Although parasites may become established in smooth, cardiac or skeletal muscles (Garnham, 1972), the parasites in the skeletal muscles are most likely to have a direct effect on the behaviour of the affected host (others affect behaviour through their effects on nutrition, see later).

Garnham (1972) noted that most parasites live in muscle tissue without causing much local reaction. Commonly, tissue pathology is observable only in newly-established infections in which muscle cells are being penetrated, infections that persist for a long time with periodic rupture of muscle cysts and invasion of new muscle fibres, or old infections with dying and degenerating parasites that result in a widespread inflammatory response (Garnham, 1972). However, extensive infections with relatively large parasites (such as the coenuri of *Taenia multiceps* or larvae of *Eustrongylides* spp.) may reduce stamina (see references in Holmes and Bethel, 1972).

Where there is muscle pathology, it should result in reduced activity of the infected host, as shown for mice infected with the nematode *Trichinella spiralis* (Rau, 1983; Rau and Putter, 1984), or kestrels (*Falco sparverius*) infected with *T. pseudospiralis* (Saumier *et al.*, 1988). The reduction in activity levels of the infected mice and kestrels is associated with the penetration of the muscle fibres by *Trichinella* and is manifest for at least 3 months following infection (Zohar and Rau, 1984, 1986; Saumier *et al.*, 1988) (but not at 1 year after infection: Zohar, 1986). Metacercariae of the 'salmon poisoning' trematode, *Nanophyetus salmincola*, also decrease the swimming ability of juvenile salmonid fishes, but only up to 96 hours following infection, the period during which migrating cercariae destroy muscle tissue and cause subcutaneous haemorrhages. Metacercariae that had completed their encystment in muscles had little effect on the swimming ability of the fish (Butler and Millemann, 1971).

Many species of microfilarial worms have a period of growth in the musculature of their insect intermediate hosts, after which they emerge and mature outside the muscle cell (Garnham, 1972). Larvae of *Brugia pahangi* have been shown to reduce the flight activity of infected *Aedes aegypti* and *A. togoi* (Townson, 1970; Husain and Kershaw, 1971). More specifically, Rowland and Lindsay (1986) showed that there is a transitory decline in the flight activity of *A. aegypti* parasitized with the microfilariae of *Brugia pahangi* when flight muscles are first invaded, and a dramatic decline in flight activity when the larvae emerge from the muscle cells (figure 2.1), coincident with major muscle damage (Becket, 1971).

Curiously, muscle invasion by some parasites has an opposite effect to the locomotory debilitation described above. Quinn *et al.* (1987) found that lemmings (*Dicrostonyx richardsoni*) infected with the muscle cysts of the protozoan *Sarcocystis rauschorum* show an increase in exploratory activity in the open field. Similarly, mice infected with *Trichinella pseudospiralis* exhibit normal exploratory

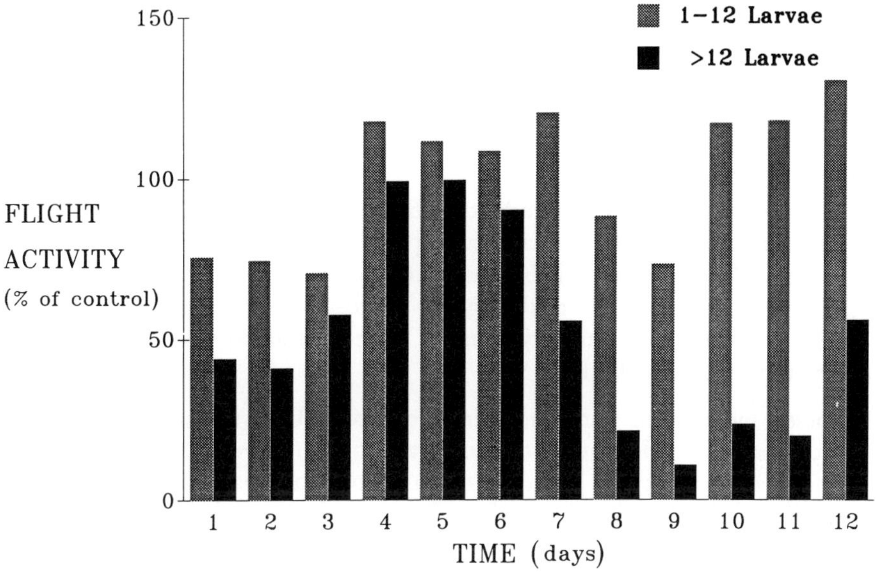

Figure 2.1. Influence of muscle pathology caused by migrating larvae of *Brugia pahangi* activity on the flight of *Aedes aegypti*. Larvae migrate into flight muscles on days 1 and 2 and out of flight muscles on days 7–9. Data on activity (as a percentage of controls) calculated from actograph scores plotted in figure 1 of Rowland and Lindsay (1986).

activity, but travel greater distances and spend less time inactive than control or lightly infected mice (Rau, 1984b), thus differing markedly from mice infected with *T. spiralis*. Presumably, such effects are due to mechanisms more subtle than malfunction of the muscles.

It may be concluded that only serious damage to muscle tissue, such as that caused by emerging microfilaria, penetrating cercariae, recent *Trichinella* infections, or inflammatory reactions to dead parasites, or infections with large numbers of large parasites, will cause observable behavioural changes. Malfunction of skeletal or flight muscles results in diminished activity. The implications of reduced activity levels may range from the inability to carry out normal maintenance activities, such as feeding, to alterations in social behaviour.

REDUCED NUTRITIONAL LEVELS

Parasites commonly reduce the nutritional levels in terms of both energy and materials of their hosts (see also Chapter 1). A nutritional deficit can be imposed by parasites found in many organs, but noticeable effects are exerted primarily by parasites in the body cavity (when the parasites are large in comparison with the size of the host), in the digestive tract (the nutrient acquiring system), or in the pulmonary or circulatory systems (the oxygen acquiring and distributing systems).

A parasitized host may incur a nutrient deficit by one or more of the following four mechanisms:

1. Parasites may cause direct or indirect nutritional drains. Parasites may compete with the host for energy or nutrients; damage host tissues, thereby stimulating costly repair responses; or otherwise stimulate energy- and nutrient-requiring host defensive responses.
2. Parasites may affect the assimilation efficiency of the affected host by altering gastrointestinal functions such as absorption or gut motility.
3. Infected hosts may eat less food (exhibit anorexia).
4. The acquistion and delivery of oxygen to the tissues can be impaired by parasites that are found in pulmonary tissue or in the circulatory system or that feed extensively on blood cells.

These mechanisms are not independent. The interactions between gastrointestinal function, appetite, nutrient or energy reserves, and immunological responses are complex and are currently attracting considerable attention (Castro, 1988; Blalock, 1989; Symons, 1989). They involve both local events (in the gastrointestinal tract) and hormonal or CNS events (Mettrick and Podesta, 1974; Blalock, 1989). In addition, the literature, particularly the veterinary literature, is replete with cases in which parasites are known to reduce productivity (weight gains, wool production, milk production), but in which the precise mechanisms are unknown (Dargie, 1987; Nansen, 1987; Symons, 1989). We shall therefore first look briefly at the effects of the four mechanisms, then assess the potential behavioural effects of reduced nutrient availability in general.

Energy or nutrient drain

Almost all parasites obtain their energy and nutrients from those of the host (or from digesta or bacteria present in the intestinal tract). Because in most hosts, at least most vertebrates, the mass of parasites present is small compared with that of their hosts, this uptake by the parasites is unlikely to be significant. Cases in which the parasites are large relative to the hosts are obvious exceptions. Most cases involve parasites in intermediate, usually invertebrate, hosts for which there is an extensive literature (e.g. cestode larvae in tubificid worms, Kennedy, 1972; *Plasmodium* in mosquitoes, Schiefer *et al.*, 1977; trypanosomes in *Glossina* flies, Bursell, 1981; trematode larval stages in snails, Lauckner, 1987).

One of the best-studied examples is that of the cestode *Schistocephalus solidus* in its intermediate host, the three-spined stickleback (*Gasterosteus aculeatus*) (Arme and Owen, 1967; Walkey and Meakins, 1970). Walkey and Meakins (1970) found that much of the energy taken in by the infected fish is redirected to the production of parasite tissue. This energy loss results in the depletion of the endogenous reserves of the host and a higher metabolic demand on the infected fish (Lester, 1971; Meakins, 1974; Meakins and Walkey, 1975), with reduced abilities to withstand anoxia (Smith and Kramer, 1987) or other environmental stresses (Pascoe and Mattey, 1977; Pascoe and Woodworth, 1980). Behavioural effects include altered habitat selection, greater time spent feeding, and reduced aversion to the risk of predation (e.g. Lester, 1971; Milinski, 1985; Godin and Sproul, 1988; see also Chapter 7).

An interesting case of nutrient drain in a colonial ant, *Leptothorax nylanderi*, was reported by Plateaux and colleagues. Ants infected with cysticercoids of the cestode *Anomotaenia brevis* are smaller, differ in colour and some aspects of body shape, are usually modified queens that rarely produce eggs, do not leave the nest and spend most of their time begging food from normal workers (Plateaux, 1972; Gabrion *et al.*, 1976). When such modified ants are housed with normal workers, the latter lay significantly fewer eggs, partly through decreased vitellogenesis and partly through retention of eggs, both attributed to the energy drain (and behavioural alterations) associated with feeding the parasitized ants (Salzemann and Plateaux, 1987). In this case, the infected ants become a nutrient drain on the entire colony.

Competition with the host for micronutrients, such as vitamins or trace minerals, is more likely to be significant. The best-known example is the uptake of vitamin B_{12} by the tapeworm *Diphyllobothrium latum*. Large worms located in the anterior intestine can absorb vitamin B_{12} secreted in the stomach prior to its reabsorption in the ileum, thus inducing a condition similar to pernicious anaemia (review in Hall, 1985). Vitamin B_{12} is apparently taken up rapidly by almost all other tapeworms (except cyclophyllideans) that have been studied; this ability is correlated with the ability to use B_{12}, probably as a cofactor associated with carbon dioxide fixation in energy metabolism (see review in Arme *et al.*, 1983). The widespread occurrence of this ability to concentrate vitamin B_{12} may indicate that parasites have similar abilities to concentrate other micronutrients not yet examined as thoroughly.

Similarly, iron deficiency anaemia (or deficiencies in blood proteins) may be produced by hookworms or other parasites that feed on blood, especially those that cause enteric haemorrhage (Hall, 1985; Holmes, 1987; Symons, 1989). Only 40–60% of the iron lost into the gut is reabsorbed (Rosenberg and Bowman, 1982).

Replacing lost blood (or other tissues) may also be a nutritional strain on the host. Mace and Davis (1972) reported that shorthorn sculpins (*Myoxocephalus scorpius*) parasitized by the leech *Malmiana nuda* expend more energy than uninfected fish. They attributed the energy loss not only to the leech's consumption of blood but also to an increase in the rate of blood and tissue formation.

Other defensive responses may be even more costly. Fever accompanying malaria can increase the basal metabolic rate of a human host by 40%, a considerable energy cost, especially if the host is malnourished (Hall, 1985). The cost of mounting an immune response has not been measured, to our knowledge, but must be high (Hall, 1985). Infections stimulate multiplication of many cell lines and extensive production of antibodies, responses massive enough to be evident by examining tiny samples of blood in capillary tubes (Levine and Wardlaw, 1988). Some parasites, such as *Trichinella spiralis* (Perrudet–Badoux and Binaghi, 1974) and *Schistosoma mansoni* (Fischer *et al.*, 1981), increase the cost by the way they attempt to evade the immune response. They non-specifically stimulate the production of cells and antibodies to a wide variety of antigens, thus reducing the capacity of the host to produce those specific to the parasite itself. Others, such as *Capillaria hepatica* (Wright, 1974) and *Fasciola hepatica* (Sukhdeo *et al.*, 1988), stimulate the immune response and feed on the cells attracted to them. (For a general discussion of situations in which immune responses are beneficial to the parasites see Damian, 1987.)

Reduced assimilation efficiency

The assimilation of nutrients by a host can be adversely affected by a wide variety of parasites, primarily those found in the gastrointestinal tract or associated glands. Reduced assimilation efficiency may be the result of poor digestion and/or malabsorption; examples of each, and the mechanisms by which parasites induce them, have been reviewed by Hall (1985) and Symons (1989). A particularly well-studied example is that of *Hymenolepis diminuta* in the laboratory rat. This tapeworm causes extensive pathology in the mucosa of the gut with villous atrophy and fusion (Martin and Holland, 1984), altered lumen pH and oxidation–reduction potential, markedly reduced absorption, and marked changes in permeability, nutritional and absorptive gradients, microfloral composition and abundance, and blood and liver chemistry (see review by Mettrick, 1980). Despite this evidence of severe local tissue damage and gut malfunction, whole-body studies indicate that digestive efficiency is not significantly affected (Mead and Roberts, 1972; Bailey, 1975; Insler and Roberts, 1976); however, Mettrick (1973) reported a decrease in efficiency.

In part, the lack of an overall effect is due to compensatory changes in the structure or function of the digestive system itself (Podesta and Mettrick, 1976). This example, plus evidence for compensatory changes in the intestine in other systems (Thomas, 1986; Stock and Holmes, 1987) suggests that the extent to which host production or behaviour is affected by reduced assimilation efficiency is likely to be small. Note, however, that these compensatory changes are costly in terms of energy and materials devoted to them (see also Chapter 1). Holmes (1987) concluded that impaired digestion and absorption in ruminants parasitized by nematodes (which provide some of the best examples of reduced assimilation) are not important causes of the poor utilization of nutrients, and that the primary cause is the increased metabolic demand on the host (including compensatory responses) as a result of the activities of the parasites.

Anorexia

Anorexia is a voluntary decrease in the food intake of an animal (Crompton, 1984), often associated with other physiological and behavioural features (Hart, 1988). The association between parasitism and host food intake has been the subject of several comprehensive reviews (Crompton, 1984; Hall, 1985; Symons, 1989). Anorexia may be caused by a wide range of protozoans and helminths (presumably through neural or hormonal means), but is most commonly associated with those that inhabit the gastrointestinal tract (through structural damage or functional alterations) (Crompton, 1984; Hall, 1985; Symons, 1989). The degree to which the host's food intake is reduced is often dependent on the density (number or biomass) of the parasite (Crompton, 1984; Symons, 1989). For example, the intakes of energy and protein of Nigerian children decreased significantly as the severity of *Ascaris* infection increased (Hussain, 1980). In addition, the degree of anorexia may be at least partially dependent on the level of protein nutrition (Abbott *et al.*, 1986). Symons (1989) concluded that some degree of anorexia is common in many parasitic infections but occurs more frequently, and in greater severity, in poorly-fed than in well-fed animals.

An anorexic animal may be lethargic, inactive or exhibit impaired physical performance (Hall, 1985). In the short-term, anorexia may be of some benefit to the parasitized host. Murray *et al.* (1978) and Hart (1988) suggested that a reduced food intake, coupled with inactivity, may allow more effective defense against infection (particularly with viruses or bacteria) or may suppress inflammation. However, long-term anorexia must reduce the ability of the host to respond to the infection by restricting nutrient (and especially protein) availability. Despite the extensive research done on anorexia, the interactions between parasites and appetite are still poorly understood, probably because of their complexity (Symons, 1989).

Altered oxygen delivery

Oxygen and nutrients are transported to the tissues by the circulatory and pulmonary systems. In some parasitic infections the hosts are capable of carrying on normal metabolic activities but when an additional metabolic demand (often a demand for energy) is imposed on the infected host, the delivery may be insufficient to supply such a demand.

Parasites that affect the oxygen-carrying capacity of the blood, such as malaria, hookworms or schistosomes, may affect the physical performance of their hosts (Hall, 1985). For example, *Plasmodium mexicanum* in fence lizards (*Sceloporus occidentalis*) markedly reduced running stamina, which is dependent on continuous oxygen delivery, but not burst running performance which is not (Schall *et al.*, 1982; Schall and Sarni, 1987). Pulmonary infections, especially in heavily parasitized hosts, can also affect host performance. Moose (*Alces alces*) with large numbers of hydatid cysts (*Echinococcus granulosus*) are more likely to be killed by hunters (Rau and Caron, 1979), and probably by wolves (Mech, 1966). Similarly, extensive damage to lung tissues due to infections with *Protostrongylus* spp. in bighorn sheep (*Ovis canadensis*) can markedly reduce their ability to run (Geist, 1971). Such pulmonary infections are probably of little consequence to resting animals but may compromise the vigour of stressed animals (Uhazy *et al.*, 1973;McNeill and Rau, 1987). Such alterations in the efficiency of oxygen delivery are probably manifested only at times of high and continuous oxygen demand, during periods of high activity or when the host is under stress.

Behavioural consequences

The assimilated energy and nutrients available to any organism are usually considered to be allocated to basic metabolism (essentially maintenance of structural and functional elements, including maintenance of body temperature for endotherms), voluntary activity, production (growth and reproduction) or storage. Although a shortage of energy or nutrients can affect any of these features, and severe shortages will affect all, less severe shortages usually affect one (or more) of the last three. The behavioural consequences depend on which features are affected, and in some cases, what mechanism is involved.

Reductions in voluntary activity (or changes in the nature of such activities) are obviously of behavioural significance; reduced activity, lethargy and depression are common symptoms of active disease. Some of these reductions appear to be due to analgesia induced by the neuroendocrine system (discussed below), some are associated with direct effects on muscles, and others appear to be due to reduced availability of energy. Examples of the latter are the 15% reduction in flight activity in *Glossina* flies infected with trypanosomes (Bursell, 1981) or the lethargy and reduced antimosquito activity in rodents with malaria (Day and Edman, 1983; Day *et al.*, 1983; Burkot, 1988). Dobson (1988) indicates that the authors of several papers (references in Dobson) 'mention that infected hosts behave in a fashion similar to that of uninfected hosts that have either recently engaged in exhausting aerobic physical activity or have been nutritionally stressed', with obvious effects on escape or foraging behaviour.

Effects on growth have been reported primarily in the veterinary literature; their behavioural implications have not been investigated. However, such effects are likely to be significant for size-related behavioural patterns, particularly if the 'growth' affected is that of 'showy' secondary sexual characteristics (Hamilton and Zuk, 1982; see Chapter 5). In this connection, the Hamilton–Zuk hypothesis appears to be based on an assumption that allocation of energy (or nutrients) to such showy characteristics is of low priority, thus reductions in 'showiness' would be early manifestations of reduced nutritional availability. That assumption deserves careful evaluation; Kodric-Brown and Brown (1984) and others provide arguments and evidence for such 'truth in advertising', but the data of Schall (1986) and Ressel and Schall (1989), who found that male lizards parasitized by blood protozoans were more brightly coloured (and resembled older males) suggest that it may not always be true.

The same 'truth in advertising' concept was applied by Kodric-Brown and Brown (1984) to energy-demanding courtship or maintenance of breeding territories; nutritional deficiencies should rapidly result in loss of status or territory. Later in the reproductive season, nutritional deficiences may reduce provisioning of eggs or embryos, or parental care. Reductions in either are likely to have significant effects on the behaviour of the juveniles. An excellent example has been provided by Rau (1985). Young mice born to or raised by mothers infected with *Trichinella spiralis* wandered from their nest boxes more frequently than those born to and raised by uninfected mothers. The pre- and postnatal effects appeared additive, because those born to and raised by infected mothers were the most severely affected. Rau suggests that these second-generation effects may have long-lasting consequences for the competitive fitness of the uninfected young. Similarly, lambs of bighorn ewes (*Ovis canadensis*) heavily infected with lungworms (*Protostrongylus* spp.) suckled less frequently and for shorter times, played less and had higher mortality rates than lambs of less heavily-infected ewes (Festa-Bianchet, 1988). In addition, rabbits (*Oryctolagus cuniculus*) infected with *Trichostrongylus retortaeformis* produced less milk and weaned fewer and smaller young than uninfected controls (Dunsmore, 1981). These second-generation effects may constitute one of the more significant ways that parasites affect the fitness of their hosts and emphasize that both females and males should assess the condition of their mates. It may well be that vigour, as expressed through behaviour, may be more important as a truthful advertisement than colour or secondary sexual ornamentation.

Because stores of nutrients, especially those of energy, are dynamic and regularly used to offset short-term deficits, it is not surprising that one of the most frequently reported symptoms of parasitic infection is a depletion of host energy stores (e.g. Walkey and Meakins, 1970; Bursell, 1981; Lemly and Esch, 1984; Schall, 1983). Such depletions should be considered symptomatic of a general nutrient depletion that must reduce the host's stamina and its capacity for sustained activity or responses to other stresses (Pascoe and Mattey, 1977; Pascoe and Woodworth, 1980). Host animals may attempt to compensate for such depletion by increasing feeding activity at the expense of predator avoidance (Milinski, 1985; Giles, 1987; for a more detailed discussion of this topic, see Chapter 4). This kind of compensation is obviously not possible if nutrient depletion is due to anorexia.

MODULATION OF CONTROL SYSTEMS

One of the most interesting modes of pathology (in the broad sense of a disturbance of normal function rather than in the narrow sense of damaged tissues) associated with parasites, and one that has caused considerable interest in recent years, is modulation of the neuroendocrine control system. This mechanism has been invoked frequently for cases in which behavioural changes are not readily explicable as the result of impaired organ function or reduced nutritional levels. For example, Bethel and Holmes (1973, 1974) suggested that acanthocephalan cystacanths alter the behaviour of their amphipod intermediate hosts by chemical means because: (1) cystacanths located remotely from the CNS elicit complex behaviour patterns, (2) cystacanths of different species in the same location can elicit different behaviour patterns, (3) only living, unencapsulated cystacanths elicit those patterns, and (4) in some cases, behavioural alterations are expressed only after the parasite has become infective, weeks after reaching its final size and location. In addition, Romig *et al.* (1980) found that metacercaria of *Dicrocoelium dendriticum* located in either the suboesophageal or supraoesophageal ganglion elicited the same altered behaviour pattern.

Two good examples of modulation have been described recently. *Gammarus lacustris* infected with cystacanths of *Polymorphus paradoxus* exhibit a modified escape response in which they swim toward the surface of the water, skim along the surface and cling to any solid object encountered (Bethel and Holmes, 1973). Helluy (1988; Helluy and Holmes, in press) demonstrated that serotonin injected into normal gammarids elicits all three aspects of the modified escape behaviour (table 2.1). Some, but not all, of the modifications in infected gammarids can be temporarily eliminated by injection of octopamine. The high haemolymph concentrations required to elicit or block these modified behaviours suggests that the neurotransmitters act within the CNS, and make it unlikely that the acanthocephalan secretes serotonin directly, given the absence, or low concentrations, of serotonin in other acanthocephalans (Budziakowski *et al.*, 1983; Terenina, 1988).

Table 2.1. Escape responses of *Gammarus lacustris* infected with cystacanths of *Polymorphus paradoxus* and/or injected with neurotransmitters.

	Uninfected	Infected
Phototaxis (Time in lighted zone)		
Controls	23	81
Injected with serotonin	52	
Injected with octopamine		76
Skimming (Frequency, in %)		
Controls	<1	17
Injected with serotonin	28	
Injected with octopamine		not done
Clinging (Frequency, in %)		
Controls	0	86
Injected with serotonin	43	
Injected with octopamine		0

Data from Helluy (1988). See Bethel and Holmes (1973) for details of responses.

The second example involves the activation of the opioid system of the host, one of the functions of which is the mediation of responses to pain or other stressful stimuli (Akil *et al.*, 1984; Amit and Galina, 1986; Bodnar, 1986; Kavaliers, 1988). Kavaliers *et al.* (1984) showed that hamsters (*Mesocricetus auratus*) infected with *Schistosoma mansoni* showed increased pain thresholds similar to those seen in hamsters given morphine; in both cases, the analgesia was blocked by an opiate antagonist, naloxone. Kavaliers and Podesta (1988) showed that the locomotor activities of hamsters infected with *S. mansoni* were increased during the prepatent period but decreased at day 40, after eggs were being produced. Both effects could be mimicked by the administration of morphine; at a low dosage (5 mg/kg) morphine increased the activity of uninfected hamsters but at a higher dosage (10 mg/kg) activity was reduced. In both cases, the increased activity could be blocked by a specific delta-opiate antagonist (delta-opiates are known to mediate activity levels: Innes and Kavaliers, 1987; Kavaliers and Innes, 1987), whereas the decreased activity could be blocked by naloxone, a specific mu-opiate antagonist (mu-opiates are known to mediate analgesia: Kavaliers, 1988). The results clearly demonstrate that both the analgesia and the level of activity in the infected hamsters are modulated through the opioid system.

Each of these cases clearly involves modulation of the neuroendocrine system of the host, but in neither case is the mechanism clearly defined. It is likely that *P. paradoxus* secretes something that activates a serotonergic system; the system could be activated by enhancing the secretion of serotonin, blocking its breakdown or reabsorption, increasing the number of serotonin receptors or increasing receptor activity. The same possibilities exist for the activation of the endogenous opioid system by *S. mansoni*. In addition, it is clear that two systems are activated by the schistosomes but it is not clear whether they involve separate mechanisms or are different stages in the same process. One of the simplest hypotheses is that the presence of the worms, and especially the tissue damage consequent to their production of eggs, continually activates the normal analgesic response, resulting in the secretion of increasing levels of opioids. Further work on these systems seems worthwhile.

These two examples suggest that parasites can modulate the neuroendocrine system through damage to the host (and activation of normal host responses to that damage) or through the active production of regulatory compounds.

Host responses to tissue damage

The most obvious mechanism whereby parasites could modify the hormone levels in their host would be through the destruction of a hormone-producing organ. Examples may include cases of parasitic castration through direct consumption of gonadal tissue (e.g. by gonad-inhabiting infections with rediae in snails), or by diverting nutrients that would otherwise be used in gonadal development, a possible mechanism for the castration of snails by sporocysts located in the digestive gland (Baudoin, 1975). Feeding and digestion involve a large number of interacting neural and hormonal controls, many of which originate in the mucosa of the stomach or duodenum. Plasma and tissue concentrations of many gastrointestinal hormones are altered by infections with helminths or protozoans (reviewed by Symons, 1989). Comparisons with pair-fed animals suggested that anorexia is responsible for a few of these changes (such as decreases in plasma insulin in several intestinal nematode and protozoan infections), and that damage to tissues in the stomach or duodenum is responsible for others (such as the decrease in plasma secretin levels in infections with *Trichinella spiralis*). Such changes can have extensive effects on digestion or feeding (see the discussion on the role of serotonin by Sukhdeo and Mettrick, 1987), although Symons (1989) points out the paucity of solid information on this subject.

A second mechanism whereby a host response to damage produced by parasites may modify neuroendocrine function is through the activation of the opioid systems (or functionally similar, non-opioid systems: see Kavaliers, 1988) of the host, as discussed above. A release of mu- and delta-opioids in response to tissue damage could be the mechanism underlying the general pattern of reduced activity, lassitude and depression that accompanies acute infections with many parasites (Hart, 1988) and that increases transmission rates for vectored parasites (Ewald, 1983).

Kavaliers and Podesta (1988) point out that opioids mediate many other behaviourally-important functions known to be affected by parasitic infections, such as feeding (Giles, 1983; Godin and Sproul, 1988), aggression (Rau, 1984a), responses to predators (Milinski, 1985; Giles, 1987) and reproduction (Edwards and Barnard, 1987). Indeed, Isseroff *et al.* (1986) speculated that opioid systems may be involved in the decreased gonadal steroid activity in mice infected with *S. mansoni*. In addition, both the immune and neuroendocrine systems contain identical opioid receptors, suggesting that opioids are the communication links allowing the immune system to be used as a sensory system and allowing neuroendocrine control of the immune system (see review by Blalock, 1989). Consequently, other interactions between parasites and endogenous opioid systems should be expected; such interactions would appear to be a fruitful field for further research.

Direct modulation through parasite secretions

Helminths, protozoans and especially parasitic arthropods are known to produce a wide variety of potentially regulatory compounds. Helminths are known to produce and secrete pheromones, including sterols, which attract conspecifics and occasionally other species (reviewed by Haseeb and Fried, 1988). Although the study of the hormone systems of parasites is in its infancy, both insect parasitoids and helminths are known to produce ecdysteroids, juvenile hormone and growth hormone (reviewed by Spindler, 1988). In addition, helminths are known to produce (and metabolize) a large number of neurotransmitters, including acetylcholine, gamma-aminobutyric acid, at least six amines, and at least fourteen peptides (reviewed by Gustafsson, 1988; see also Smart, 1988; Magee *et al.*, 1989). Most of these compounds regulate the parasite's own metabolism and activities; the extent to which they may also be secreted into and regulate those of the host is not known (but see Thorndyke and Whitfield, 1987; for a review of hormonal interactions involving insect parasitoids see also Beckage, 1985; Thompson discusses the relationship between infection, endocrinology and behaviour in more detail in Chapter 3).

There are at least three ways that parasite secretions can modulate neuroendocrine function: through affecting mediator (hormone or neurotransmitter) metabolism, through secretion of mediators and through immunomodulation.

A number of gastrointestinal nematodes are known to produce acetylcholinesterases, and presumably secrete them into the gut lumen (see review by Lightowlers and Rickard, 1988). The amounts produced are very large, particularly in *Nematodirus battus* (Lee and Martin, 1976); the synthesis of the enzyme must impose a high metabolic cost on the parasite. Several theories on the benefits of these secretions have been proposed. Lee (1970) suggested that these enzymes were used as 'biochemical holdfasts', suppressing local nerve transmission and acetylcholine-induced peristalsis, thus allowing the nematodes to maintain their positions. More recently, Philipp (1984) has suggested that they may be more important as inhibitors of acetylcholine-stimulated mucus secretion by goblet cells (one of the major defensive mechanisms of the gut: see Miller, 1987). In addition, Stibbs and Seed (1976) suggested that *Trypanosoma brucei gambiense* may leak tyrosine aminotransferase into the serum of parasitized animals, diverting tyrosine from the catecholamine pathway and lowering catecholamine release in the central or autonomic nervous systems, resulting in changes in motor activity and sleep. No such reduction in catecholamine synthesis was found by Stibbs and Curtis (1987); they did find that catecholamines were metabolized more rapidly, plus a significant drop in brain serotonin levels, consistent with reduced rates of release of serotonin in infected animals. These examples indicate that one mechanism whereby parasites could affect behaviour is by their metabolites deactivating or interfering with the release of regulatory compounds.

Some cases of parasitic castration appear to be associated with perturbations of the endocrine system. A wide variety of trematodes castrate freshwater snails, inhibiting reproductive activity, depleting energy reserves and increasing growth rates (Loker, 1979; others referred to earlier). Joosse and Van Elk (1986) suggested that the trematodes prevent or inhibit secretion of gonadotrophic hormones or alter gonadal sensitivity to such hormones, inhibiting maturation of the gonad in young snails or resulting in gonadal atrophy in mature snails. Similarly, gonadotrophin-secreting cells in the pituitary are damaged when plerocercoids of *Ligula intestinalis* are present in the coelom (Orr, 1966) of rudd (*Scardinius erythrophthalmus*): this leads to an inhibition of gamete development at a stage known to be dependent on pituitary gonadotrophins (Arme, 1968). Although Arme (1968) suggested the production of an anti-gonadotrophin by the parasite, later work showed no evidence, at least of a steroid hormone (Arme *et al.*, 1982). The mechanism is not known, but it is clear that the parasite does prevent the production of gonadotrophins.

A more common mechanism may be the production and secretion of the regulatory compounds themselves. The most extensively studied systems involve the growth factors produced by plerocercoids of the cestodes *Spirometra mansonoides* and *S. erinacei* in rodents and other intermediate hosts (table 2.2). Many of the physiological properties of the parasite growth factors, particularly that of *S. mansonoides*, are similar, but not identical, to those of human growth hormone, and consequently, infected animals become grossly overweight (Mueller, 1974). (The mechanisms involved are discussed comprehensively in Chapter 3.)

Other systems in which chemical mediation of host neuroendocrine functions is likely involve parasitic castration (see reviews by Reinhard, 1956; Wülker, 1964; Kuris, 1974; Baudoin, 1975; Minchella *et al.*, 1985). Some cases are likely to involve the secretion of hormone-analogues by the parasite. One example is the secretion of a juvenile-hormone analogue by the microsporidian *Nosema* in insects (Fisher, 1964). Another probable example is the parasitic barnacle *Sacculina granifera*, which alters the behaviour of male and female sand crabs to resemble that of an

Table 2.2. Comparison of the activities of plerocercoid growth factors from *Spirometra mansonoides* and *S. erinacei* with those associated with growth hormone.

	Growth hormone	Plerocercoid growth factors	
		S. mansonoides	*S. erinacei*
Growth-related activities			
Weight gain	Increases	Increases	Increases
Skeletal growth	Increases	Increases	Increases
Somatomedin activity	Increases	Increases	Increases
Production of endogenous growth hormone	Decreases	Decreases	?
Metabolic activities			
Insulin-like			
In hypophysectomized rats	Yes	Yes	No
In normal rats	No	Yes	No
Lipogenic	No	Yes	No
Anti-insulin-like			
Lipolytic	Yes	No	Suppressed
Diabetogenic	Yes	No	Yes
Other endocrine processes			
Regulation of receptors for prolactin and oestrogen	Increases	Decreases	?
Onset of puberty	Normal	Delayed	?
Growth rate of hormone-sensitive mammary cancer	Increases	Decreases	?

Modified from Phares (1987) by inclusion of data from Hirai *et al.* (1983), Shiwaku *et al.* (1983), Tsuboi and Hirai (1986) and Hirai *et al.* (1987).
?, data not available.

ovigerous female crab (Bishop and Cannon, 1979). The ovaries of parasitized females are atrophied and infected males do not initiate courtship displays and are not aggressive. The most striking alterations of the crab's behaviour are in its seaward migration, digging and occupation of depressions, abdominal flapping and brood-sac grooming (all those of ovigerous females, but not other crabs). Bishop and Cannon attribute these behavioural changes to a feminizing hormone that mimics the female hormone of the crab. (See Reinhard (1956) and Bishop and Cannon (1979) for similar conclusions of earlier workers on similar systems.) The fact that only certain of the female secondary sexual characteristics are developed in such cases suggests a very specific action of the hormone (Reinhard, 1956).

Modulation of the immune system provides some of the best examples of parasites releasing compounds that influence the metabolism and activities of the host (although details are still far from clear). Many parasites, helminths in particular, persist for relatively long periods of time despite the active responses of the host's immune system. As Lightowlers and Rickard (1988, p.S132) remark: 'The complexity of the host's immune armamentarium has been matched by the evolution of a myriad of techniques by which parasites evade and subvert these immune responses.' The 'immune armamentarium' is regulated by a complex system of chemical communication, including substances elaborated by the neuroendocrine system and by the cells of the immune system itself (see reviews by Behnke, 1987; Blalock, 1989). Therefore, among the techniques available to parasites is the secretion of compounds that inhibit, rather than sensitize, the immune system. Indeed, parasites are known to induce suppressor cell populations or suppress lymphocyte activity by interfering with the chemical communication between lymphocytes or between lymphocytes and macrophages (for details see reviews by Behnke, 1987; Lightowlers and Rickard, 1988). Given the connections between the immune system and the neuroendocrine system, it is possible that immunomodulators may also have effects on behaviour, but no such actions have been described as yet.

Behavioural consequences

Unlike the fairly predictable consequences of organ malfunction or reduced nutritional availability, known or suspected cases of modulation of control systems suggest that this mode of pathology can lead to a bewildering array of highly-specific behavioural changes. These changes may be grouped in various ways. We have followed Moore (1984) and Dobson (1988) in using 'motility', and Helluy and Holmes (in press) in using 'habitat shifts' (which may be influenced by altered responses to light, gravity, humidity, temperature, colour or mechanical stimuli), and 'altered escape behaviour' (in which the sensitivity, orientation or duration of the escape manoeuvre may be altered). Other categories of behaviour altered by parasites are learning, exploratory behaviour and social behaviour (including dominance, aggression, courtship and mating, and parental care). Examples of each are presented in table 2.3. Note that a connection between host behaviour and neuroendocrine modulation has been established in very few of these cases (those marked [a] in table 2.3).

Thus, there is some direct evidence but considerably more suggestive evidence for parasites altering the behaviour of their hosts by modulation of neuroendocrine control systems. Clearly, this is an area that could benefit from more detailed work, combining behavioural studies with studies on the cellular and molecular biology of these control systems.

ADAPTIVE SIGNIFICANCE OF PATHOLOGY

The modes of pathology we have covered, the kinds of behaviour we expect each to produce and the adaptive significance of each (in terms of whether it enhances or reduces genetic fitness) are summarized in table 2.4. The table also indicates whether or not the pathology is dependent on parasite density, and whether or not an infected host can compensate (behaviourally) for the damage produced by the parasite.

The clearest pattern shown in table 2.4 is that only two kinds of pathology are potentially advantageous to the host, and both of them only over a short term. The most obvious case would be that involving the analgesic effect of mu-opioids. With relatively minor damage, the analgesic effect may facilitate fairly normal or increased activity (or may even stimulate an increased feeding response) sufficient

Table 2.3. Behavioural alterations known or suspected to be caused by parasites modulating the neuroendocrine system of their hosts.

Parasite	Host	Habitat shift	Escape responses	Learning	Motility	Exploratory activity	Social behaviour	References
Direct transfer								
Rabies virus	Canids, mammals		X		I		X	Macdonald, 1980
Entomophthora spp.	Ants	X	X		D			Spindler *et al.*, 1986
Vector-borne								
Dirofilaria immitis	Mosquito (vector)				I			Berry *et al.*, 1988
Trypanosoma brucei	Voles (host)				D[a]			Stibbs and Curtis, 1987
Plasmodium spp.	Mice (host)		X					Day and Edman, 1983
Adult parasites								
Schistosoma mansoni	Mice, hamsters			D	D[a]		X	Kershaw *et al.*, 1959, Strech *et al.*, 1960; Kavaliers *et al.*, 1984; Isseroff *et al.*, 1986; Kavaliers and Podesta, 1988
Sacculina granifera	Crab	X			X		X	Bishop and Cannon, 1979
Larval parasites (transferred by ingestion with intermediate host)								
Protozoans								
Sarcocystis rauschorum	Lemming					I		Quinn *et al.*, 1987
Digenea								
Parvatrema affinis	Clam	X			I			Swennon, 1969; Hulscher, 1973; Swennon and Ching, 1974
Brachylecithum mosquensis	Ants	X	X		D		X	Carney, 1969
Dicrocoelium spp.	Ants	X	X		D		X	Spindler *et al.*, 1986; Hohorst and Graefe, 1961; Romig *et al.*, 1980
Leucochloridium spp.	Snails	X	X					Wesenberg-Lund, 1931; Lewis, 1977
Microphallus papillorobustus	Amphipods	X	X					Helluy, 1983; Helluy, 1984
Plagiorchis noblei	Mosquito larvae	X						Webber *et al.*, 1987

Cestoda							
Anomotaenia brevis	Ant	X	X	D	D	X	Plateaux, 1972; Grabrion *et al.*, 1976
Ligula intestinalis	Freshwater fish					X[a]	Orr, 1966; Arme and Owen, 1967
Acanthocephala							
Acanthocepualus spp.	Isopods	X		I			Seidenberg, 1973; Camp and Huizinga, 1979; Muzzall and Rabalais, 1975
Corynosoma constrictum	Amphipod	X	X				Bethel and Holmes, 1973
Echinorhynchus truttae	Amphipod	X		I			Lehmann, 1967
Moniliformis moniliformis	Cockroach	X		I			Moore, 1983b; Wilson and Edwards, 1986
Plagiorhychus cylindratus	Isopod	X		I			Moore, 1983a
Polymorphus paradoxus	Amphipod	X	X[a]				Bethel and Holmes, 1973; Helluy and Holmes, in press
Pomphorhynchus laevis	Amphipod	X					Kennedy *et al.*, 1978
Nematoda							
Dispharynx nasuta	Isopod	X					Moore and Lasswell, 1986
Trichinella spiralis	Mice					X	Rau, 1984a; Edwards, 1988; Edwards and Barnard, 1987
T. pseudospiralis	Mice			I			Rau, 1984b
Toxocara canis	Mice			I	I		Hay and Aiken, 1984; Hay *et al.*, 1986

[a] Cases of known modulation. Only cases that cannot be explained by other pathological mechanisms are included. X, altered; I, increased; D, decreased.

Table 2.4. Summary of the principal modes of pathology that influence behaviour, the principal behavioural consequences and the adaptive value of the pathology for host and parasites.

Pathology	Behaviour	Dependent on density of parasite?	Adaptive Significance: Host	P1	P2	P3	P4	Can host compensate behaviourally for damage?
Organ malfunction								
Sense organs, CNS	Reduced efficiency, incoordination	Yes	–	–	+	+, –	+	Yes, No
Musculature	Reduced motor activity, reduced stamina	Yes	–	–	+	–	+	Yes
Nutrient deprivation								
Anorexia	Reduced feeding, reduced activity	Yes	a	?	–	–	+	No
Energy drain, reduced assimilation, impaired oxygen delivery	Reduced activity or stamina	Yes	–	–	+	–	+	Yes
Neuroendocrine modulation								
Response to damage (analgesic opioids)	Continued (reduced*) activity	Yes	a	–, +	+, –	?	+	?
Direct secretion	Habitat shifts	No	–	+	+	+	+	No
	Escape responses	No	–	–, +	+	–, +	+	No
	Learning	Yes?	–	?	?	?	?	No
	Motility	Yes, No	–	?	+	+	+	No
	Exploratory activity	Yes, No	–	?	+	?	?	No
	Social behaviour	Yes, No	–	+	+	?	?	No

*At high levels. P1, adult parasite; P2, larva that must be ingested with intermediate host; P3, vector-borne parasite, pathology in vector; P4, vector-borne parasite, in vertebrate host. +, advantageous; –, disadvantageous; ?, significance unclear; a, advantageous in short-term, disadvantageous in long-term.

to compensate for the damage. With greater amounts of damage, the increased levels of opioids could reduce activity levels (the equivalent of the 'bed rest' prescribed for human illness), allowing available resources to be marshalled for defence or repair, which may or may not be adequate to compensate for the damage. Short-term anorexia could assist this process (see Hart, 1988, for further discussion). Long-term reductions in activity or long-term anorexia, however, would be disadvantageous.

For an individual host, the results of all other pathological processes are intrinsically disadvantageous. However, there appear to be two groups of processes. In the first, organ malfunction and nutrient deprivation, the effects on fitness are dependent on the extent of the damage (for this discussion, 'damage' will include reduction in nutrients). So long as adequate nutrient supplies are available, and the parasitized individual can compete for them, minor damage or reductions in the share of the nutrients that go to the host, rather than its parasites, can probably be compensated for by increased feeding. Where increased feeding can be done without increased exposure to predators, the resulting loss in fitness may be very small.

Relatively minor damage may even be offset by occasional circumstances under which the presence of parasites may be advantageous. One class of examples are situations in which parasites may be used as weapons of competition (Barbehenn, 1969), such as the meningeal worm, *Parelaphostrongylus tenuis*, which (as discussed earlier) is relatively non-pathogenic in white-tailed deer but usually lethal in moose, wapiti (*Cervus elaphus*) or caribou (*Rangifer tarandus*). Infections in white-tailed deer have been implicated in restricting moose to deep snow areas in New Brunswick (Kelsall and Prescott, 1971), and in the failure of several attempted reintroductions of wapiti or caribou (Anderson and Prestwood, 1979). Other examples are summarized by Holmes (1979, 1982). A second class of examples includes those in which the parasites provide a micronutrient inadequately available in the diet. Ground squirrels (*Spermophilus richardsonii*) fed a pyridoxine-deficient diet *ad libitum* gained weight when infected with *Trypanosoma otospermophili*, but lost weight when uninfected (Munger and Holmes, 1988; see Lincicome (1971) for other examples). In addition, a complex of relatively well-adapted parasites may protect against invasion by potentially more pathogenic parasites. Wild mute swans (*Cygnus olor*) have a well-developed fauna of specialist cestodes (Czaplinski, 1975) and the cestodes *Dicranotaenia coronula* and *Diorchis stefanskii* are rare and immature parasites of them. However, in a zoo, where the swans lacked their normal cestode fauna, they acquired large and pathogenic numbers of these cestodes (Kotecki, 1970).

The major feature that unites these modes of pathology is that their significance to the host is likely to be a complex function of the amount of damage done and the ability of the host to compensate for that damage. The amount of damage is likely to be a function of the number of parasites present; because most parasites are overdispersed or clumped among hosts, relatively few hosts will be severely affected. The ability to compensate is likely to be a function of size and general well-being. At the host population level, differences in these parameters will almost certainly mean that the net effect of the parasites will be to increase the variance in 'condition' in the population. Steinwascher (1979a,b) provides an excellent example. The yeast *Candida humicola* acts as a mutualistic symbiont in large *Rana clamitans* tadpoles increasing growth rates and thereby the probability of metamorphosis, but as a parasite in smaller tadpoles it decreases growth rates. Many selection pressures, including predation and mate choice, act on that variance, as indicated in other chapters in this book. It may well be that increasing the variance in the host population is the most important role parasites play in the evolution of their hosts.

Parasites that actively modulate the neuroendocrine system of their hosts appear to be quite different in their significance to the hosts. Their effects are usually not dependent on the number (or biomass) of the parasites (a generalization evident at the time of the review by Reinhard, 1956), but may be critically dependent on the state of development of the parasite (e.g., Bethel and Holmes, 1974; Bishop and Cannon, 1979). Wherever adequately studied, the behavioural effects appear to be beneficial to the parasites (discussed more fully below), not the hosts (Dobson, 1988; but see Chapter 7). There is some variation among host individuals in the extent to which altered behaviours are expressed, but the determinants and significance of that variation are not clear (Helluy and Holmes, in press). The net effect of this class of behavioural pathology appears to be decreased fitness of the host. At the host population level, this creates two classes of hosts, those infected (with lower fecundities, higher predation rates, or both) and those uninfected. Analyses of models by Dobson (1988) suggest that low-density host populations may be destabilized by the effects of the parasites.

There are apparently few ways that hosts can compensate for this kind of damage. Minchella (1985) reviewed evidence for the two types of facultative compensation shown by snails to infection with trematodes. Many adult snails (or those close to the size at which they would normally begin reproduction) responded to infection by trematode miracidiae with a rapid burst of reproduction. Longer-lived snails responded to parasitic castration by growth (gigantism), which increases life expectancy and the possibility of outliving the trematode infection.

Longer-term genetic compensation is suggested by the data of Tokeson and Holmes (1982). The development of larval *Polymorphus marilis* is inhibited by water temperatures below about 10 °C; many overwinter as small, undeveloped larvae that do not become large enough to castrate their intermediate host, *Gammarus lacustris*, until the host has matured one clutch of eggs. A similar situation was described by McPhail and Peacock (1983): most plerocercoids of *Schistocephalus solidus* remained small and relatively non-pathogenic until after the host sticklebacks (*Gasterosteus aculeatus*) had reproduced, which takes from 1–4 years in different populations. These timing features obviously reduce the loss of fitness of infected individuals but it is not clear whether they are host adaptations or fortunate side-effects of adaptations to climatic conditions (in the former) or physiological dependence on host hormones (in the latter).

For the parasites, influences on fitness are essentially determined by effects on reproductive life span or, especially, on transmission. Three patterns are apparent.

The most obvious is that almost any kind of damage done to the host is generally advantageous to larval parasites that must be ingested with their intermediate host for the cycle to be continued. Most original authors and reviewers (Holmes and Bethel, 1972; Helluy, 1983; Moore, 1984; Dobson, 1988; but see Chapter 7), conclude that reduced stamina, debilitation, altered behaviour or the behavioural compensation for loss of function enhances transmission. Predator tests done in the laboratory and field observations generally support that conclusion. Some authors and reviewers also comment on the most obvious cost to the parasites, that of the increased susceptibility of infected intermediate hosts to predation by animals in which the parasite cannot mature. It would appear that this cost would be considerably less for the relatively specific behavioural alterations than for the pathology that produces debilitation. In this regard, it is interesting that Dobson (1988, p.159) noted that: 'the most highly evolved and sophisticated modifications of host behaviour occur in the host with which the parasite has been associated for the longer length of evolutionary time.' He contrasted the subtle manipulations of host physiology in the invertebrate hosts of acanthocephalans, in which the invertebrate is considered to be the original host (Conway Morris and Crompton, 1982) with the relatively simple reductions in survival and fecundity in the intermediate hosts of cestodes, in which the invertebrate host is considered to have been added later in evolution (Llewellyn, 1965). Note, however, that the effects of the cestode *Anomotaenia brevis* on its ant intermediate host (discussed above) appear to be comparable with, or more complex than, the effects of acanthocephalans on their hosts.

The modelling study of Dobson (1988) indicated that the enhanced transmission associated with the behavioural changes increases the basic rate of reproduction of the parasite, which decreases the size of the host populations (both intermediate and definitive hosts) needed to maintain the parasite in an endemic state. Dobson points out that such mechanisms will therefore be more advantageous for parasites colonizing patchy, small host populations than for those continuously present in large host populations. The data available are inadequate to test this hypothesis.

In contrast, damage to the host appears almost always disadvantageous to adult parasites. Parasites that are actively reproducing face conflicting pressures – greater exploitation of host resources may allow increased reproduction, but also increases damage to the host, resulting in a higher probability of death of both host and parasite. The evolutionary consequences of the consequent tradeoffs have been modelled by Anderson and May (1982); in both model and nature, examples often lead to intermediate levels of pathogenesis (as in myxomatosis, Ross, 1982).

The best examples we could find of damage to the host that was advantageous to an adult parasite involve sacculinid barnacles. These barnacles often castrate their host, and some (such as *Sacculina granifera* in table 2.1.) also feminize the host, eliciting brood-caring behaviour directed at the parasite's brood sac. The parasitized crab also migrates with the normal brooding females, so that both crab larvae and those of the barnacles are released in the same area (Bishop and Cannon, 1979). Other examples of altered behaviour that aids transmission of the propagules produced by adult parasites are the vertical migration of insects infected

with the fungus *Entomophthora* spp. (table 2.1), and the scattering of eggs, cysts or bacteria by diarrhoea, coughing or vomiting (see brief discussion of direct life cycle parasites in Dobson, 1988).

Other examples of pathology favourable to adult parasites involve situations in which the immunopathological conditions are advantageous to the parasites. The best-known example involves the formation of granulomas around the eggs of *Schistosoma mansoni*; these granulomas constitute one of the major modes of pathology, especially when they occur in the liver, spleen or other organs. However, the granulomas also aid in the passage of the eggs from the site of oviposition to the intestinal lumen. Doenhoff *et al.* (1985) showed that immunosuppressed mice excreted significantly fewer eggs in their faeces. Other examples are reviewed by Damian (1987).

Vector-borne parasites face tradeoffs in both the vertebrate host and the vector. However, damage to the vertebrate host is generally advantageous, in part because it reduces the effectiveness of anti-vector behaviour (e.g. Day and Edman, 1983) and in part because rapid reproduction provides the high densities of infective stages necessary for efficient infection of the vector (Ewald, 1983). In contrast, damage to the vector, either through induced nutritional deficiencies or tissue damage (by filarid larvae emerging from muscles or by rupturing malarial oocysts) can affect the ability of the vector to find and infect new vertebrate hosts, and is therefore generally disadvantageous. Damage to the vector 'may be the best compromise the parasite has achieved under the circumstances', and may be offset by other kinds of damage, such as damage to labral sensilla or blockage of the food canal, that enhance transmission (Rowland and Boersma, 1988). As a result, with vector-borne parasites, damage in the vertebrate host is advantageous but damage in the invertebrate host involves major tradeoffs. This pattern is the direct opposite of that for parasites transmitted by ingestion of infected intermediate hosts.

The net result is that relatively few of the behavioural changes associated with parasites appear likely to be host adaptations (anorexia, analgesia, behavioural compensation for reduced efficiency or nutrient depletions), many appear likely to be parasite adaptations (most cases of neuroendocrine modulation, debilitation of vertebrate hosts of vector-borne parasites and of intermediate hosts of parasites transmitted by ingestion of the intermediate host), and a large number appear likely to be the side-effects of the way parasites in general live.

ACKNOWLEDGEMENTS

Preparation of this manuscript was supported by a grant from the Natural Sciences and Engineering Research Council of Canada. Part of the work was done while JCH was on sabbatical leave at the University of Exeter; JCH thanks the Department of Biological Sciences for space and facilities. We thank R. Bates, J. Goldberg, C. R. Kennedy, W. M. Samuel, M. Stock and an anonymous reviewer for constructive criticisms of earlier drafts.

REFERENCES

Abbott, E. M., Parkins, J. J. and Holmes, P. H., 1986, *Veterinary Parasitology*, **20**, 275–289.
Akil, H., Watson, S. J., Lewis M. E., Khacchaturian, H. and Walker, M. J., 1984, *Annual Review of Neuroscience*, **7**, 223–256.
Amit, Z. and Galina, H., 1986, *Physiological Reviews*, **66**, 1091–1120.
Anderson, R. C., 1964, *Pathologica Veterinaria*, **1**, 289–322.
Anderson, N., Hansky, J. and Titchen, D. A., 1981, *Parasitology*, **82**, 401–410.
Anderson, R. M. and May, R. M., 1982, *Parasitology*, **85**, 411–426.
Anderson, R. C. and Prestwood, A. K., 1979, in *Diseases and Parasites of White-Tailed Deer* edited by W. R. Davidson, F. A. Hayes, V. F. Nettles and F. E. Kellogg (Washington, DC: US Department of the Interior, US Fish and Wildlife Service) pp.266–317.
Arme, C., 1968, *Biological Bulletin*, **134**, 15–25.
Arme, C., Bridges, J. F. and Hoole, D., 1983, in *Biology of the Eucestoda*, Vol. 2 edited by C. Arme and P. W. Pappas (London, New York: Academic Press), pp. 499–538.
Arme, C., Griffiths, D. V. and Sumpter, J. P., 1982, *Journal of Parasitology*, **68**, 169–171.
Arme, C. and Owen, R W., 1967, *Parasitology*, **57**, 301–314.
Bailey, G. N. A., 1975, *International Journal for Parasitology*, **5**, 609–613.
Barbehenn, K. R., 1969, *Biotropica*, **1**, 29–35.
Baudoin, M., 1975, *Evolution*, **29**, 335–352.
Beckage, N. E., 1985, *Annual Review of Entomology*, **30**, 371–413.
Behnke, J., 1987, *Advances in Parasitology*, **26**, 1–71.
Becket, E. B., 1971, *Parasitology*, **63**, 365–372.
Berry, W. J., Rowley, W. A. and Christensen, B. M., 1988, *Journal of Parasitology*, **74**, 970–974.
Bethel, W. M. and Holmes, J. C., 1973, *Journal of Parasitology*, **59**, 945-956.
Bethel, W. M. and Holmes, J. C., 1974, *Journal of Parasitology*, **60**, 272–274.
Bishop, R. K. and Cannon, L. R. G., 1979, *Journal of Fish Diseases*, **2**, 131–144.
Blalock, J. E., 1989, *Physiological Reviews*, **69**, 1–32.
Bodnar, R. J., 1986, *Annals of the New York Academy of Sciences*, **467**, 345–360.
Budziakowski, M. E., Mettrick, D. F. and Webb, R., 1983, *Journal of Neurobiology*, **14**, 313–325.
Burkot, T. R., 1988, *Parasitology Today*, **4**, 156–162.
Bursell, E., 1981, *Parasitology*, **82** (Suppl), 107–113.
Butler, J. A. and Millemann, R. G., 1971, *Journal of Parasitology*, **57**, 860–865.
Camp, J. W. and Huizinga, H. W., 1979, *Journal of Parasitology*, **65**, 667–669.
Carney, W. P., 1969, *American Midland Naturalist*, **82**, 605–611.
Castro, G. A., 1988, *Journal of Parasitology*, **74**, 906–914.
Conway Morris, S. and Crompton, D. W. T., 1982, *Biological Reviews*, **57**, 85–115.
Crompton, D. W. T., 1984, *Federation Proceedings*, **43**, 239–245.
Crowden, A. E. and Broom, D. M., 1980, *Animal Behaviour*, **28**, 287–294.
Czaplinski, B., 1975, *Acta Parasitologica Polonica*, **23**, 305–327.
Dailey, M. D. and Stroud, R., 1978, *Journal of Wildlife Diseases*, **14**, 503.
Dailey, M. D. and Walker, W. A., 1978, *Journal of Parasitology*, **64**, 593–596.
Dargie, J. D., 1987, *International Journal for Parasitology*, **17**, 453–463.
Damian, R., 1987, *Journal of Parasitology*, **73**, 3–13.
Day, J. F., Ebert, K. M. and Edman, J. D., 1983, *Journal of Medical Entomology*, **20**, 120–127.
Day, J. F. and Edman, J. D., 1983, *Journal of Parasitology*, **69**, 163–170.
Delyamure, S. L., 1955, in *Helminthofauna of Marine Mammals (Ecology and Phylogeny)* edited by K. I. Skryjabin (Moscow: Izdatel'stvo Akademii Nauk SSSR).
Despommier, D., 1976, in *Ecological Aspects of Parasitology* edited by C. R. Kennedy (Amsterdam, Oxford: North-Holland Publishing Company), pp. 269–285.
Dobson, A. P., 1988, *The Quarterly Review of Biology*, **63**, 139–165.
Doenhoff, M. J., Hassounah, O. A. and Lucas, S. B., 1985, *Immunology Today*, **6**, 203–206.
Dolinsky, Z. S., Burright, R. G. and Donovick, P. J., 1981, *Science*, **213**, 1142–1144.
Donovick, P. J., Dolinsky, Z. S, Perdue, V. P. and Burright, R. G., 1981, *Brain Research Bulletin*, **7**, 317–323.
Dunsmore, J. D., 1981, in *Proceedings of the Worldwide Furbearer Conference* edited by J. A. Chapman and D. Pursley (Falls Church, Maryland: R. Donnelly and Sons), pp. 654–669.
Dunsmore, J. D., Thompson, R. C. A. and Bates, J. A., 1983, *International Journal for Parasitology*, **13**, 517–521.

Edwards, J. C., 1988, *Animal Behaviour*, **36**, 529–540.

Edwards, J. C. and Barnard, C. J., 1987, *Animal Behaviour*, **35**, 533–540.

Erasmus, D. A., 1958, *Parasitology*, **48**, 312–335.

Ewald, P. W., 1983, *Annual Review of Ecology and Systematics*, **14**, 465–485.

Festa-Bianchet, M., 1988, *Animal Behaviour*, **36**, 1445–1454.

Fischer, E., Camus, C., Sentoro, F. and Capron, A., 1981, *Clinical and Experimental Immunology*, **46**, 89.

Fisher, F. M., 1964, *Biological Bulletin*, **126**, 235–252.

Gabrion, C., Plateaux, L. and Quentin, C., 1976, *Annales de Parasitologie*, **51**, 407–420.

Garnham. P. C. C., 1972, in *The Structure and Function of Muscle* edited by G. H. Bourie (New York and London: Academic Press), pp. 239–286.

Geist, V., 1971, *Mountain Sheep: A Study in Behaviour and Evolution* (Chicago: University of Chicago Press), p.383.

Geraci, J. R. and St. Aubin, D. J., 1987, *International Journal for Parasitology*, **17**, 407–421.

Giles, N., 1983, *Animal Behaviour*, **31**, 1192–1194.

Giles, N., 1987, *Journal of Fish Biology*, **30**, 631–638.

Godin, J. G. J. and Sproul, C. D., 1988, *Canadian Journal of Zoology*, **66**, 2360–2367.

Gustafsson, M. K. S., 1988, in *Parasitology in Focus* edited by H. Mehlhorn (Berlin, Heidelberg: Springer-Verlag), pp. 295–310.

Hall, A., 1985, *Progress in Food and Nutrition Science*, **9**, 227–256.

Hamilton, W. D. and Zuk, M., 1982, *Science*, **218**, 384–387.

Hart, J. A., 1988, *Neuroscience and Biobehavior Reviews*, **12**, 123–137.

Haseeb, M. A. and Fried, B., 1988, *Advances in Parasitology*, **27**, 170–207.

Hay, J. and Aitken, P. P., 1984, *Annals of Tropical Medicine and Parasitology*, **78**, 145–155.

Hay, J., Aitken, P. P. and Graham, D. J., 1984a, *Zeitschrift für Parasitenkunde*, **70**, 575–588.

Hay, J., Aitken, P. P., Hair, D. M. and Hutchison, W. M., 1984b, *Annals of Tropical Medicine and Parasitology*, **78**, 611–618.

Hay, J., Aitken, P. P., Hutchison, W. H. and Graham, D. J., 1983a, *Annals of Tropical Medicine and Parasitology*, **77**, 261–277.

Hay, J. and Hutchison, W. H., 1983, *Annals of Tropical Medicine and Parasitology*, 77, 437–439.

Hay, J., Hutchison, W. M. and Aitken, P. P., 1983b, *Annals of Tropical Medicine and Parasitology*, 77, 543–544.

Hay, J., Hutchison, W. M., Aitken, P. P. and Graham, D. J., 1983c, *Annals of Tropical Medicine and Parasitology*, **77**, 483–495.

Hay, J., Kendall, A. T., Aitken, P. P. and Arnott, M. A., 1986, *Annals of Tropical Medicine and Parasitology*, **80**, 531–533.

Helluy, S., 1983, *Revue d'Ecologie (La Terre et la Vie)*, **38**, 211–223.

Helluy, S., 1984, *Annales de Parasitologie*, **59**, 41–56.

Helluy, S. M., 1988, *On the mechanisms by which a larval parasite* (*Polymorphus paradoxus*, *Acanthocephala*) *alters the behaviour of its intermediate host* (*Gammarus lacustris*, *Crustacea*), PhD Thesis, University of Alberta, Edmonton, Alberta.

Helluy, S. M. and Holmes, J. C., 1990, *Canadian Journal of Zoology*, **68**, (in press).

Hirai, K., Shiwaku, K., Tsuboi, T., Torii, M., Nishida, H. and Yamane, Y., 1983, *Zeitschrift für Parasitenkunde*, **669**, 489–499.

Hirai, K., Tsuboi, T., Torri, M. and Nishida, H., 1987, *Parasitology Research*, **74**, 183–187.

Hohorst, W., 1964, *Zeitschrift für Parasitenkunde*, **22**, 105–106.

Hohorst, W. and Graefe, G., 1961, *Naturwissenschaften*, **48**, 229–230.

Holmes, J. C., 1979, in *Host-Parasite Interfaces* edited by B. B. Nickol (New York: Academic Press), pp. 27–46.

Holmes, J. C., 1982, in *Population Biology of Infectious Diseases* edited by R. M. Anderson and R. M. May (Berlin: Springer-Verlag), pp. 37–51.

Holmes, P. H., 1987, *International Journal for Parasitology*, **17**, 443–451.

Holmes. J. C. and Bethel, W. M., 1972, *Zoological Journal of the Linnean Society of London*, **51**, (Supplement), 123–149.

Howard, E. B., Britt, J. O. and Matsumato, G., 1983, in *Parasitic Diseases* (Florida: CRC Press, Inc.), pp. 119–232.

Hulscher, B. L., 1973, *Netherlands Journal of Sea Research*, **6**, 141–156.

Husain, A. and Kershaw, W. E. 1971, *Transactions of the Royal Society of Tropical Medicine and Hygiene*, **65**, 617–619.

Hussain, M. A., 1980, *The Nigerian Journal of Nutrition Science*, **1**, 5–15.

Hutchison, W. M., Aitken, P. P. and Wells, B. W. P., 1980a, *Annals of Tropical Medicine and Parasitology*, **74**, 145–150.

Hutchison, W. M., Bradley, M., Cheyne, W. M., Wells, B. W. P. and Hay, J., 1980b, *Annals of Tropical Medicine and Parasitology*, **74**, 337–345.

Innes, D. G. L. and Kavaliers, M., 1987, *Canadian Journal of Zoology*, **65**, 2504–2512.

Insler, G. D. and Roberts, L. S., 1976, *Experimental Parasitology*, **39**, 351–357.

Isseroff, H., Sylvester, P. W. and Held, W. A., 1986, *Molecular and Biochemical Parasitology*, **18**, 401–412.

Joosse, J. and Van Elk, R., 1986, *Experimental Parasitology*, **62**, 1–13.

Kavaliers, M., 1988, *Brain Research Bulletin*, **21**, 923–931.

Kavaliers, M. and Innes, D., 1987, *Brain Research*, **425**, 49–56.

Kavaliers, M. and Podesta, R., 1988, *Canadian Journal of Zoology*, **66**, 2653–2657.

Kavaliers, M., Podesta, R. B., Hirst, M. and Young, B., 1984, *Life Sciences*, **35**, 2365–2373.

Kelsall, J. P. and Prescott, W., 1971, *Moose and Deer Behaviour in Snow in Fundy National Park, New Brunswick* (Ottawa: Canadian Wildlife Service, Report Series No.15), p. 25.

Kennedy, C. R., 1972, *Parasitology*, **64**, 485–499.

Kennedy, C. R., Broughton, P. F. and Hine, P. M., 1978, *Journal of Fish Biology*, **13**, 265–275.

Kershaw, W. E., Laytham, G. W. H. and Dickerson, G., 1959, *Annals of Tropical Medicine and Parasitology*, **53**, 504–506.

Kodric-Brown, A. and Brown, J. H., 1984, *The American Naturalist*, **124**, 309–323.

Kotecki, N. R., 1970, *Acta Parasitologica Polonica*, **17**, 329–355.

Kummeneje, K., 1974, *Nordisk Veterinaermedicin*, **26**, 456–458.

Kuris, A. M., 1974, *Quarterly Review of Biology*, **49**, 129–148.

Lauckner, G., 1987, *International Journal for Parasitology*, **17**, 391–398.

Lee, D. L., 1970, *Tissue and Cell*, **2**, 225–231.

Lee, D. L. and Martin, J., 1976, in *Biochemistry of Parasites and Host-Parasite Relationships* edited by H. van den Bossche (Amsterdam: Elsevier/North-Holland Biomedical Press), pp. 311–318.

Lehmann, U., 1967, *Zeitschrift für Morphologie und Okologie der Tiere*, **60**, 227–274.

Lemly, A. D. and Esch, G. W., 1984, *Journal of Parasitology*, **70**, 475–492.

Lester, R. J. G., 1971, *Canadian Journal of Zoology*, **49**, 361–366.

Lester, R. J. G. and Huizinga, H. W., 1977, *Canadian Journal of Zoology*, **55**, 64–73.

Levine, R. A. and Wardlaw, S. C., 1988, *American Scientist*, **76**, 592–598.

Lewis, P. D., 1977, *Proceedings of the Montana Academy of Sciences*, **37**, 70–81.

Lightowlers, M. W. and Rickard, M. D., 1988, *Parasitology*, **96**, S123–S166.

Lincicome, D. R., 1971, in *Aspects of the Biology of Symbiosis* edited by T. C. Cheng (Baltimore: University Park Press), pp. 139–227.

Livesey, J. L., Molyneux, D. H. and Jenni, L., 1980, *Acta Tropica*, **37**, 151–161.

Llewellyn, J., 1965, in *Evolution of Parasites* edited by A. E. R. Taylor (Oxford: Blackwell Scientific Publications), pp. 47–78.

Loker, G. S., 1979, *Journal of Invertebrate Zoology*, **34**, 138–144.

Mace, T. F. and Davis, C. C., 1972, *Oikos*, **23**, 336–343.

Macdonald, D. W., 1980, *Rabies and Wildlife: A Biologist's Perspective* (Oxford: Oxford University Press), p. 151.

Mackenzie, C. D., Gatrill, A. J., Luty, A. J., Manyasi, D. G., Moore, C., Sayers, G. and Sulaimon, S., 1987, *Parasitology*, **94**, S9–S28.

Magee, R. M., Fairweather, I., Johnston, C. F., Halton, D. W. and Shaw, C., 1989, *Parasitology*, **98**, 227–238.

Martin, J. and Holland, C., 1984, *Journal of Helminthology*, **58**, 93–99.

McNeill, M. A., and Rau, M. E., 1987, *Journal of Wildlife Diseases*, **23**, 418–421.

McPhail, J. D. and Peacock, S. D., 1983, *Canadian Journal of Zoology*, **61**, 901–908.

Mead, R. W., and Roberts, L. S., 1972, *Comparative Biochemistry and Physiology*, **41**, 749–760.

Meakins, R. H., 1974, *Polskie Archiwum Hydrobiologii*, **21**, 455–466.

Meakins, R. H. and Walkey, M., 1975, *Journal of Fish Biology*, **7**, 817–824.

Mech, L. D., 1966, *The Wolves of Isle Royale* (Washington, DC: US Government Printing Office), p. 210.

Mettrick, D. F., 1973, *Canadian Journal of Public Health*, **64**, 70–82.

Mettrick, D. F., 1980. In *Biology of the Tapeworm* Hymenolepis diminuta, edited by H. P. Arai (New York: Academic Press), pp. 281–356.

Mettrick, D. F. and Podesta, R. B., 1974, *Advances in Parasitology*, **12**, 183–279.
Milinski, M., 1985, *Behaviour*, **93**, 203–216.
Miller, H. R. P., 1987, *Parasitology*, **94**, S77-S100.
Minchella, D. J., 1985, *Parasitology*, **90**, 205–216.
Minchella, D. J., Leathers, B. K., Brown, K. M. and McNair, J. N., 1985, *The American Naturalist*, **126**, 843–854.
Molyneux, D. H. and Jefferies, D., 1986, *Parasitology*, **92**, 721–736.
Moore, J., 1983a, *Ecology*, **64**, 1000–1015.
Moore, J., 1983b, *Journal of Parasitology*, **69**, 1174–1176.
Moore, J., 1984, *The American Naturalist*, **123**, 572–577.
Moore, J. and Lasswell, J., 1986, *Journal of Parasitology*, **72**, 186–189.
Mueller, J. F., 1974, *Journal of Parasitology*, **60**, 3–14.
Munger, J. C., and Holmes, J. C., 1988, *Canadian Journal of Zoology*, **66**, 222–227.
Murray, J., Murray, A. and Murray, N., 1978, *Perspectives in Biology and Medicine*, **22**, 134–142.
Muzzall, P. M. and Rabalais, F. C., 1975, *Proceedings of the Helminthological Society of Washington*, **42**, 116–118.
Nansen, P., 1987, *International Journal for Parasitology*, **17**, 425–433.
O'Connor, G. R., 1976, in *Ecological Aspects of Parasitology* edited by C. R. Kennedy (Amsterdam, Oxford: North-Holland Publishing Company), pp. 327–347.
Olson, L. J. and Rose, J. E., 1966, *Experimental Parasitology*, **19**, 77–84.
Orr, T. S. C., 1966, *Nature*, **212**, 736.
Pascoe, D. and Mattey, D., 1977, *Zeitschrift für Parasitenkunde*, **51**, 179–186.
Pascoe, D. and Woodworth, J., 1980, *Zeitschrift für Parasitenkunde*, **62**, 159–163.
Perrudet-Badoux, A. and Binaghi, R. A., 1974, *Immunology*, **26**, 1217–1233.
Phares, C. K., 1987, *Parasitology Today*, **3**, 346–349.
Philipp, M., 1984, *Transactions of the Royal Society for Tropical Medicine and Hygiene*, **78**, 138–139.
Plateaux, L., 1972, *Annales des Sciences Naturelles, Zoologie Paris*, **14**, 203–220.
Podesta, R. B. and Mettrick, D. F., 1976, *Canadian Journal of Zoology*, **54**, 794–803.
Quinn, S. C., Brooks, R. J. and Cawthorn, R. J., 1987, *Journal of Parasitology*, **73**, 265–271.
Radabaugh, D. C., 1980a, *Journal of Fish Biology*, **16**, 621–628.
Radabaugh, D. C., 1980b, *Journal of Parasitology*, **66**, 183–184.
Rau, M. E., 1983, *Parasitology*, **86**, 311–318.
Rau, M. E., 1984a, *Parasitology*, **88**, 371–373.
Rau, M. E., 1984b, *Parasitology*, **88**, 415–419.
Rau, M. E., 1985, *Journal of Parasitology*, **71**, 774–778.
Rau, M. E. and Caron, F., 1979, *Canadian Journal of Zoology*, **57**, 2466–2468.
Rau, M. E. and Putter, L., 1984, *Parasitology*, **89**, 579–583.
Rees, G., 1955, *Parasitology*, **45**, 295–312.
Reinhard, E. G., 1956, *Parasitology*, **5**, 79–107.
Ressel, S. and Schall, J. J., 1989, *Oecologia*, **78**, 158–164.
Ridgway, S. H. and Dailey, M. D., 1972, *Journal of Wildlife Diseases*, **8**, 33–43.
Romig, T., Lucius, R. and Frank, W., 1980, *Zeitschrift für Parasitenkunde*, **63**, 277–286.
Rosenberg, J. H. and Bowman, B. B., 1982, *Review of Infectious Diseases*, **4**, 763–767.
Ross, J., 1982, in *Animal Disease in Relation to Animal Conservation. Symposium of the Zoological Society of London* edited by M. A. Edwards and U. McDonnell (London: Academic Press), pp. 77–91.
Rossignol, P. A. and Rossignol, A. M., 1988, *Parasitology*, **97**, 363–372.
Rowland, M. and Boersma, E., 1988, *Parasitology*, **97**, 221–227.
Rowland, M. W. and Lindsay, S. W., 1986, *Physiological Entomology*, **11**, 325–334.
Salzemann, A. and Plateaux, L., 1987, in *Chemistry and Biology of Social Insects* edited by J. Eder and H. Rembold (Munich: Verlag J. Peperny), p.45.
Saumier, M. D., Rau, M. E. and Bird, D. M., 1988, *Canadian Journal of Zoology*, **66**, 1685–1692.
Schall, J. J., 1983, *Parasitology*, **87**, 1–6.
Schall, J. J., 1986, *Journal of Herpetology*, **20**, 318–324.
Schall, J. J., Bennett, A. F. and Putnam, R. W., 1982, *Science*, **217**, 1057–1059.
Schall, J. J. and Sarni, G. A., 1987, *Copeia*, **1987**, 84–93.
Schiefer, B. A., Ward, R. A. and Eldridge, B. F., 1977, *Experimental Parasitology*, **41**, 397–404.
Seidenberg, A. J., 1973, *Journal of Parasitology*, **59**, 957–962.
Shariff, M., Richards, R. H. and Sommerville, C., 1980, *Journal of Fish Diseases*, **3**, 455–465.

Shiwaku, K., Hirai, K., Torii, M. and Tsuboi, T., 1983, *Parasitology*, **87**, 447–453.
Smart, D., 1988, *International Journal for Parasitology*, **18** , 485–492.
Smith, R. S. and Kramer, D. L., 1987, *Canadian Journal of Zoology*, **65**, 1862–1865.
Spindler, E. -M., Zahler, M. and Loose-Frank, B., 1986, *Zeitschrift für Parasitenkunde*, **72**, 689–692.
Spindler, K. D., 1988, in *Parasitology in Focus* edited by H. Mehlhorn (Berlin, Heidelberg: Springer-Verlag), pp. 465–476.
Sprent, J. F. A., 1955, *Parasitology*, **45**, 31–40.
Steinwascher, K., 1979a, *Ecology*, **60**, 884–890.
Steinwascher, K., 1979b, *Ecology*, **60**, 1172–1183.
Stibbs, H. H. and Curtis, D. A., 1987, *Annals of Tropical Medicine and Parasitology*, **81**, 673–679.
Stibbs, H. H. and Seed, J. R., 1976, *Experimental Parasitology*, **39**, 1–6.
Stock, T. M. and Holmes, J. C., 1987, *Journal of Parasitology*, **73**, 1116–1123.
Stretch, R. G. A., Laytham, G. W. H. and Kershaw, W. E., 1960, *Annals of Tropical Medicine and Parasitology*, **54**, 487–492.
Sukhdeo, M. V. K. and Mettrick, D. F., 1987, *Advances in Parasitology*, **26**, 73–144.
Sukhdeo, M. V. K., Sangster, N. C. and Mettrick, D. F., 1988, *International Journal for Parasitology*, **18**, 509–512.
Summers, B., Cypress, R. H., Dolinsky, Z. S., Burright, R. G. and Donovick, P. J., 1983, *Brain Research Bulletin*, **10**, 547—550.
Swennon, C., 1969, *Netherlands Journal of Sea Research*, **4**, 376–379.
Swennon, C. and Ching, H. L., 1974, *Netherlands Journal of Sea Research*, **8**, 108–115.
Symons, L. E. A., 1989, *Pathophysiology of Endoparasitic Infection* (Sydney: Academic Press), p. 331.
Szidat. L., 1969, *Journal of the Fisheries Research Board of Canada*, **26**, 753–786.
Terenina, N. B., 1988, *Parazitologiya*, **22**, 476–480.
Thevenaz, Ph. and Hecker, H., 1980, *Acta Tropica*, **37**, 163–175.
Thomas, V. G., 1986, *Canadian Journal of Zoology*, **64**, 251–254.
Thorndyke, M. C. and Whitfield, P. J., 1987, *General and Comparative Endocrinology*, **68**, 202–207.
Tokeson, J. P. E. and Holmes, J. C., 1982, *Journal of Parasitology*, **68**, 112–119.
Townson, H., 1970, *Annals of Tropical Medicine and Parasitology*, **64**, 411–420.
Tsuboi, T. and Hirai, K., 1986, *Parasitology*, **93**, 143–151.
Uhazy, L. S., Holmes, J. C. and Stelfox, J. G., 1973, *Canadian Journal of Zoology*, **51**, 817–824.
Vercruysse, J., Taraschewski, H., and Voigt, W. P., 1988, in *Parasitology in Focus* edited by H. Mehlhorn (Berlin, Heidelberg: Springer-Verlag), pp. 477–537.
Walkey, M. and Meakins, R. H., 1970, *Journal of Fish Biology*, **2**, 361–372.
Webber, R. A., Rau, M. E. and Lewis, D. J., 1987, *Canadian Journal of Zoology*, **65**, 1340–1342.
Wesenberg-Lund, C., 1931, *Kongelige Danske Videnskabernes Selskabes Naturvidenskabelige og Mathematiske Afhandlinger*, **4**, 89–142.
Wilson, K., and Edwards, J., 1986, *Animal Behaviour*, **34**, 942–945.
Witting, P. A., 1979, *Zeitschrift für Parasitenkunde*, **61**, 29–51.
Wright, K. A., 1974, *Canadian Journal of Zoology*, **52**, 1215–1220.
Wülker, W., 1964, *Experimental Parasitology*, **15**, 561–597.
Zohar, A. S., 1986, *Behavioural Changes in* Trichinella spiralis-*Infected Mice*, MSc Thesis, McGill University, Montreal.
Zohar, A. S. and Rau, M. E., 1984, *Journal of Parasitology*, **70**, 927–930.
Zohar, A. S. and Rau, M. E., 1986, *Journal of Parasitology*, **72**, 464–466.

3. Physiological alterations during parasitism and their effects on host behaviour

S. N. Thompson

BEHAVIOUR AND INTIMATE ASSOCIATIONS

This chapter examines the physiological alterations that occur during the host relationships of multicellular animal parasites and attempts to establish how these influence or interact with behaviour. Behaviour is important in defining many symbioses (Debary, 1879) because intimate associations (Whitfield, 1979) are often categorized on the basis of an outcome that is observable, and reflected in the behaviour of the symbionts. For example, the mutualistic nature of the relationship between anemonefish and sea anemones is readily apparent by observing the behaviour of the fish relative to the anemone (Fricke, 1975). Indeed, the behaviour can only be understood within the context of the relationship (Whitfield, 1979). Nevertheless, behaviour of the symbionts, although interactive, is separate and distinct. Such is often not the case with parasitic relationships during which the host's behaviour may be influenced in an intimate way expressed by the host–parasite complex.

Behaviour is important during the establishment of parasitism. For a successful relationship to occur the parasite must come into frequent contact with a host and the behaviour of each must favour such contact (Price, 1980; Schmidt and Roberts, 1981; see, however, Chapter 9). Physiological interaction, however, is quickly established. The host's immunological response and the parasite's counter-response for self preservation are known to play key roles (Davies *et al.*, 1980; Grossman *et al.*, 1986; Damian, 1987). As the association develops, physiological alterations in the host may affect host behaviour in a manner that changes the relationship of the host–parasite complex with the environment. Patterns of inherent behaviour associated with the life history of the unparasitized host may be temporarily or permanently changed. By the very nature of parasitism such change is commonly manifested in a weakening of the host. By influencing host behaviour, physiological alteration may determine the fate of the relationship. Thus, the role of physiological alteration becomes implicit in Dogiel's (1964) ecologically-oriented definition of parasites as '. . . those animals which use other living animals as their environment and sources of food at the same time relinquishing to their hosts, partly or completely, the task of regulating their relationships with the external environment.'

PARASITISM AND PHYSIOLOGICAL INTEGRATION

Organisms interact at many levels during symbiotic associations. Metabolic, physiological and ecological factors play interactive roles that determine the nature of the relationship. Because of their complex and varied character symbiotic relationships have traditionally been categorized as mutualistic, commensalistic or parasitic. Parasitic associations are those during which one symbiont, the parasite, feeds at the expense of another, the host, causing damage (Henry, 1966). A more precise definition of parasitism has been the subject of extensive discussion (Goff, 1982; Lewin, 1982). Smyth (1976) presented a global description of the 'ideal' parasite as one that: (1) is adapted to the physiochemical conditions of the host, (2) utilizes host nutrient in a manner compatible with host survival, (3) minimizes the host immune (defense) response, and (4) has a life cycle synchronous with that of the host. This description serves to point out that many factors influence the success or outcome of parasitic symbioses. It is clear, however, that Smyth's parasite might be considered commensalistic (Van Beneden, 1876) or, under some circumstances, mutualistic. The characteristics of various relationships overlap thus emphasizing the dynamic nature of symbiosis.

The importance of metabolic and physiological interaction in parasitism was implied by Smyth (1962) who stated that a parasite exhibits direct or indirect metabolic dependence on the host. This view was expanded by MacInnis (1976) who defined parasitism as an intimate association during which one partner, the parasite, is dependent upon a minimum of one gene or its product from the other interacting species, the host, for survival. This requirement is often met through alterations to the host's physiology at the nutritional, metabolic and/or endocrinological levels in a manner that allows survival of the parasite in a living host (see also Chapter 2). The result is the formation of an interacting or integrated host–parasite complex. The fate of the host may be permanently altered (Read, 1970), but the new balance may not necessarily be detrimental to the host in the short-term. That is, in an 'ideal' host–parasite relationship physiological homeostasis is initially maintained.

The internal environment of vertebrate animals is remarkably constant. Many physiological parameters vary only within narrow and precise limits. Constant internal conditions, however, are less characteristic of invertebrates. Emphasizing that the maintenance of function rather than a specific set of internal conditions is the critical factor in physiological adaptation, Mangum and Towle (1977) proposed the term enantiostasis to encompass 'a counterbalanced state of opposing conditions . . . when the effect of change in one chemical or physical property of the internal milieu is opposed by a change in another. Thus, the internal milieu is unstable, but the net effect of the flux on a particular physiological system is stability, or a tendency toward it.'

Functional stability characterizes all living systems. Physiological regulation maintains appropriate conditions that allow optimal function of the organism. When exposed to stress such as may occur during parasitic infection, organisms compensate through physiological adjustments. Most often these alterations reflect the normal functioning of the host's regulatory mechanisms, although parasitism may affect the regulatory process itself. Failure of the host–parasite complex to maintain homeostasis results from a failure of the mechanism of integration (Read, 1970) and leads to disease. Little is known, however, of the mechanisms of pathogenicity and the manner by which the integrative character of the host–parasite association is modified toward that end (Castro, 1980).

NUTRITIONAL ALTERATION

Nutrition plays an important role in all symbiotic associations. In the case of parasitism, the parasite must compete with the host for nutrients and the host must be adequately nourished for the parasite to complete its development. The competition is usually a complex process with the parasite's nutrition being intimately associated with the physiology of the host (Von Brand, 1979). The factors involved are interactive; that is, parasitic infection influences the host's nutrition and the nutritional status of the host influences the parasite (Read, 1950; Crompton and Nesheim, 1982; Bundy and Golden, 1987). Nutritionally-related effects of infection are highly variable and may include modified host food consumption and growth pattern, altered digestive function and assimilation as well as changes in metabolic conversion efficiency. Decreased food consumption is commonly observed during parasitic infection (Crompton, 1984; Symons, 1985) and host nutritional status, particularly malnutrition, may exacerbate the pathological effects of many infections (Crouch, 1982; Keusch 1982; Pawlowski, 1984).

Studies on vertebrate hosts

Inappetence and host growth

The relationship between host growth, food consumption and malnutrition during infection by several gastrointestinal parasites has been well documented. Crompton *et al.* (1981) reported that well-nourished rats infected with the nematode *Nippostrongylus brasiliensis* displayed lower food intake and growth than their uninfected counterparts. No difference in growth was evident between infected rats and pair-fed controls. When rats were undernourished and fed *ad libitum* on diets low in protein, food consumption was even lower and weight loss greater in infected rats than was observed in the experiments with well-nourished individuals. Undernourished infected rats also displayed greater weight loss than undernourished pair-fed controls. Thus, the effect of infection with malnutrition was synergistic (Gordon, 1976; Beisel, 1982) resulting in a combined effect greater than that expected from either factor alone. The effect of *N. brasiliensis* infection on the pattern of feeding in rats was described by Ovington (1985). Recently, Ash *et al.* (1985a,b) described the effects of infection and malnutrition on serum levels of amino acids and hormones in relation to protein synthesis and growth. Although reduced food consumption may be solely responsible for the poor growth observed during many infections, metabolic changes affecting conversion efficiencies may also play a role and influence growth patterns (Symons, 1985).

The basis for inappetence or anorexia during parasitic infection is unknown. Crompton (1984) outlined the environmental and physiological factors known to initiate, modulate and terminate normal feeding behaviour. These include external olfactory and visual signals, internal mechanical and hormonal factors associated with intestinal function as well as postabsorptive signals such as levels of various serum components. Some of these factors are known to vary during parasitic infection but little is known concerning their potential influence on appetence and nutrition during parasitic infection. Physiological alterations associated with nutrition, however, may be expressed through subtle modifications in behaviour. Keymer *et al.* (1983a) reported that rats may learn to avoid food containing *N. brasiliensis*. Rats were demonstrated to discriminate between two equally-preferred but differently flavoured diets. When fed one diet and inoculated subcutaneously with parasite larvae rats preferred to feed on the other diet when given a choice. In contrast, rats that were not inoculated displayed no preference and ate equal amounts of both diets. Although the food avoidance observed was not sufficient to account for the entire reduction of food intake by *N. brasiliensis*-infected rats, learned taste aversion may play a significant role in influencing nutrition and inappetance during some host–parasite relationships.

Anorexia is seldom considered other than a pathological response to infection (Crompton, 1984; Symons, 1985). Murray *et al.* (1978), however, suggested that the loss of appetite and resultant decreased food intake may play a beneficial role during the initial stages of infection by limiting nutrient availability to the parasite. Subsequent studies demonstrated that mortality of anorexic mice infected with the bacterium *Listeria monocytogenes* was increased following forced feeding (Murray and Murray, 1979). Learned taste aversion such as that described above (Keymer, 1983a) may represent a behavioural mechanism that ensures that the benefits of food deprivation accrue to the host.

Nutrition and gastrointestinal pathology

The precise manner in which infection and nutrition interact is in some cases difficult to establish because of direct physical damage caused by the parasite. Many clinical workers have emphasized the role of intestinal damage in nutritional pathology (Martin, 1981; Holmes, 1987; Rosenberg and Bowman, 1984). Pathological damage might be expected to result in altered food intake with inappetance reflecting a protective feedback mechanism. During some parasitic infections, however, intestinal pathology may have limited pathophysiological significance. Domingo and Warren (1969), for example, described extensive pathological damage to the muscularis and serosa of the intestinal wall of mice during formation of granulomatous tissue around the eggs of *Schistosoma mansoni*. The villi and mucosal epithelia cells, however, appeared unaffected and no difference in absorption was observed between infected and uninfected mice when absorption was measured over the entire small intestine. Localized malabsorption occurred in rats during infection with the intestinal phase of the nematode, *Trichinella spiralis*, and infected rats displayed extensive intestinal inflammation (Castro *et al.*, 1967). Castro *et al.* (1979) examined the interaction of growth, decreased food consumption, intestinal pathology and basal metabolism. When fed *ad libitum* infected rats grew at low rates. Individuals fed intraduodenally or intravenously, however, gained weight normally. Thus, despite the observed pathological effects, decreased growth was related to low food consumption rather than an overall depression of digestion, malabsorption or basal metabolic alteration.

Pathology and intensity of infection

Nutrition provides an ideal example of the critical role intensity of infection plays in the physiological alterations characterizing many host–parasite associations. Despite the prevalence of parasitism in a host population the intensity of infection generally determines to what extent the host will be physiologically altered and may ultimately become diseased. In most cases, naturally acquired parasitic infections are not intense and are often asymptomatic. Although the precise relationship between intensity and pathogenic processes is unknown, intensity clearly influences the expression and interaction of the physiological alterations occurring during infection. Keymer *et al.* (1983b) reported that the above described effects of *N. brasiliensis* on rat food intake were dose-dependent. Moreover, the relationship between host malnutrition and worm burden was synergistic. During schistosomiasis, nutrition and infection may interact synergistically or antagonistically depending upon the severity of infection, stage of parasite development and nutritional state (Akpom, 1982). Knauft and Warren (1969) demonstrated that during 'moderate' infection severe calorie deprivation suppressed parasite maturation and egg production and ameliorated the pathological effects of *Schistosoma mansoni*. Calorie deprivation, however, had limited effects on the host. In contrast, protein deficiency had severe effects on both host and parasite, leading to a synergistic pathological state and extreme host liver cell necrosis. Despite having negative effects on the host and parasite, malnutrition reduced mortality from schistosomiasis (Akpom and Warren, 1975).

Studies on invertebrate hosts

Food intake, growth and conversion efficiency in insect hosts

Studies of parasitic associations involving invertebrate hosts have also demonstrated alterations in host nutrition. The nutritional interactions of hymenopterous insect parasites with their insect hosts are of particular interest because the parasite biomass often represents a significantly greater proportion of the host–parasite complex than is normally observed in relationships involving vertebrate hosts. Thus, it is essential that the parasites development and nutrition be carefully regulated to ensure the survival of the host during parasite development (Slansky, 1986). Most hymenopterous parasites develop within the haemocoel of the host larva and are parasitic only during their immature larval stages. The developing larvae obtain their nutrition from the haemolymph or 'blood' of the host. In the case of solitary parasites, the mature larvae generally consume the host's internal tissues prior to pupation. Host death is imminent.

Investigations of several hymenopterous parasite species have demonstrated reduced host food consumption immediately following parasitization and during the early stages of parasite development (Thompson, 1983a). Thompson (1982a,b) reported that food consumption by *Trichoplusia ni* was sharply reduced following parasitization by the ichneumonid *Hyposoter exiquae*. Moreover, this effect was accompanied by reduced host growth and delayed development. Additional study with pair-fed animals demonstrated that while the altered growth and development pattern could be accounted for by the reduction in food consumption, parasitized hosts displayed a greater rate of assimilation than unparasitized pair-fed animals (Thompson, 1983b). Both the percentage assimilation and growth rate were greater in parasitized insects than in pair-fed controls. Parasitized hosts appeared to compensate for decreased food consumption but pair-fed controls were unable to do so. The net conversion efficiency of parasitized insects, however, was equivalent to pair-fed animals but lower than that of *ad libitum*-fed controls.

In contrast to the effects of parasitization by *H. exiquae* and other solitary parasites on host nutrition, Cloutier and Mackauer (1979, 1981) reported that the aphid *Acyrthosiphon pisum* displayed increased rates of feeding and growth when parasitized by the solitary hymenopteran, *Aphidius smithi*. The precise effects observed were related in a complex manner to the host's stage of development when parasitized as well as the development stage of the parasite (figure 3.1(a)). The rate of assimilation of *A. pisum* was elevated following parasitization but percentage assimilation and net conversion efficiency were unaltered. Cloutier (1986) presented a hypothetical pictorial model of food consumption and utilization in parasitized *A. pisum* (figure 3.1(b)). The model emphasizes that host feeding is regulated by nervous input from peripheral food receptors, receptors associated with the alimentary canal as well as by signals such as haemolymph trehalose levels that reflect the host's nutritional state. It was suggested that the parasite influences host feeding by affecting the concentration of signal nutrients in the haemolymph. The nutrient in the host's haemolymph is utilized proportionately by the host tissues and the developing parasite. As the parasite grows, its use of haemolymph nutrient increases relative to that of the host. Likewise, total weight gain reflects that of the host as well as the parasite, including stored waste products. Weight loss by the parasite results from energy metabolism or respiration which when added to that of the host makes up the total respiratory loss. Host excretion comprizes both unabsorbed food as well as excreted end products of host metabolism.

(a)

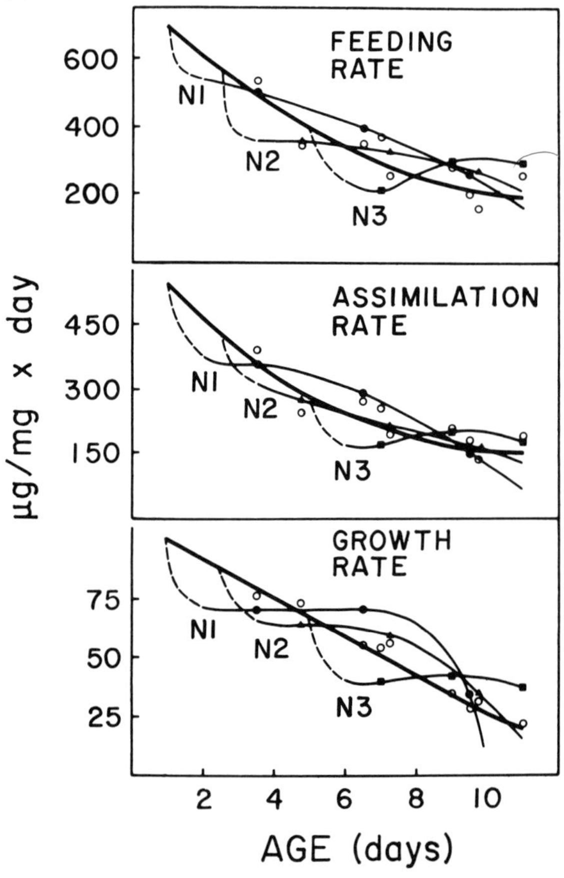

Figure 3.1. Effects of the insect parasite, *Aphidius smithi*, on the nutrition of its aphid host, *Acrythosiphon pisum*. (a) Effects on feeding, assimilation and growth rates with age of *A. pisum* when parasitized in the first (N1), second (N2) and third (N3) instars. Heavy line is unparasitized control. After Cloutier and Mackauer (1979). (b) Model of feeding and food utilization in parasitized *A. pisum*. ←——, biomass transfer and ← – – –, regulatory input Gh and Gp, nutrient converted to host and parasite biomass, respectively; Rh and Rp, haemolymph nutrient metabolized and respired by host and parasite, respectively; Sh, secretion of unabsorbed food and honeydew by the host; Sp, accumulated exretory products of the parasite which are retained until pupation and contribute to the biomass of the host–parasite complex. Boxes show quantifiable biomass in different compartments of the host—parasite complex.
After Cloutier (1986).

(b)

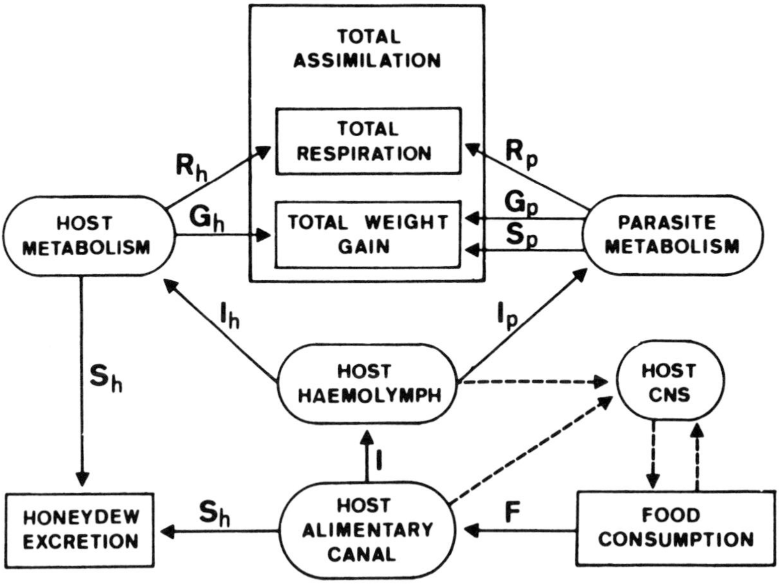

Figure 3.1. Cont'd.

Alterations in net conversion efficiency have been demonstrated in insect hosts parasitized by gregarious hymenopterous parasites. Slansky (1978) reported that larvae of *Pieris rapae* parasitized by the braconid *Apanteles glomeratus* displayed increased growth rates and higher conversion efficiencies. Host food consumption was unaffected. It was further demonstrated that parasite larvae consumed no more than 60% of the food assimilated by the host thus ensuring survival of the host. Beckage and Riddiford (1983) demonstrated that successful development of *A. congregatus* in *Manduca sexta* was closely related to the nutritional status of the host as well as to parasite burden. Although parasitized insects displayed reduced food consumption, starvation of host larvae at the time of the parasite's first larval ecdysis prevented further development of the parasite. When the host was starved at progressively later stages of parasite development, an increasing proportion of parasites completed development and emerged. Thus, host nutrition played a critical role in ensuring the success of this parasitic relationship. Additional investigation demonstrated that the biomass of the host–parasite complex was positively correlated to the numbers of parasites developing in the host. Moreover, the host contribution to the total biomass of the host–parasite complex was greater in heavily parasitized hosts than in lightly parasitized individuals. The latter result was consistent with the finding of Führer and Keja (1976) who reported that the rate of assimilation in *P. brassicae* parasitized by *A. glomeratus* was increased with parasite load.

Numerous studies with insects have demonstrated an inverse relationship between feeding and locomotion (Bernays and Simpson 1982). Beckage and

Templeton (1986) reported that feeding and locomotion of terminal stage parasitized *M. sexta* gradually diminished and ceased prior to emergence of *A. congregatus*. Thus, the normal behavioural response of *M. sexta* to reduced nutrient intake, that is, increased locomotion and food searching, was absent during parasitism.

Growth and conversion efficiency in molluscan hosts

Alterations in nutritionally-related physiology have also been reported in parasitized invertebrates serving as intermediate hosts. Increased growth or 'gigantism' has been reported during many parasitic associations of molluscs and digenean parasites (Wright, 1969; Malek and Cheng, 1974). In reference to the large size of trematode-parasitized *Lymnaea* spp., Wesenberg-Lund (1934; cited in Wright, 1969) stated that the host was 'eating for two'. Davis and Farley (1973) suggested that higher efficiency of food conversion was responsible for the accelerated growth of *Littorina saxatilis* following parasitization by the trematode *Cryptocotyle lingua*.

Few studies have been made on the nutritional physiology of snail hosts to assess the impact of trematode infection on nutrient assimilation and energy conversion. Thompson and Mejia-Scales (1989) recently examined the effects of *S. mansoni* infection on the nutrition of *Biomphalaria glabrata* when snails were fed an alginate medium containing the alga *Spirulina*. Growth was reduced in snails with patent infections of *S. mansoni*, but food consumption and assimilation were unaffected. Similar results were observed in uninfected snails fed media with reduced nutrient levels supporting the hypothesis that infection produces an artificial state of host starvation, as discussed below (Becker, 1980a). Infected snails displayed sharply reduced efficiencies of food conversion but conversion efficiency was unaffected by food deprivation. The decreased conversion efficiency in infected *B. glabrata* indicated that a significant portion of the food energy consumed by the host was directly or indirectly utilized by the parasite for development and reproduction.

HORMONAL ALTERATION

Hormonal interactions characterize many host–parasite relationships. The host's endocrine function often plays a critical role in the formation and success of the host–parasite complex.

Studies on invertebrate hosts

Developmental alterations in insect hosts

The development of many parasites of invertebrate hosts is synchronized with that of the host, and parasite development may be directly affected by host hormones or by biochemical alterations in the host brought about by hormonally-triggered events (Thompson, 1983a; Beckage, 1985; Lawrence, 1986). In other cases, hormones may mediate many of the host responses observed during parasitic infection. Host responses may result from direct perturbations in the host's normal endocrine function or from changes in the host's hormonally-regulated activities brought about by production of hormones or hormone-like substances by the parasite (Thompson, 1983a; Beckage, 1985).

Parasitic associations involving hymenopterous insect parasites demonstrate the diversity of hormonally-mediated effects on the host. These parasites often delay the development and metamorphosis of their insect hosts. Altered host hormone levels have been demonstrated to be responsible in many cases, as, for example, during the host associations of *Apanteles* spp. Smith and Smilowitz (1976) reported arrested larval development of *Pieris rapae* following parasitization by *Apanteles glomeratus*. Endocrine involvement was suggested by Schopf (1984) who reported elevated levels of juvenile hormone in parasitized insects. Moreover, based on the results of ligature experiments it was concluded that elevated hormone titres were due to secretion by the parasite during ecdysis. The small size of the ecdysial glands of parasitized larvae suggested that reduced host ecdysteroid secretion may also occur and explain the absence of pupation by parasitized individuals.

Beckage and Riddiford (1978) reported similar developmental effects in *Manduca sexta* larvae following parasitization by *A. congregatus* (figure 3.2). Beckage and Riddiford (1982) demonstrated that a high juvenile hormone level in the presence of ecdysteroid in the host's haemolymph suppressed the host's development during parasite emergence, and low ecdysteroid and 20-hydroxyecdysone moulting hormone concentrations following emergence caused failure of host moulting thereafter. Recently, Beckage and Templeton (1986) suggested that the conversion of ecdysteroid to 20-hydroxyecdysone was low as a result of depressed ecdysone 20-monooxygenase levels in the midgut and fat body of parasitized larvae (figure 3.2(b) and (c)). A low level of haemolymph juvenile hormone esterase activity contributed to the high juvenile hormone levels in parasitized larvae (Beckage and Riddiford, 1982) (figure 3.2(d)).

Altered hormone levels may also be responsible for the absence of wandering behaviour in parasitized *M. sexta* (Beckage and Riddiford, 1978, 1982). In normal unparasitized larvae ecdysteroid in the absence of juvenile hormone results in pupal commitment and induces wandering during the last day of the terminal stage prior to pupation (Bollenbacher, 1988) (figure 3.2). The high juvenile hormone and low ecdysteroid levels reported by Beckage and Riddiford (1982) in parasitized larvae were, therefore, consistent with an absence of wandering. Changes in neural control over locomotor activity, however, may also contribute. During the feeding phase of the terminal instar, locomotor activity is stimulated by the suboesophageal ganglion and the brain exerts an inhibitory influence. In contrast, locomotor activity during the wandering phase is regulated by the brain. Beckage and Templeton (1986) were able to induce wandering in parasitized larvae by removing the brain prior to parasite emergence, thus suggesting that locomotor control had not yet shifted to that organ. It was concluded that intact parasitized larvae remain

(a)

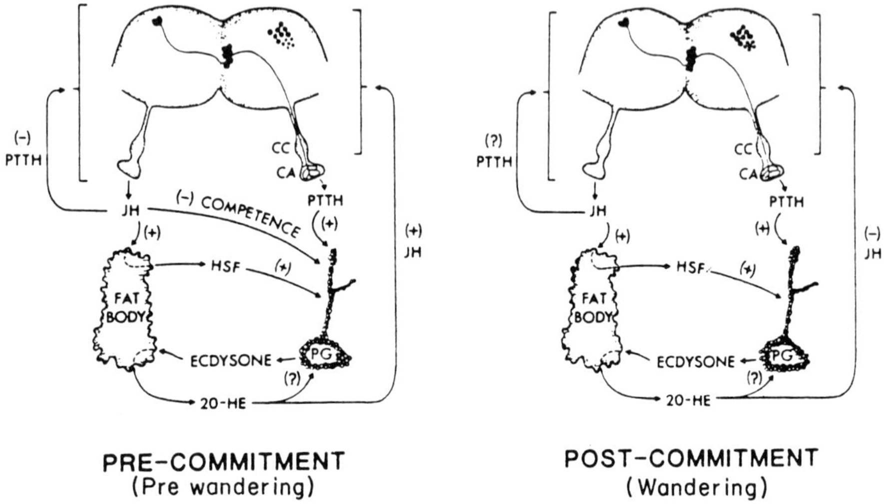
(–) PTTH
CC
CA
PTTH
JH
(–) COMPETENCE
(+)
(+)
HSF
(+)
FAT BODY
ECDYSONE
PG
(?)
20-HE
(+) JH
PRE-COMMITMENT
(Pre wandering)
(?) PTTH
CC
CA
PTTH
JH
(+)
(+)
HSF
(+)
FAT BODY
ECDYSONE
PG
(?)
20-HE
(–) JH
POST-COMMITMENT
(Wandering)

(b)

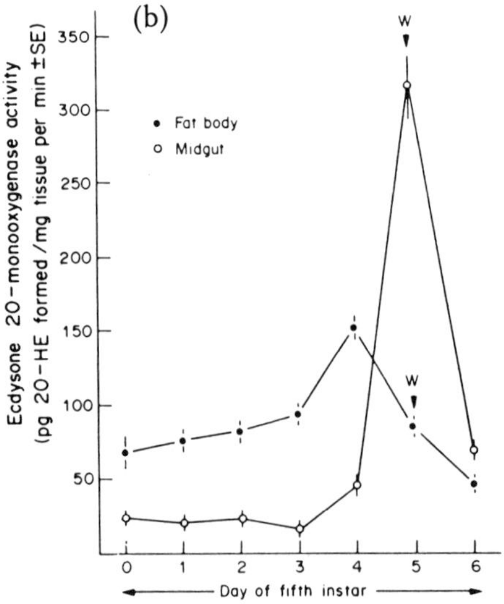
Ecdysone 20-monooxygenase activity
(pg 20-HE formed/mg tissue per min ±SE)
350
300
250
200
150
100
50
• Fat body
○ Midgut
W
W
0 1 2 3 4 5 6
Day of fifth instar

(c)

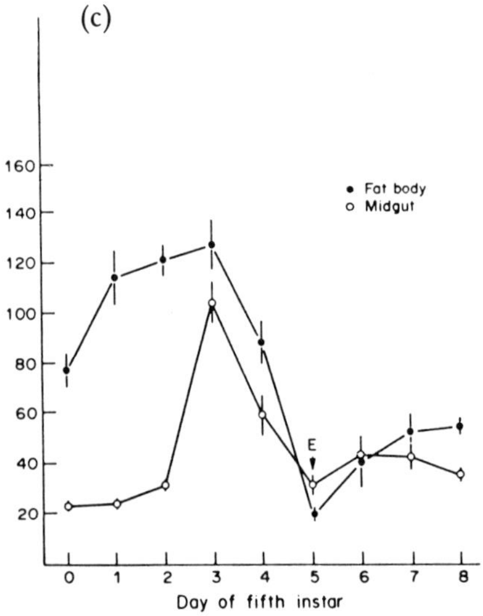
160
140
120
100
80
60
40
20
• Fat body
○ Midgut
E
0 1 2 3 4 5 6 7 8
Day of fifth instar

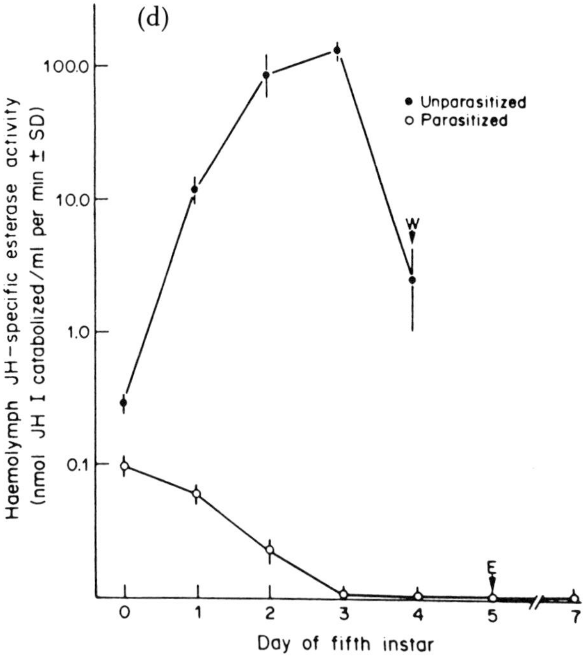

Figure 3.2. Pictorial representation of the endocrine system and hormonal control of larval–pupal development of the insect *Manduca sexta*. (a) Pictorial representation of the endocrine interactions of precommitted (prewandering) feeding last instar larvae and larvae committed (wandering) to pupation. PTTH, prothoracicotropic hormone; JH, juvenile hormone; HSF, haemolymph stimulatory factor; 20-HE, 20-hydroxyecdysone; CC, corpora cardiaca; CA, corpora allata; PG, prothoracic gland. After Bollenbacher (1988). Larvae parasitized by *Apanteles congregatus* fail to reach the commitment stage. (b) Activity of ecdysone 20-monooxygenase in the tissues of unparasitized last instar *M. sexta* larvae. (c) Activity of ecdysone 20-monooxygenase in the tissues of last instar host larvae parasitized by *A. congregatus*. (d) Effect of parasitization on haemolymph juvenile hormone esterase activity of last instar host larvae. After Beckage and Templeton (1986).

capable of active locomotion and absence of locomotion is not due to disruption of neuromuscular coordination or parasitic lesions in the musculature. Although the basis for the altered feeding and locomotor responses of *M. sexta* parasitized by *A. congregatus* are not yet clear, further study may implicate previously unknown chemical mediators of insect activity. Davenport and Evans (1984) recently demonstrated that increased locomotion in response to starvation in *Schistocerca gregaria* was correlated with haemolymph octopamine levels.

Delayed larval development and failure to pupate by *T. ni* parasitized by *H. exiguae* also appears to be hormonally based. Smilowitz (1974) and Iwantsch and Smilowitz (1975) demonstrated that activation of the ecdysial gland and subsequent release of ecdysteroid prior to pupation failed to occur in parasitized larvae. It was suggested that inhibition of ecdysiotropin secretion may have been responsible. Thompson (1982a) reported that host juvenile hormone esterase activity present at the normal time of moulting from the fourth to fifth instar was absent in parasitized insects and presumed to be delayed. It was concluded that this resulted in the maintenance of high juvenile hormone titres at the normal time of larval moulting thus delaying the process.

In contrast to the above results demonstrating delayed development and absence of pupation in parasitized hosts, Jones *et al.* (1981) reported that *T. ni* parasitized by the hymenopteran *Chelonus curvimaculatus* displayed cocoon-spinning behaviour and initiated pupation prematurely following the penultimate fourth larval stadium rather than after the fifth. Food consumption of parasitized insects was less than that of controls, but control and parasitized hosts reached the fourth stadium in the same time. Fourth instar parasitized *T. ni*, however, were smaller than the critical size required to complete pupation. Although larvae initiated metamorphosis, precocious pupation was not successful. Similar effects on development of *T. ni* were later reported following parasitization by *C. insularis* (Jones, 1985a). Subsequent elaborate and detailed studies on these host–parasite relationships demonstrated that developmental events in penultimate larvae including corpora allata activity, juvenile hormone esterase activity, *in vivo* rates of juvenile hormone metabolism and biochemical alterations in haemolymph composition all followed the pattern observed in normal last instar larvae initiating pupation (Jones and Sreekrishna, 1984; Jones, 1985b). Premature metamorphosis appeared to result directly from a decrease in the critical size parameter that signalled attainment of the last instar. Lack of successful pupation resulted from alteration of normal hormonal events in the host including a lack of sufficient prepupal ecdysteroids that in turn resulted from low levels of juvenile hormone synthesis (Jones, 1985a). Recently, Jones *et al.* (1986) reported that suppressed juvenile hormone titres in parasitized larvae resulted in both low ecdysteroid levels and a decreased conversion of ecdysone of 20-hydroxyecdysone.

Reproductive efficiency and parasitic castration in molluscan hosts

Altered reproductive development, activity and behaviour are commonly observed in parasitized invertebrate hosts and such changes often appear to involve hormonal physiology. The most definitive studies on the endocrine basis for altered reproductive activity in parasitized invertebrates have been carried out on the molluscan intermediate hosts of digenean parasites. Reduction or elimination of egg production often termed 'parasitic castration' is a well-known phenomenon (Pan, 1965; Meuleman, 1972; Malek and Cheng, 1974; see also Chapters 1, 2 and 7) and often accompanies the increased growth observed in parasitized hosts, as discussed above. Thus, Rothschild (1936; cited in Wright, 1969) suggested that overgrown snails infected with larval trematodes were 'eunuchs'; their gigantic size reflecting possible alterations in the hormonal physiology of the host. This conclusion was initially supported by the knowledge that during many host–parasite associations direct tissue destruction by the developing parasite was not observed.

Extensive studies on the effects of infection by the avian schistosome *Trichobilharzia ocellata* on the host's endocrine function in relation to reproductive

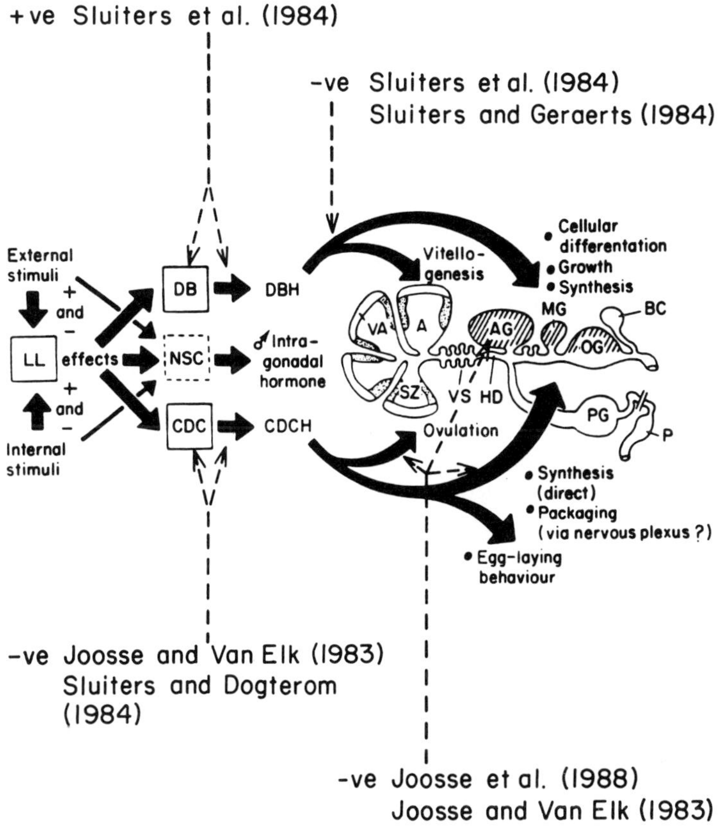

Figure 3.3. Pictorial representation of the endocrine system and hormonal control of reproduction in the snail *Lymnaea stagnalis*. CDCH, caudodorsal cell hormone; CDC, caudodorsal cells; DB, dorsal bodies; LL, lateral lobes; NSC, neurosecretory cells; A, acinus of the ovotestis; AG, albumin gland; BC, bursa copulatrix; HG, hermaphrodite duct; MG, muciparous gland; OG, oothecal gland; P, penis; PG, prostate gland; SZ, spermatogenic zone; VA, vitellogenic area; VS, vesiculae seminales. After Geraerts and Joosse (1984). The figure has been modified to show those functions and targets altered during parasitism by the trematode, *Trichobilharzia ocellata*.

activity have been carried out on *Lymnaea stagnalis* (figure 3.3). Sluiters *et al.* (1980) reported that this hermaphroditic species failed to oviposit following infection, and growth of the gonad as well as gamete production were sharply reduced. Moreover, differentiation of both male and female accessory sex organs was retarded (Sluiters, 1981; Joosse and Van Elk, 1983). Maturation of the female reproductive organs in *L. stagnalis* is controlled by gonadotropic dorsal body hormone, secreted by the dorsal bodies attached to the perineurium of the cerebral ganglia. The hormone stimulates vitellogenesis of oocytes in the ovotestis as well as growth, cellular differentiation and synthetic activity of the female accessory organs. Sluiters *et al.* (1984) demonstrated an increased activity of the dorsal bodies of infected snails but the action of the dorsal body hormone on the reproductive system was impeded (Sluiters and Geraerts, 1984a,b). Sluiters and Dogterom (1984) subsequently reported that infected snails had reduced levels of a second gonadotropic hormone, caudodorsal cell hormone, that stimulates ovulation and oviposition and is produced in the neuroendocrine caudodorsal cells of the cerebral ganglia. Moreover, infected snails were less responsive to injections of hormone. Mellink and Van den Bovenkamp (1983) reported that oviposition was delayed when snails were injected with haemolymph from parasitized animals and De Jong-Brink *et al.* (1988a) suggested that the parasite produces a factor that interferes with the activity of the reproductive hormones.

Further study of infected *L. stagnalis* demonstrated a depressed activity of calfluxin, a gonadotropic hormone regulating the development and activity of the albumin gland (De Jong-Brink *et al.*, 1988a,b). The albumin gland is a secondary sex organ producing the nutritive fluid containing galactogen and protein for nourishment of the developing embryo. Calfluxin regulates calcium influx into mitochondria at a specific time during oocyte migration through the reproductive tract. The effect on calfluxin activity was related to the presence of a heat-stable pronase-labile agent, schistosomin, associated with the secondary or daughter sporocyst stage of the parasite. Recently, Joosse *et al.* (1988) reported that galactogen synthesis in isolated albumin glands was reduced markedly during incubation in haemolymph of infected snails. The authors concluded that the agent(s) responsible has a peptide nature and probably competes with the dorsal body hormone for the same receptors in the albumin gland. It is not clear at this time, however, whether the responsible agent(s) is a product of the parasite itself or whether the parasite induces the host to produce the agent in response to parasitization.

Studies of several other trematode–host relationships have also suggested a relationship between parasitic castration, neuroendocrine pathology and a 'toxic' factor from the parasite (Venkata Rama Krishna and Sunder Simha, 1980). Recent investigations have demonstrated that a factor in the parasitic larvae of *Zoogonus lasius* influenced reproductive development of *Ilyanassa obsoleta*. Parasitization of this snail species results in inhibition of gametogenesis and parasitic castration. Although *Z. lasius* has a redial larval stage causing host tissue destruction, Cheng *et al.* (1973) concluded that physical damage was not responsible for the observed effects and suggested that a factor originating from the parasite was involved. Subsequently, Pearson and Cheng (1985) demonstrated *in vitro* that an extract of the parasite sporocyst inhibited spermatogenesis in excised testes. Sullivan *et al.* (1985) suggested that such a factor may act through antagonism of the neurosecretions essential for gamete formation.

Reproductive efficiency and sexual character in insect hosts

Parasitic castration is also a common occurrence in parasitized insects and crustaceans. This effect by strepsipterous parasites of homopterous and hymenopterous insect hosts was described over a century ago. Of related interest are apparent sexual inversions and changes in external and internal sex characteristics often observed in parasitized hosts. The latter changes in hymenopterous hosts result in intersexes, individuals possessing secondary sexual characteristics of both sexes. Wüker (1964) emphasized that sexual changes are highly variable, not uniform, and true intersexes are rare. The changes undoubtedly have significant behavioural implications, but little is known concerning intersex behaviour and the physiological basis for the changes has yet to be established. Parasitic castration itself is not likely to be responsible for alterations in secondary sex characters because alterations sometimes occur in the absence of castration. Moreover, insect gonads do not appear to produce sex hormones. The potential contribution of alterations in the hosts' neuroendocrine physiology brought about by parasitization or resulting from secretions of the parasite was discussed by Wüker (1974). Nutritional stress that brings about expression of genes for the opposite sex has been suggested as a possible factor (Wüker, 1964; Askew, 1971).

Studies on vertebrate hosts

In contrast to extensive work on parasite-induced alterations in endocrine physiology of invertebrates, less is known concerning vertebrate hosts. However, as indicated in Chapter 2, the endocrine regulated physiology of some vertebrate hosts is markedly altered by a chemical factor produced by the plerocercoid larval stage of the tapeworm *Spirometra* spp. (Mueller, 1980). Mueller (1963) first reported that infection by spargana of *S. mansonoides* resulted in abnormal growth of the paratenic mouse host. Obesity was initially thought responsible, but it was later demonstrated that skeletal and muscular growth were stimulated (Mueller, 1965). Similar growth was also reported in other infected rodents. Ruegamer and Mueller (1973a) reported accelerated growth in infected rats even when hosts were maintained on nutritionally deficient diets. Moreover, growth of both thyroidectomized and hypophysectomized rats was stimulated by infection (Ruegamer and Mueller, 1973a,b). Phares and Ruegamer (1973) demonstrated that the growth effect was due to parasite secretion of a growth factor with properties similar to pituitary growth hormone. Indeed, Steelman *et al.* (1970) demonstrated that the action of the growth factor on the tibial cartilage of hypophysectomized rats was identical to that of growth hormone. Recently, Phares and Booth (1987) reported that antihuman growth hormone antibodies cross-react with the parasite growth factor and Phares and Watts (1988) demonstrated that the growth factor displaces growth hormone and binds to growth hormone receptors on cultured human lymphocytes (figure 3.4). *In vivo*, the secretion of growth hormone and thyroxine are reduced and immune responsiveness may be depressed (Sharp *et al.*, 1982).

Meyer *et al.* (1965) reported that *S. mansonoides*-infected hamsters had an increased carcase fat content. Similar observations were made by Phares and Carroll (1977, 1978). Recently, Phares and Carroll (1984) reported an increased activity in lipogenic enzymes and suggested that the parasite factor exhibits a chronic insulin-like activity on lipid metabolism. This effect of the parasite factor is in contrast to that of growth hormone which causes a depletion of adipose tissue stores. It should also be noted that infection has no insulin-like effect on blood sugar. Indeed,

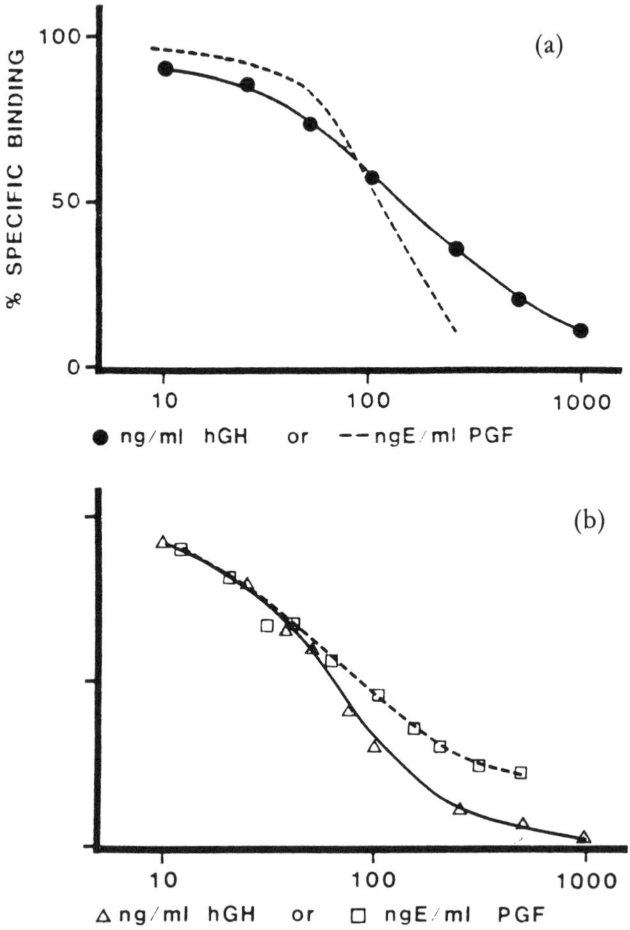

Figure 3.4. Binding competition between human growth hormone (hGH) and plerocercoid growth factor (PGF) from *Spirometra mansonoides*. The binding curves were analyzed independently by logit-log regression. (a) Competitive displacement of [^{125}I]hGH from human lymphocytes by hGH and PGF. The ED_{50}s (estimated dose required to displace 50% of the ^{125}I for hGH and PGF were 129 and 118 ng/ml, respectively. (b) Competitive displacement of [^{125}I] hGH from an anti-hGH monoclonal antibody by increasing doses of hGH and PGF. The ED_{50}s for hGH and PGF in this assay were 58 and 85 ng/ml, respectively. After Phares and Cox (1987.)

infected mice displayed elevated fasting blood sugar levels (Sudan *et al.*, 1965). Because parasite infection was previously demonstrated to elevate somatomedin levels in rats (Garland *et al.*, 1971) and the latter hormone is known to mimic insulin, Phares and Carroll (1984) concluded that increased somatomedin activity may be the basis for the observed effects on lipogenesis.

Phares and Cox (1987) carried out molecular hybridization studies and demonstrated significant homology between sequences of *S. mansonoides* plerocercoid DNA and that of the human growth hormone gene. It was concluded that a gene for growth hormone was incorporated into the genome of *S. mansonoides* during evolution and was subsequently modified to direct the synthesis of a factor with the ability to stimulate growth but which has lost the anti-insulin/diabetogenic characteristics of growth hormone.

Additional studies on the basis for the growth effects observed during *Spirometra* infection and the potential involvement of the host's endocrine system have been carried out on mice infected with *S. erinacei*. Although the effects of the latter species on its host differ in several ways from those reported for *S. mansonoides* (Hirai *et al.*, 1983; see table 2.2 in Chapter 2 for details) it has been demonstrated to enhance growth in rodents in a manner similar to growth hormone and the secretion of a parasite factor mimicking growth hormone was suggested (Hirai *et al.*, 1978; Shiwaku and Hirai, 1982; Hirai *et al.*, 1986). Shiwaku *et al.* (1983) hypothesized that the parasite stimulates the growth hormone releasing factor and somatostatin–growth hormone–somatomedin axis or a secretion of the larva has intrinsic somatomedin activity. The authors reported increased growth and elevated serum somatomedin activity in dwarf mice genetically lacking growth hormone. Because the genetic defect could not be corrected by hormone treatment, it was concluded that the parasite secretes a growth hormone-like or somatomedin-like substance. Subsequently, Shiwaku *et al.* (1986) confirmed that plerocercoids induce the mitosis of costal cartilage cells by an enhancement of serum somatomedin activity.

Hirai *et al.* (1988) recently examined the effects of *S. erinacei* plerocercoid infection on thyroid hormone levels in mice. Infection of propylthiouracil-induced hypothyroid mice resulted in growth equivalent to that observed in uninfected normal mice. T4 levels were reduced by infection, and it was, therefore, concluded that the growth factor secreted by the parasite was responsible for the observed growth stimulation.

Mueller (1980) considered the hypothesis that accelerated growth and large size of mice infected by the plerocercoid of *S. mansonoides* might prove adaptive to the parasite by reducing host speed and manoeuverability, thus improving the probability of predation by the definitive host, the cat. It was pointed out that the water snake *Natrix* spp. rather than the mouse is the natural host, and, moreover, that the effects of parasite-derived growth factor have not been observed in the snake. The adaptive significance for the physiological alterations resulting from growth factor activity is, therefore, difficult to reconcile with the normal life cycle of the parasite.

METABOLIC ALTERATION

Studies on vertebrate hosts

Parasitic infection often results in altered host metabolism. In vertebrate hosts such alterations generally reflect pathological responses.

Protein metabolism in sheep and guinea-pigs

Altered protein metabolism in relation to anorexia and gastrointestinal pathology has been extensively studied during infections of guinea-pigs and sheep with the intestinal nematode *Trichostrongylus colubriformis*. Several workers have reported inappetence during infection with this parasite species and further that the degree of anorexia was closely related to parasite burden (Steel *et al.*, 1980, 1982). Depressed food consumption and skeletal muscle protein synthesis were reported in infected hosts and pair-fed uninfected controls (Symons and Jones, 1971, 1975). Based on incorporation of injected radioactive leucine per unit body weight a sharp elevation in hepatic protein synthesis was observed in infected hosts. Pair-fed sheep also displayed increased synthesis of liver proteins (Symons and Jones, 1975).

Further studies with infected guinea-pigs demonstrated that increased liver protein synthesis occurred in response to loss of serum protein into the intestine (Symons *et al.*, 1974). Such loss did not occur in uninfected hosts fed nutritionally inadequate diets. Symons and Jones (1978) reported a greater depression of muscle protein synthesis and rate of weight loss in relation to decreased food intake together with a greater increase in liver protein synthesis in infected guinea-pigs when compared with pair-fed animals, indicating that anorexia was not solely responsible for the metabolic alterations observed. Symons and Jones (1983) examined protein synthesis in guinea-pig intestine. Protein synthesis in infected animals per unit body weight was 25% greater in the small intestine and 70% greater in the large intestine than demonstrated in controls. Synthesis by pair-fed animals was less than that observed in infected or control animals and the effect of infection, therefore, was not attributable to anorexia alone.

Jones and Symons (1982) utilized constant infusion of radioactive tyrosine to monitor protein synthesis in individual organs and the whole body of lambs. Whole body protein synthesis per unit body weight was similar in *T. colubriformis*-infected lambs and *ad libitum*-fed controls. Pair-fed sheep displayed reduced synthesis. Based on protein intake, however, infected lambs had higher rates of protein synthesis than controls and pair-fed animals had rates similar to controls. Protein synthesis was increased in the liver and depressed in skeletal muscle in infected lambs. In contrast to the results of previous studies, protein synthesis in liver and skeletal muscle was unaffected in pair-fed animals. This difference in results was subsequently discussed by Symons (1985). Based on the above findings, as well as analyses demonstrating significantly reduced plasma albumin levels in infected but not pair-fed animals, it was concluded that enteric loss of plasma was responsible for the increased liver protein synthesis observed during infection. Symons (1985) later concluded that the increase in whole body protein synthesis per unit protein consumed observed in infected lambs was the result of increased synthesis of plasma proteins as well as intestinal protein, the latter observed in studies of infected guinea-pigs.

It is evident from the studies of Symons and colleagues outlined above that the interactions between food consumption, infection, metabolism and pathogenesis during nematode infection is complex (figure 3.5). Although some effects of

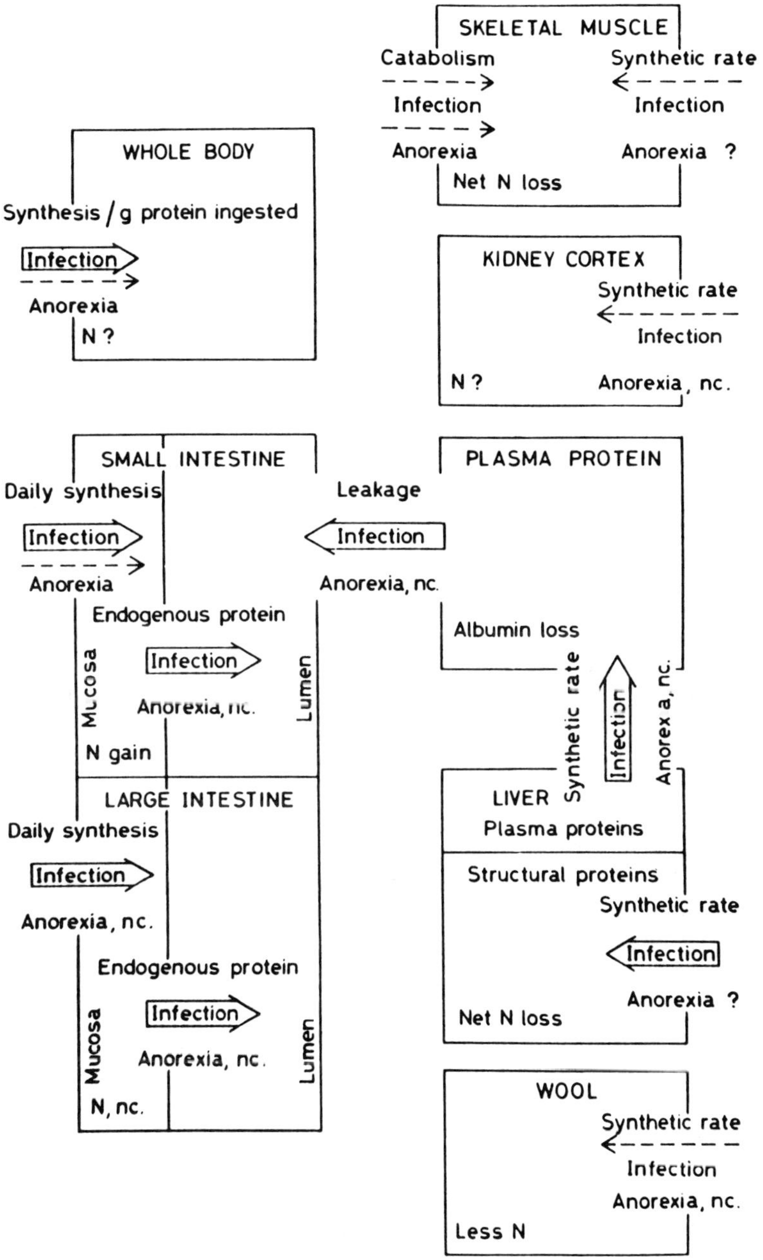

Figure 3.5. Schematic summary of the pathological effects of infection by the nematode *Trichostrongylus colubriformis* and anorexia on protein synthesis in the sheep. ⟸, increases; ← – – –, decreases; nc, no change; N, nitrogen. After Symons (1985)

infection were distinct from those of anorexia, the latter was an important factor. The effects of infection on metabolism were clearly related to the enteric leakage of plasma and protein that characterizes infection by *T. colubriformis* as well as other intestinal nematodes. Similar studies during infection with *Ostertagia* spp. were discussed by Holmes (1987).

Studies on invertebrate hosts

Alterations in metabolism and their interactions with nutrition and hormonal physiology are well characterized in host–parasite relationships involving invertebrate hosts. In contrast to the effects of parasitism on the metabolism of vertebrate hosts, those observed in invertebrates often appear less associated with pathological effects.

Carbohydrate metabolism in insect hosts

Metabolic alterations during the host relationships of several hymenopterous insect parasites have been well characterized. The most exhaustive investigations have been made with the solitary insect parasite, *Hyposoter exiguae*, in association with its host, *Trichoplusia ni* (figure 3.6). Thompson and Yamada (1984a) evaluated the overall effect of parasitism in *T. ni* by examining the levels of adenylates and adenylate energy charge (ATP + 1/2 ADP)/(ATP + ADP + AMP) in the fat body. The adenylate nucleotide pool is of prime importance in cellular energy metabolism and adenylate ratios, particularly the energy charge ratio, play a critical role in metabolic regulation. Under normal physiological conditions the energy charge is maintained within a very specific range and numerous studies have suggested that this reflects the overall energetic state of cells and tissues. Measurement of the adenylate nucleotide pool as an indicator of energy status and host viability has been suggested as a means for evaluating the overall nature of host response to parasitic infection (Yamada and Sherman, 1980). Drastic changes may reflect metabolic stress and pathogenesis. In the fat body of parasitized *T. ni* larvae, the ATP level was reduced and AMP elevated. The energy charge, however, was only slightly altered following parasitization and it was concluded that development of *H. exiguae* is closely integrated with host physiology and occurs independent of pathological effects on host metabolism.

Sharply elevated levels of carbohydrate were observed in *T. ni* following parasitization (Thompson, 1982a). Trehalose levels were increased twofold within 24 hours and this was accompanied by a marked elevation in fat body glycogen stores. Metabolic studies demonstrated a several-fold increase in fructose 1,6-bisphosphatase activity in the fat body (Thompson and Binder, 1984). Moreover, haemolymph levels of free amino acids were depleted (Thompson, 1986a) and uric acid levels increased (Thompson and Cohen, 1984). Thompson (1986a) concluded that the effects of parasitism on carbohydrate reserves resulted from an increase in the rate of gluconeogenesis from amino acids. Increased gluconeogenesis, however, was not demonstrated *in vivo*.

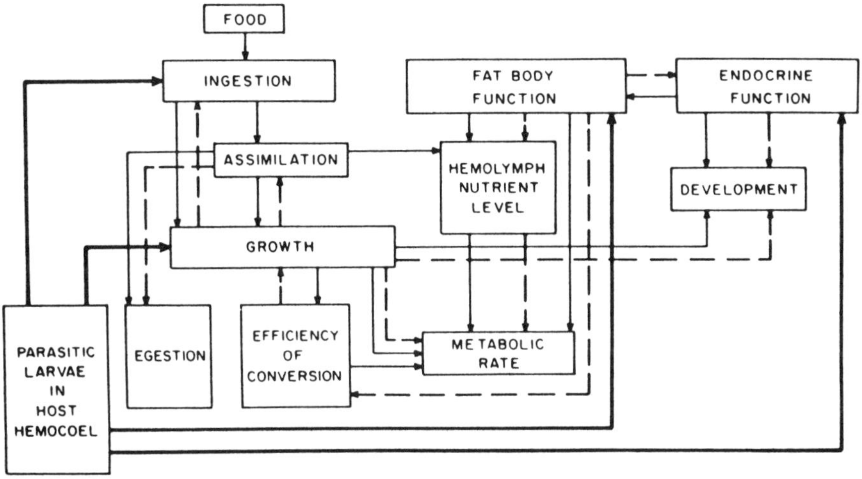

Figure 3.6. Schematic summary of the effects of the parasite *Hyposoter ex iguae* on the nutrition and physiology of its host, *Trichoplusia ni*. ←—, direct host–parasite interaction; ←—, physiological interactions in the host; ←---, indirect host–parasite interaction. After Thompson (1983a).

Thompson (1985b) reported that fat body fructose 1,6-bisphosphatase was less sensitive to AMP inhibition in *T. ni* after parasitization by *H. exiguae* explaining the enzyme's elevated activity coincident with increased AMP. Decreased sensitivity of fructose 1,6-bisphosphatase to AMP inhibition is known to contribute to the glucogenic effect of the hormone glucagon in mammals and Thompson (1986a) suggested that the effects of *H. exiguae* on metabolism of *T. ni* may be hormonally based. Whether the metabolic alterations observed may be correlated with changes in development and developmental hormone titres is unknown.

The effects of dietary carbohydrate level on haemolymph trehalose in *T. ni* were reported by Thompson (1986b). On a carbohydrate-free medium parasitized larvae had significantly higher haemolymph trehalose levels than unparasitized controls. When dietary sucrose was increased from 0 to 4, 8 and 12 g/l haemolymph trehalose was unaffected in controls but was elevated in parasitized insects. Because dietary carbohydrate is known to inhibit gluconeogenesis, it was concluded that this metabolic process may be abnormally regulated in response to dietary carbohydrate in *T. ni* parasitized by *H. exiguae*.

Carbohydrate and protein metabolism in molluscan hosts

Comprehensive studies have been made on the effects of parasitism on the metabolism of schistosome-infected snails. In contrast to the effects of parasitization by many insect parasites on the carbohydrate levels of their hosts, a characteristic feature of schistosome infection is the depletion of snail tissue carbohydrate levels. Cheng and Lee (1971) demonstrated a sharp decline in haemolymph glucose levels in *B. glabrata* and depletion of glycogen reserves in the digestive gland-gonad was reported by Christie *et al.* (1974). Mohamed and El Fiki (1980) subsequently demonstrated similar effects of *S. mansoni* infection in *Biomphalaria alexandrina*. The metabolic bases for altered glycogen levels in infected *B. glabrata* were investigated by Schwartz and Carter (1982), who demonstrated that the activities of glycogen synthase and phosphorylase were altered during infection in a manner consistent with observed changes in tissue carbohydrate levels.

Depletion of haemolymph protein and amino acid levels in schistosome-infected *B. glabrata* was described by Gilbertson *et al.* (1967), Lee and Cheng (1972), and Anteson and Williams (1975). Christie and Michelson (1975) reported that aminotransferase activities were decreased coincident with declining protein levels in the digestive gland-gonad. The effects of these alterations on nitrogenous excretion were examined by Becker and Schmale (1975) who reported a significant elevation of haemolymph urea in infected snails. Becker and Schmale (1975, 1978) concluded that urea production was increased in response to elevated haemolymph ammonia levels produced by protein degradation. Subsequently, Schmale and Becker (1977) demonstrated that the Krebs-ornithine cycle was active in *B. glabrata* and that the activities of the individual enzymes were increased following infection with *S. mansoni*.

S. mansoni infected *B. glabrata* maintained adenylate levels and ratios equivalent to those of uninfected individuals demonstrating that the short-term effects of infection on metabolism were not pathogenic (Thompson and Yamada, 1984b). Many of the above investigations have demonstrated that host starvation in invertebrates results in similar effects to those of infection. For example, Trede and Becker (1982) and Stanislawski *et al.* (1979) reported that amino acid and glucose levels were depleted in a similar manner by infection and starvation.

Similarly, Christie *et al.* (1974) reported that the glycogen depletion observed during infection also occurred with starvation. Becker (1980a) evaluated these and other findings and concluded that the effects of 5 days starvation on the metabolites of *B. glabrata* were comparable with those observed following 30 days of infection by *S. mansoni*. Nutritional studies employing semi-synthetic diets have further demonstrated the similarity between the effects of starvation and schistosome infection on *B. glabrata* (Stanislawski and Becker, 1979).

Several investigations have been made on the effects of schistosome infection on parameters associated with the metabolic rate of *B. glabrata*. Lee and Cheng (1971a,b) demonstrated that oxygen consumption was positively correlated with heartbeat rate and that both were elevated during infection with *S. mansoni*. Subsequently, Meakins (1980) reported that increased oxygen consumption by infected snails was related to sporocyst migration and reproductive activity in the digestive gland. Williams and Gilbertson (1983) attempted to correlate heartbeat rate and locomotion with feeding. They demonstrated that feeding activity increased heartbeat rate and locomotor activity. When fed *ad libitum*, infected snails fed more often than controls and had higher heartbeat rates during both day and night. Controls, however, fed more often during the night, while infected individuals fed as often during the day as in the night. When deprived of food for

24 hours, infected individuals had slightly higher heartbeat rates than controls only during the day and also had greater night-time than daytime heartbeat rates. There was no difference in heartbeat rate between feeding infected snails and uninfected snails that were starved prior to feeding. When feeding, infected snails displayed less nocturnal locomotor activity than controls, but no difference in locomotor activity was observed in snails deprived of food. Starvation also increased the night-time to daytime ratio of locomotor activity of infected snails.

The effect of infection by *S. mansoni* on the metabolic rate of *B. glabrata* was also examined by Becker and Lamprecht (1977). No difference in energy conversion was demonstrated between infected and uninfected snails based on calorimetric analyses. Becker (1980b) also reported that heat production of infected and uninfected individuals was the same. However, when differences in locomotor acitivity were taken into consideration, infected snails had significantly higher heat production. It was concluded that infected *B. glabrata* have a higher basal metabolic rate, but reduce their locomotor activity to compensate and maintain a constant rate of energy conversion.

THE ADAPTIVE SIGNIFICANCE OF PHYSIOLOGICAL ALTERATION

Dawkins (1982) effectively promulgated the adaptationist view that a parasitized host represents an 'extended phenotype' of the parasite. The host is modified by the parasite for the purpose of ensuring parasite survival and development. Although the overall effects of parasitism on the host are, by definition, detrimental to the host and beneficial for the parasite, the critical role of the host in determining the nature and outcome of the association is frequently neglected. At the genetic level both parasite and host genomes play an active role in influencing the outcome of parasitism (Anderson and May, 1982), and hence, in evolutionary terms an 'arms race' develops (see Chapter 1).

The host–parasite relationship evolves as a result of constant action and reaction by the host as well as the parasite. During establishment of the association, alterations in host physiology likely reflect responses to various physiological and metabolic stresses induced by parasitism. Those responses then provide a basis upon which natural selection could act allowing the parasite species to exploit the modified physiological state of the host. Sprent (1969) described this evolutionary process as 'adaptation tolerance' and discussed it in relation to the immunological interactions of the relationship. In that case the host defense response results in a strong selection pressure on the parasite, favouring the emergence of a less immunogenic form, while at the same time there occurs a corresponding selection pressure on the host to modify its responsiveness (Wakelin, 1976). When viewed in the broader context of physiological integration those host physiological responses considered 'dull pathological by-products of infection' (Dawkins, 1982) or 'indirect stress effects' (Jones *et al.*, 1986) take on new significance. They become the 'fodder' for the dynamic stimulus-response nature of host–parasite co-evolution.

Nowhere have the relative contributions of parasite and host to the physiological manifestations of parasitism been debated more than in studies of trematode-infected snails. The argument was recently elaborated by Joosse and Van Elk (1986) who stated: 'Giant growth, depletion of energy stores, and inhibition of reproductive activity are striking effects of many trematode parasites on their intermediate snail host. Two hypotheses have been put forward to explain these phenomena: (1) host and parasite compete for energy-rich and other essential nutrients, with the parasite as the winner, and (2) the parasite intervenes in the endocrine control of reproduction of the snail.' These hypotheses, however, are not mutually exclusive, and there is supportive evidence for both.

Sluiters (1981) demonstrated that the effects of *T. ocellata* on growth, gonad development and egg laying of *L. stagnalis* occurred during the pre-patent development of the parasite, within 1 to 2 weeks following infection. Meuleman (1972) reported similar results with *B. pfeifferi* infected with *S. mansoni*. Thus, the effects were evident before significant nutrient demand by the parasite was realized, and Joosse and Van Elk (1983) were justified in concluding that nutrient demand was an unlikely explanation.

McClelland and Bourns (1969) related the larger size, increased longevity and absence of reproductive activity in infected *L. stagnalis* to the energetic cost of egg *versus* parasite production. They suggested that growth and longevity were governed by the nutritional drain resulting from production of eggs or cercariae and suggested that it was energetically less expensive for the host to produce parasite than egg biomass. They concluded that reproductive inhibition served as a 'resource management device' for the parasite. Gigantism was viewed as a mere side-effect, a non-adaptive consequence of parasitism. In contrast, Minchella (1985) related similar events to survival of the host, and suggested that long-lived infected snails respond to infection by using energy normally consumed during snail reproduction for increased growth and/or parasite production. Growth of the host was considered a survival strategy whereby host reproduction is delayed until the infection has run its course, or the host is self-cured. In this scenario the host is conserving energy despite the activities of the parasite. Minchella (1985) and McClelland and Bourns (1969), however, suggested that the trematode caused reproductive inhibition in the host through indirect hormonal control. Yoshino and Lodes (1988) recently reported that excretory-secretory products of larval schistosomes modulate the synthesis and secretion of a variety of polypeptides in circulating haemocytes of *B. glabrata*.

In contrast to the inhibitory effects of infection on reproductive activity of *L. stagnalis* during the pre-patent period of *T. ocellata* development (Sluiters, 1981), Minchella and LoVerde (1981) demonstrated that *S. mansoni*-infected *B. alexandrina* and *B. glabrata* displayed a significant increase in egg laying during the pre-patent period followed by a decline at patency. The authors termed this fecundity compensation and suggested it was a host response to expected future loss in reproductive success. Recently, Thornhill *et al.* (1986) examined growth and reproduction in immature non-laying *B. glabrata* during 5 weeks following exposure to *S. mansoni*. Exposed snails included both parasitized and non-parasitized individuals. Exposed snails showed a significant increase in the number of eggs laid, whether they were parasitized or not. Growth of both parasitized and non-parasitized snails was also greater than controls. It was suggested that a hormonal factor from the miracidial stage may be responsible for inducing sexual maturation in exposed snails and it was emphasized that normal development of the parasite within the snail was not required for stimulation of oviposition. Schwanbek *et al.* (1986) reported that *S. mansoni*-infected *B. glabrata*, fed diets of higher nutritive value than provided by the usual lettuce diet, continued to lay eggs throughout the patent period and, indeed, displayed an astonishing increase in egg laying as the infection progressed. These investigations strongly suggest that nutritional factors are important to host response during infection.

The host's role in life-history variation in molluscs infected by trematodes was outlined by Minchella (1985), who stated: 'The ultimate goal of digenetic trematodes is not simply the infection of molluscan hosts, but rather the continuation of the parasite life-cycle . . . The many examples of host manipulation by parasites combined with the apparent lack of specific host counteradaptations has sometimes led to an over-emphasis on the power of parasite genes to control host behaviour.'

It seems paradoxical that little is known of the physiological bases for many behavioural changes that appear to have adaptive consequences for the host–parasite relationship. The best examples are behaviours arising from infection of the intermediate hosts of acanthocephalan parasites (Holmes and Bethel, 1972; Moore, 1984a). Every intermediate host association of acanthocephalans thus far examined involves a change in host behaviour. The changes may be survival strategies for both host and parasite (see Moore, 1984b) but appear adaptive for the parasite in that the intermediate host becomes more easily accessible for consumption by the definitive host (Bethel and Holmes, 1973, 1977). Because acanthocephalans produce similar behavioural alterations in a great variety of intermediate hosts, Moore (1984b) hypothesized that the ability to modify host behaviour was present in the common ancestor of the parasites and may be shared by their descendants (see Chapter 7). She suggested that: 'The parasite may simply elicit pre-existing host behaviours under inappropriate conditions', rather than inducing novel behavioural patterns. Thus, the physiological bases for the modified behaviour observed in infected hosts may be elusive because in many cases they merely reflect subtle modifications of existing mechanisms.

Crompton (1970), however, discussed the physiological basis for the lack of upstream movement in parasitized *Gammarus pulex*, the amphipod intermediate host of *Echinorhynchus truttae*. Based on glucose uptake and oxygen consumption estimated for another acanthocephalan species, *Polymorphus minutus*, which infects *G. pulex* it was reported that hosts infected by a single cystacanth would have to increase food consumption and ventilation rate to supply the parasite's metabolic demands and maintain normal glucose and oxygen levels in the haemolymph. Crompton concluded that the metabolic stress induced by parasitism weakened the host to the point where it failed to maintain its normal migratory behaviour. Hosts were thus unable to move upstream and were more readily preyed upon by the definitive host, trout.

CONCLUSION

Physiological alterations are common during the host associations of most parasites. Their nature varies widely. Some changes result in pathological consequences that lead to disease, while others appear benign. As discussed throughout this chapter, the significance of many host responses in the evolution of the association has proven difficult to assess. Successful parasitism ultimately leads to survival of the parasite at the expense of the host. Nevertheless, an attempt was made here to de-emphasize the common view that parasitism is a symbiotic association during which the parasite controls its own destiny with the host a passive or altruistic partner. Several recent studies suggest that the genes of some parasites may interact directly with the host's genome and be expressed by alterations in host physiology. The effects of parasitism then can be understood readily in the context of parasite success alone. As Dawkins (1982) pointed out, however, for changes in host biology to occur, there must be a pool of genetic variability with regard to a character altered in order for natural selection to have favoured it. Regardless of the level considered, genetic or behavioural, host and parasite interaction becomes the framework for the evolution of the host–parasite relationship.

REFERENCES

Akpom, C. A., 1982, *Reviews of Infectious Diseases*, **4**, 776–782.
Akpom, C. A. and Warren, K. S., 1975, *Journal of Infectious Diseases*, **132**, 6–14.
Anderson, R. M and May, R. M., 1982 , *Parasitology*, **85**, 411–426.
Anteson, R. K. and Williams, J. F., 1975, *Journal of Parasitology*, **61**, 149–151.
Ash, C. P. J., Crompton, D. W. T., and Lunn, P. G., 1985a *Parasitology*, **90**, 157–168.
Ash, C. P. J., Crompton, D. W. T., and Lunn, P. G., 1985b, *Parasitology*, **91**, 359–368.
Askew, R. R., 1971, *Parasitic Insects* (American Elsevier: New York).
Beckage, N. E., 1985, *Annual Review of Entomology*, **30**, 371–413.
Beckage, N. E. and Riddiford, L. M., 1978, *Entomologia Experimentalis et Applicata*, **23**, 139–151.
Beckage, N. E. and Riddiford, L. M., 1982, *General and Comparative Endocrinology*, **47**, 308–322.
Beckage, N. E. and Riddiford, L. M., 1983, *Physiological Entomology*, **8**, 231–241.
Beckage, N. E. and Templeton, T. J., 1986, *Journal of Insect Physiology*, **32**, 299–314.
Becker, W., 1980a, *Zeitschrift für Parasitenkunde*, **63**, 101–111.
Becker, W., 1980b, *Journal of Comparative Physiology*, **135B**, 101–105.
Becker, W. and Lamprecht, I., 1977, *Zeitschrift für Parasitenkunde*, **53**, 297–305.
Becker, W and Schmale, H., 1975, *Comparative Biochemistry and Physiology*, **51A**, 407–411.
Becker, W. and Schmale, H., 1978, *Comparative Biochemistry and Physiology*, **59B**, 75–79.
Beisel, W. R., 1982, *Reviews of Infectious Diseases*, **4**, 746–755.
Bernays, E. A. and Simpson, S. J., 1982, *Advances in Insect Physiology*, **26**, 119–126.
Bethel, W. M. and Holmes, J. C., 1973, *Journal of Parasitology*, **59**, 945–956.
Bethel, W. M. and Holmes, J. C., 1977, *Canadian Journal of Zoology*, **55**, 110–115.
Bollenbacher, W. E., 1988, *Journal of Insect Physiology*, **34**, 941–947.
Bundy, D. A. P. and Golden, M. H. N., 1987, *Parasitology*, **95**, 623–635.

Castro, G. A., 1980, In *The Host-Invader Interplay* edited by H. Van den Bossche (Amsterdam: Elsevier/North Holland), pp. 457–467.

Castro, G. A., Olson, C. J., Baker, R. D., 1967, *Journal of Parasitology*, **53**, 595–612.

Castro G. A., Copeland, E. M., Dudrick, S. J., and Ramaswamy, K., 1979, *American Journal of Tropical Medicine and Hygiene*, **28**, 500–507.

Cheng, T. C. and Lee F. O., 1971, *Journal of Invertebrate Pathology*, **18**, 395–399.

Cheng, T. C., Sullivan, J. T., and Harris, K. R., 1973, *Journal of Invertebrate Pathology*, **21**, 183–190.

Christie, J. D. and Michelson, E. H., 1975, *Comparative Biochemistry and Physiology*, **50B**, 233–236.

Christie, J. D., Foster, W. B., and Stauber, L. A., 1974, *Journal of Invertebrate Pathology*, **23**, 55–62.

Cloutier, C., 1986, *Journal of Insect Physiology*, **32**, 263–267.

Cloutier, C. and Mackauer, M., 1979, *Canadian Journal of Zoology*, **57**, 1605–1611.

Cloutier, C and Mackauer, M., 1981, *Proceedings of the IX International Congress of Plant Protection*, **1**, 99–100.

Crompton, D. W. T., 1970, *An Ecological Approach to Acanthocephalan Physiology* (Cambridge University Press: Cambridge).

Crompton, D. W. T., 1984, *Federation Proceedings*, **43**, 239–245.

Crompton, D. W. T., and Nesheim, M. C., 1982, *Bioscience*, **32**, 677–680.

Crompton, D. W. T., Walters, D. E., and Arnold S., 1981, *Parasitology*, **82**, 23–38.

Crouch, P. A. R., 1982, *Papua New Guinea Medical Journal*, **25**, 43–49.

Damian, R. T.., 1987, *Journal of Parasitology*, **73**, 3–13.

Davenport, A. P., and Evans, P. D., 1984, *Physiological Entomology*, **9**, 263–274.

Davis, D. S. and Farley, J., 1973, *Parasitology*, **66**, 191–197.

Davies, A. J. S., Hall J. G., Targett G. A. T., and Murray, M., 1980, *Journal of Parasitology*, **66**, 705–721.

Dawkins, R., 1982, In *The Extended Phenotype*: *The Gene as the Unit of Selection* (Oxford: W. H. Freeman), pp. 209–227.

Debary, A., 1879, *Die Erscheinung der Symbiose* (Strassberg: Karl J. Trüber).

De Jong-Brink, M., Elsaadany, M. M., and Boer, H. H., 1988a, *Experimental Parasitology*, **65**, 91–100.

De Jong-Brink, M., Elsaadany, M. M., and Boer, H. H., 1988b, *Experimental Parasitology*, **65**, 109–118.

Dogiel, V. A., 1964, *General Parasitology* (London: Oliver and Boyd).

Domingo, E. O. and Warren, K. S., 1969, *Gastroenterology*, **56**, 231–240.

Fricke, H. W., 1975, *Society for Experimental Biology Symposium*, **29**, 581–594.

Führer, E. and Keja, Th.D., 1976, *Entomologia et Experimentalis Applicata*, **19**, 287–300.

Garaerts, W. P. M. and Joosse, J., 1984, In *The Mollusca* edited by A. S. M. Saleuddin and K. M. Wilbur (New York: Academic Press), Vol. 1, pp. 141–207.

Garland, J. T., Ruegamer, W. R. and Daughaday, W. H., 1971, *Endocrinology*, **88**, 924–927.

Gilbertson, D. E., Etges, F. J. and Ogle, J. D., 1967, *Journal of Parasitology*, **53**, 565–568.

Goff, L. J., 1982, *Bioscience*, **32**, 255–256.

Gordon, H. E., 1976, In *Nutrition in Preventative Medicine* edited by G. H. Beaton and J. M. Bengoa (Geneva: WHO), pp. 193–209.

Grossman, Z., Greenblatt, C. L., and Cohen, I. R., 1986, *Journal of Theoretical Biology*, **121**, 129–139.

Henry, S. M., 1966, Forward, In *Symbiosis* edited by S. M. Henry (New York: Academic Press), Vol. I, pp. ix–xi.

Hirai, K., Nishida, H., Shiwaku, K., and Okuda, H., 1978, *Japanese Journal of Parasitology*, **27** 527–533.

Hirai, K., Shiwaku, K., Tsuboi, T., Torii, M., Nishida, H. and Yamane, Y., 1983, *Zeitschrift für Parasitenkunde*, **69**, 489–499.

Hirai, K., Tsuboi, T., and Torii, M., 1986, *Japanese Journal of Parasitology*, **35**(Suppl): 108.

Hirai, K., Tsuboi, T., and Torii, M., 1988, *Parasitology Research*, 74, 262–266.

Holmes, J.C. and Bethel, W. M., 1972, In *Behavioural Aspects of Parasite Transmission* edited by E. U. Canning and C. A. Wright, *Zoological Journal of the Linnean Society*, Supplement 1, pp. 123–149.

Holmes, P. H., 1987, *Parasitology*, **94**, S29–S51.

Iwantsch, G. and Smilowitz, Z., 1975, *Journal of Insect Physiology*, **21**, 1151–1157.

Jones, D., 1985a, *Journal of Comparative Physiology*, **B155**, 235–240.

Jones, D., 1985b, *Journal of Comparative Physiology*, **B155**, 583–590.

Jones, D. and Sreekrishna, S., 1984, In *Insect Neurochemistry and Neurophysiology* edited by A. B. Borkovec and T. J. Kelly (New York: Plenum), pp. 389–391.

Jones, D., Jones, G., and Hammock, B. D., 1981, *Physiological Entomology*, **6**, 387–394.

Jones, D., Jones, G., Rudnicka, M., Click, A., Reck-Malleczewen, V., and Iwaya, M., 1986, *Journal of Insect Physiology*, **32**, 315–328.

Jones, W. O. and Symons, L. E. A., 1982, *International Journal for Parasitology*, **12**, 295–301.
Joosse, J. and Van Elk, R., 1983, In *Molluscan Neuro-endocrinology* edited by J. Lever and H. H. Boer (Amsterdam: North Holland), pp. 118–120.
Joosse, J. and Van Elk, R., 1986, *Experimental Parasitology*, **62**, 1–13.
Joose, J., Van Elk, R., Mosselman, S., Wortelboer, H., and Van Diepen, J. C. E., 1988, *Parasitology Research*, **74**, 228–234.
Keusch, G. T., 1982, *Reviews of Infectious Diseases*, **4**, 901–907.
Keymer, A., Crompton, D. W. T., and Sahakian, B. J., 1983a, *Parasitology*, **86**, 455–460.
Keymer, A., Crompton, D. W. T., and Walters, D. E., 1983b, *Parasitology*, **86**, 461–475.
Knauft, R. F. and Warren K. S., 1969, *Journal of Infectious Diseases*, **120**, 560–575.
Lawrence, P. O., 1986, *Journal of Insect Physiology*, **32**, 295–298.
Lee, F. O. and Cheng, T. C. 1971a, *Experimental Parasitology*, **30**, 393–399.
Lee, F. O. and Cheng, T. C. 1971b, *Journal of Invertebrate Pathology*, **18**, 412–418.
Lee, F. O. and Cheng, T. C, 1972, *Experimental Parasitology*, **31**, 203–216.
Lewin, R. A., 1982, *Bioscience*, **32**, 254, 256–260.
McClelland, G. and Bourns, T. K. R, 1969, *Experimental Parasitology*, **24**, 137–146.
MacInnis, A. J. 1976, In *Ecological Aspects of Parasitology* edited by C. R. Kennedy (Amsterdam: North Holland), pp. 3–20.
Malek, E. A. and Cheng, T. C, 1974, In *Medical and Economic Malacology* (New York: Academic Press), pp. 204–241.
Mangum C. P. and Towle, D. W. 1977, *American Scientist*, **65**, 67–75.
Martin, J. 1981, *Parasitology*, **82**, 38.
Meakins, R. H., 1980, *Comparative Biochemistry and Physiology*, **66A**, 137–140.
Mellink, J. J. and Van den Bovenkamp, W., 1983, *Tropical and Geographical Medicine*, **35**, 212.
Meuleman, E. A., 1972, *Netherlands Journal of Zoology*, **22**, 355–427.
Meyer, F. D., Kimura, S. and Mueller, J. F., 1965, *Journal of Parasitology*, **51**, 57–58.
Minchella, D. J., 1985, *Parasitology*, **90**, 205–216.
Minchella, D. J. and LoVerde, R. T., 1981, *American Naturalist*, **118**, 876–881.
Mohamed, A. M. and El Fiki, S. A., 1980 *Journal of the Egyptian Society of Parasitology*, **10**, 65–72.
Moore, J., 1984a, *Scientific American*, **250**, 108–115.
Moore, J., 1984b, *American Naturalist*, **123**, 572–577.
Mueller, J. F., 1963, *Annals of the New York Academy of Science*, **113**, 217–233.
Mueller, J. F., 1965, *Journal of Parasitology*, **51**, 523–531.
Mueller, J. F., 1980, In *Growth and Growth Factors: Proceedings of the International Symposium on Growth and Growth Factors* edited by K. Shizume and K. Takano (Tokyo: University of Tokyo Press), pp. 193–201.
Murray, M. J. and Murray, A. B., 1979, *American Journal of Clinical Nutrition*, **32**, 593–596.
Murray, J., Murray, A. and Murray, N., 1978, *Perspectives in Biology and Medicine*, **22**, 134–142.
Ovington, K. S., 1985, *Parasitology*, **91**, 157–167.
Pan, C. T. 1965, *American Journal of Tropical Medicine and Hygiene*, **14**, 931–976.
Pawlowski, Z. S., 1984, *Federation Proceedings*, **43**, 256–260.
Pearson, E. J. and Cheng, T. C., 1985, *Journal of Invertebrate Pathology*, **46**, 239–246.
Phares, C. K. and Booth, B. J .M., 1987, *Endocrinology*, **121**, 1839–1844.
Phares, C. K. and Carroll, R. M., 1977, *Journal of Parasitology*, **63**, 690–693.
Phares, C. K. and Carroll, R. M., 1978, *Journal of Parasitolovy*, **64**, 401–405.
Phares, C. K. and Carroll, R. M., 1984, *Journal of Helminthology*, **58**, 25–30.
Phares, C. K. and Cox, G. H., 1987, In *Molecular Paradigms for Eradicating Helminthic Parasites* edited by A. J. MacInnis (New York: Alan R. Liss), pp. 391–405.
Phares, C. K. and Ruegamer W. R., 1973, *Proceedings of the Society of Experimental Biology and Medicine*, **142**, 374–377.
Phares, C. K. and Watts, D. J., 1988, *Journal of Parasitology*, **74**, 896–898.
Price, P., 1980, *Evolutionary Biology of Parasites* (Princeton: Princeton University Press).
Read, C. P., 1950, *Rice Institute Pamphlet*, **37(2)**, 1–94.
Read, C. P., 1970, *Parasitism and Symbiology* (New York: Ronald Press).
Rosenberg, I. H. and Bowman, B. B., 1984, *Federation Proceedings*, **43**, 246–250.
Ruegamer, W. R. and Mueller, J. F., 1973a, *Journal of Nutrition* **103**, 1496–1501.
Ruegamer, W. R. and Mueller, J. F., 1973b, *Proceedings of the Society of Experimental Biology and Medicine*, **143**, 133–137.
Schmale, H. and Becker, W., 1977, *Comparative Biochemistry and Physiology*, **58B**, 321–330.

Schmidt, S. P. and Roberts, L. S., 1981, *Foundations of Parasitology* (St. Louis:C. V. Mosby).
Schopf, A., 1984, *Entomologia Experimentalis et Applicata*, **36**, 265–272.
Schwanbek, A., Becker, W. and Rupprecht, H., 1986, *Zeitschrift für Parasitenkunde*, **72**, 365–373.
Schwartz, C. F. W. and Carter, C. W. M., 1982, *Journal of Parasitology*, **68**, 236–242.
Sharp, S. E., Phares, C. K. and Heidrick, M. L., 1982, *Journal of Parasitology*, **68**, 993–998.
Shiwaku, K. and Hirai, K., 1982, *Japanese Journal of Parasitology*, **31**, 183–193.
Shiwaku, K., Hirai, K., Torii, M., and Tsuboi, T., 1983, *Parasitology*, **87**, 447–453.
Shiwaku, K., Hirai, K., Tsuboi, T., and Torrii, M., 1986, *Japanese Journal of Parasitology*, **35**, 411–417.
Slansky, F., 1978, *Environmental Entomology*, **7**, 179–185.
Slansky, F., 1986, *Journal of Insect Physiology*, **32**, 255–261.
Sluiters, J. F., 1981, *Zeitschrift für Parasitenkunde*, **64**, 303–319.
Sluiters, J. F., Brussaard-Wust, C. C. M., and Meuleman, E. A., 1980, *Zeitschrift für Parasitenkunde*, **63**, 13–26.
Sluiters, J. F., and Dogterom, G. E., 1984, *Zeitschrift für Parasitenkunde*, **70**, 485–489.
Sluiters, J. F., and Geraerts, W. P. M., 1984a, *Proceedings Koninklijke Nederlandse Akademie van Wetenschappen*, **C87**, 231–233.
Sluiters, J. F., and Geraerts, W. P. M., 1948b, *Zeitschrift für Parasitenkunde*, **70**, 477–484.
Sluiters, J. F., Roub os, E. W., and Joosse, J., 1984, *Zeitschrift für Parasitenkunde*, **70**, 67–72.
Smilowitz, Z. 1974, *Annals of the Entomological Society of America*, **67**, 317–320.
Smith, C. L. and Smilowitz, Z., 1976, *Entomologia Experimentalis et Applicata*, **19**, 189–195.
Smyth, J. D., 1962, *An Introduction to Animal Parasitology*, (London: English Universities Press).
Smyth, J. D., 1976, *An Introduction to Animal Parasitology* (London: Hodder and Stoughton).
Sprent, J. F. A., 1969, In *Immunity to Parasitic Animals* edited by G. J. Jackson, R. Herman and I. Singer (Amsterdam: North Holland), pp.3–62.
Stanislawski, E. and Becker, W., 1979, *Comparative Biochemistry and Physiology*, **63B**, 527–533.
Stanislawski, E., Becker, W., and Muller, G., 1979, *Comparative Biochemistry and Physiology*, **63B**, 477–482.
Steel, J. W., Jones, W. O., and Symons, L. E. A., 1982, *Australian Journal of Agricultural Research*, **33**, 131–140.
Steel, J. W., Symons, L. E. A., and Jones, W. O., 1980, *Australian Journal of Agricultural Research*, **31**, 821–838.
Steelman, S. L., Morgan. E. R., Cuccaro, A. J., and Glitzer, M. S., 1970, *Proceedings of the Society of Experimental Biology and Medicine*, **133**, 269–273.
Sudan, E. H., Williams, J. S., Meroney, F. C., and Mueller, J. F., 1965, *Journal of Parasitology*, **51**, 532–536.
Sullivan, J. T., Cheng, T. C., and Howland, K. H., 1985, *Transactions of the American Microscopical Society*, **104**, 154–171.
Symons, L. E. A., 1985 *Advances in Parasitology*, **24**, 103–133.
Symons, L. E. A. and Jones, W. O., 1971, *Experimental Parasitology*, **29**, 230–241.
Symons, L. E. A. and Jones, W. O., 1975, *Australian Journal of Agricultural Research*, **26**, 1063–1074.
Symons, L. E. A. and Jones, W. O., 1978, *Experimental Parasitology*, **44**, 194–202.
Symons, L. E. A. and Jones, W. O., 1983, *International Journal for Parasitology*, **13**, 309–312.
Symons, L. E. A., Jones, W. O. and Steel, J. W., 1974, *Experimental Parasitology*, **35**, 492–502.
Thompson, S. N., 1982a, *Parasitology*, **84**, 491–510.
Thompson, S. N., 1982b, *Journal of Parasitology*, **68**, 936–941.
Thompson, S. N., 1983a, *Comparative Biochemistry and Physiology*, **74B**, 183–211.
Thompson, S. N., 1983b, *Parasitology*, **87**, 15–28.
Thompson, S. N., 1985a, *Comparative Biochemistry and Physiology*, **81B**, 21–42.
Thompson, S. N., 1985b, *Journal of Parasitology*, **71**, 117–119.
Thompson, S. N., 1986a, *Journal of Insect Physiology*, **32**, 287–293.
Thompson, S. N., 1986b, *Parasitology*, **92**, 25–30.
Thompson, S. N. and Binder, B. F., 1984, *Journal of Parasitology*, **70**, 644–651.
Thompson, S. N. and Cohen, A. C., 1984, *Comparative Biochemistry and Physiology*, **78A**, 785–786.
Thompson, S. N. and Mejia-Scales, V., 1989, *Journal of Parasitology*, **75**, 329–332.
Thompson, S. N and Yamada, K. A., 1984a, *Journal of Invertebrate Pathology*, **44**, 46–51.
Thompson, S. N. and Yamada, K. A., 1984b, *Molecular and Biochemical Parasitology*, **13**, 323–331.
Thornhill, J. A., Jones, J. T. and Kusel, J. R., 1986, *Parasitology*, **93**, 443–450.
Trede, G. and Becker, W., 1982, *Comparative Biochemistry and Physiology*, **73B**, 405–409.
Van Beneden, P., 1876, *Animal Parasites and Messmates*, (New York: Appleton-Century).

Venkata Rama Krishna, G. and Sunder Simha, S., 1980, *Current Science*, **49**, 160–162.
Von Brand, T., 1979, In *Biochemistry and Physiology of Endoparasites* (Amsterdam: Elsevier/North Holland Biomedical Press), pp. 321–390.
Wakelin, D., 1976, In *Ecological Aspects of Parasitology*, edited by C. R. Kenneday (Amsterdam: North-Holland), pp. 115–141.
Whitfield, P. J., 1979, *The Biology of Parasitism* (Baltimore: University Park Press).
Williams, C. L. and Gilbertson, D. E., 1983, *Journal of Parasitology*, **69**, 671–676.
Wright, C. A., 1969, *Helminthological Abstracts*, **35**, 207–224.
Wüker, W., 1964, *Experimental Parasitology* **15**, 561–597.
Wüker, W., 1974, In *Intersexuality in the Animal Kingdom*, edited by R. Reimboth (Berlin: Springer Verlag), pp. 121–134.
Yamada, K. A. and Sherman, I. W., 1980, *Molecular and Biochemical Parasitology*, **1**, 187–198.
Yoshino, T. P. and Lodes, MJ., 1988, *Journal of Parasitology*, **74**, 538–547.

4. Parasites and host decision-making

M. Milinski

INTRODUCTION

When hungry an animal will look for food, when thirsty it will try to drink, when threatened by a predator it will try to avoid being eaten. However, what should the animal do when it is hungry, thirsty and under risk of predation at the same time? Behavioural ecologists assume that natural selection has designed animals to choose the behavioural option which maximizes its fitness, often equated with its survival in order to reproduce. If two behaviours are mutually exclusive such as feeding and drinking, the animal should feed if the net benefit from feeding is greater than that from drinking and *vice versa*. Sometimes the animal is able to fulfil different needs to a certain degree simultaneously, e.g. feeding at a reduced rate allows for some vigilance. This trade-off should be the best compromise for fulfilling both needs. The hungrier ther animal is, the more weight it should give to feeding at the expense of vigilance and *vice versa* (Sibly and McFarland, 1976: Milinski and Heller, 1978). Animal decision-making appears to be well adapted for the economic solution of everyday problems and takes into account differences in costs and benefits of alternative options that are so small that even well trained ethologists are repeatedly astounded by its fine-tuned precision (for recent reviews of decision rules see McFarland, 1985: Krebs and Davies, 1987; Houston *et al.*, 1988).

Thus, behavioural ecologists usually look at animal decision-making from a functional point of view. Such considerations specify what animals ought to do to make the best decisions under particular circumstances, but they do not say anything about the mechanisms that animals might employ to attain these objectives (McFarland, 1985). To say that an animal makes decisions implies no conscious intent but merely refers to the fact that the animal takes one behavioural alternative rather than another (McFarland, 1985). Of course there must be causal mechanisms with which animals reach decisions. These can include preprogrammed rules, stochastic processes, feedback loops, hormones, and so forth.

A parasite may either change variables like energy reserves, oxygen demand, etc. necessary for the mechanisms of host decision-making or the parasite may interfere with these mechanisms directly, e.g. by cutting feedback loops or changing hormone levels (see Chapter 3). However, we do not need to understand how the parasite alters the decision-making processes in order to investigate host decision-making from a functional point of view, we just need to ask for the fitness consequences of a particular behavioural option for both the host and the parasite. I will primarily discuss in this chapter possible fitness consequences of the host's decision-making, i.e. of the host taking one behavioural option rather than another.

PROBLEMS IN THE EVOLUTION OF HOST DECISION-MAKING

One definition of a parasite may be 'an organism which is not only in continuous, intimate association with another organism – the host . . ., but is also metabolically dependent, directly or indirectly, on it to some degree' (Smyth, 1985) and therefore a parasite necessarily places constraints on its host which minimally will involve utilizing nutrients and space. These constraints may increase the host's needs as well as its costs for a number of behavioural actions (see Chapter 1) and therefore change, as a mere consequence of their existence, physiological variables and thus the outcome of the host's decision-making. If the costs and benefits of alternative options are influenced differently by the presence of a parasite, the compromize behaviour of a parasitized animal will differ from that of a healthy one.

Given that the host cannot get rid of the parasite, the new compromise should be the best option which allows the host to maximize its survival in order to reproduce, taking the inevitable constraints placed by the parasite into account. From the parasite's point of view, this may be the best the host can do to guarantee its availability as a resource for the parasite, especially in more persistent host–parasite associations. However, this is by no means a cooperative venture. The host simply does the best it can. Although this somewhat automatic response to constraints placed by the parasite seems to be adaptive, no coevolution between host and parasite needs to have occurred.

Animals have a number of conditional strategies to cope with constraints which do not differ qualitatively from many of the constraints placed on them by parasites. For example, an animal may give more weight to feeding at the expense of vigilance for predators because it has an increased need for energy caused by a period of unfavourable weather. It may use the same decision rule when the energy drain is caused by a parasite. In this way, many of the responses in an animal's behavioural repertoire need not be tailored specifically to parasites to efficiently compensate for parasitic constraints. Therefore, for the evolution of such 'adaptive' changes of host behaviour, it is irrelevant whether the host is still able to reproduce or to increase its inclusive fitness when parasitized. This can be seen easily if we consider the necessary conditions for the evolution of a specific response of the host to a parasitic constraint, e.g. to increase vigilance when the parasite renders its host more conspicuous. A mutant in the host population coding for the specific response (or an initial step to it) will increase in frequency only if it has a higher fitness, i.e. more offspring, than the wild type. If, however, the host is unable to reproduce when infected by a specific parasite, the favourable mutant would not produce any offspring (although it is more vigilant) in this case and therefore would have no selective advantage over the wild type. Nevertheless, the host can use appropriate conditional strategies evolved by selective forces other than those caused by a parasite and appear to behave adaptively. Of course, if the host is still able to reproduce, the use of such conditional strategies should increase its fitness. I believe that this kind of behavioural change in parasitized animals is the most common one and examples will be discussed in this chapter.

If the best compromise for the host is different from that which would maximize the parasite's fitness, e.g. because the parasite needs its present host to be eaten by the next one in order to complete its life-cycle, there is selection pressure on the parasite for direct manipulation of the host's decision-making and, if the host is still able to reproduce, on the host for developing specific counter-strategies. In this case both the host and the parasite compete for the host's behaviour. Under these conditions coevolution between host and parasite is possible and an evolutionary arms race can occur (see Chapter 1). A parasite which castrates its host would prevent the host species from evolving any specific adaptations countering parasitic manipulation. However, one would not expect that parasites castrate their hosts in order to prevent them from developing counter-adaptations. This would require group selection because castrating an individual host would not influence the probability that this individual has a mutation for a specific counter-adaptation, e.g. increased vigilance, which would nevertheless adversely affect the castrating parasite if it exists. Castration would only prevent this individual host from leaving offspring with the counter-adaptation. This would benefit the next parasite generation, castraters and non-castraters unselectively. Thus, castraters would not have more offspring than non-castraters and the genetic basis for castrating would not spread through the parasite population. A possible functional explanation for observed parasitic castration of hosts is, for example, that energy otherwise spent producing offspring can be used by the parasite instead (Baudoin, 1975; Smith Trail, 1980). The existence of direct manipulation of host behaviour (see Chapter 7) is sometimes difficult to prove because its effect on the host may differ only quantitatively from the behavioural change caused by parasitic constraints.

PARASITIZED ENERGY AND HOST DECISION-MAKING

Since the three-spined stickleback has been used by a number of investigators to study the influence of parasites on host decision-making, it seems appropriate to use this species for discussing some of the problems which may also be inherent in many other host–parasite systems. Three-spined sticklebacks (*Gasterosteus aculeatus*) make a trade-off between feeding and avoiding predation when there is a large predatory fish waiting close to a patch containing *Tubifex* worms (a preferred food). How close should a stickleback dare to approach the predator in order to have a good meal? The sticklebacks prefer the worms farthest away form the predator (figure 4.1) and feed more slowly than when the predator is nearby (Milinski, 1985). This compromise between feeding and avoiding predation should be different for sticklebacks which have a greater need for food. They should be less choosy in order to feed faster. For the stickleback's decision to shift the compromise towards feeding, it does not matter functionally whether it has become hungrier because of a shortage of food or because a parasite lying in the fish's body cavity has utilized some of its host's energy.

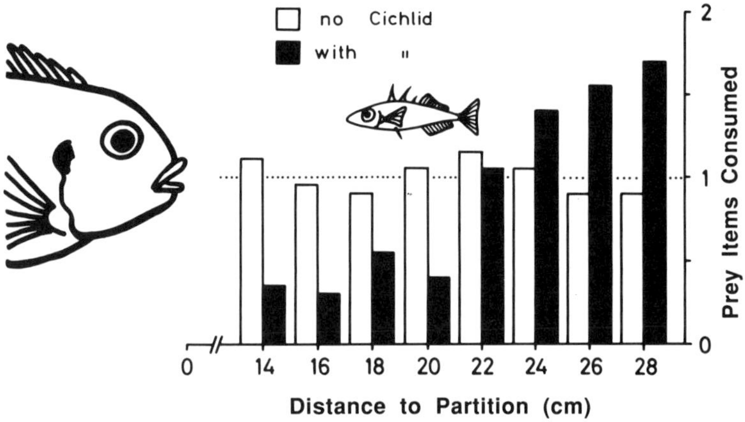

Figure 4.1. Mean number of *Tubifex* worms consumed by unparasitized sticklebacks at different distances from a watchful fish predator behind a glass partition. Each of 23 fish was allowed to eat eight worms. After Milinski (1985).

Three-spined sticklebacks are frequently parasitized by plerocercoids of the tapeworm *Schistocephalus solidus*, for which copepods are the first intermediate hosts and fish-eating birds the definitive hosts (Wootton, 1976). The plerocercoid lies free in the body cavity of the fish and grows, converting energy more efficiently than its host until after about 3 months it becomes infective to a bird (Walkey and Meakins, 1970). In the stickleback the plerocercoid nearly doubles its weight every 4 days (Orr and Hopkins, 1969) and can exceed the net weight of its host (personal observation). Despite the high infection intensities recorded there is little evidence of acute symptons of disease (Arme and Owen, 1967), and when kept together in a tank, parasitized and unparasitized sticklebacks do not obviously differ in their movements. Since an increasing drain of energy seems to be the predominant effect of *S. solidus* on its stickleback host (Walkey and Meakins, 1970), this host–parasite system may be an ideal one for studying whether energetic constraints placed by parasites affect host decision-making. We would predict that with increasing energy drain caused by the growing plerocercoid parasitized sticklebacks should change the trade-off between feeding and predator avoidance towards feeding.

Giles (1983) frightened three-spined sticklebacks which varied in the intensity of infection by plerocercoids of *S. solidus*. He suddenly lowered a model heron's head over the test tank so that the tip of the bill hit the water surface. There was a negative correlation between recovery time from the heron stimulus and the sticklebacks' burden of *S. solidus* larvae, and only parasitized fish resumed feeding in the 5 minutes following the stimulus. Furthermore, when sticklebacks with a high burden of *S. solidus* were offered *Tubifex* worms in the presence or absence of a predatory cichlid (Milinski, 1985), they fed at all distances equally often irrespective of the presence of the cichlid (figure 4.2). Infected fish did not interrupt their feeding or even raise their spines when the predator tried to attack them from behind a glass partition. Also their feeding rate did not differ significantly in the two situations. So, as expected, fish with a greater need for food accept a higher risk of predation in order to feed quickly. Nevertheless, it is astonishing that these heavily parasitized fish did not even raise a spine when a predator tried to attack them. The fact that they can react to a small prey item which is further away than the distance at which they did not react to the large predator, seems to indicate that vision is not directly impaired by *S. solidus* as is the case for *Diplostomum spathaceum* parasitizing the eyes of dace, *Leuciscus leuciscus* (Crowden and Broom, 1980).

If the parasite influences stickleback decision-making only by causing an energy drain, one would expect that parasitized fish should become more fearful when they are more satiated. However, even when they had *ad libitum* access to food prior to the trial, they continued searching for food during the trial without any indication of fear. Were they still so hungry that feeding had absolute priority or did the parasite manipulate the stickleback's decision-making more directly, e.g. by suppressing the fear response or the ability to recognize predators? The plerocercoids in this experiment were probably already at a stage that would be

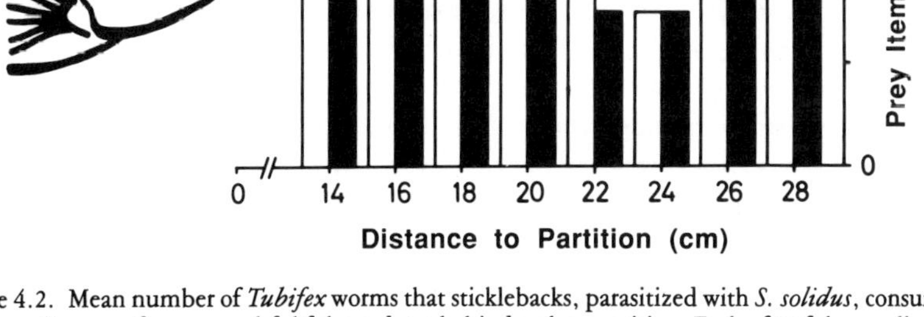

Figure 4.2. Mean number of *Tubifex* worms that sticklebacks, parasitized with *S. solidus*, consumed at different distances from a watchful fish predator behind a glass partition. Each of 12 fish was allowed to eat eight worms. After Milinski (1985).

infective to a definitive host. Therefore these plerocercoids stood to gain if their stickleback hosts were eaten by a potential definite host which is normally a bird. There are many other cases in which parasites manipulate host behaviour such that the probability of their transmission to the next host is enhanced (Holmes and Bethel, 1972; Moore, 1984; see Chapter 7).

Is the increased need for energy sufficient to explain the altered anti-predator behaviour of parasitized sticklebacks? The mature plerocercoid distends the fish's body so much that it is hard to imagine that there is space for food at all. Perhaps these fish are unable to become satiated. To test this hypothesis, parasitized and unparasitized sticklebacks of similar standard length were fed waterfleas of a particular size class until the fish could no longer swallow further items (Milinski, 1985). The healthy fish consumed twice as many prey items as the parasitized ones. The parasite occupies the space which the host needs to process the food for both animals. Although parasitized sticklebacks consume more food than healthy ones per day, *S. solidus* plerocercoids are still able to produce a negative imbalance in their host's energy budget (Walkey and Meakins, 1970). This permanent energy deficit and the constraint of the reduced gut capacity probably force the fish to try to feed almost continuously. Therefore, we do not need to invoke direct manipulation of anti-predator behaviour in this case. The stickleback's decision-making may take into account that there is always a high demand for energy.

Cysts of the microsporidian sporozoan *Glugea anomala* are also frequently found in and on three-spined sticklebacks (Arme and Owen, 1967; Canning and Lom, 1986). This parasite also utilizes the fish's energy and therefore sticklebacks harbouring *G. anomala* should take more risks in terms of predation than uninfected fish in order to feed quickly as in the case of the fish parasititized by *S. solidus*. However, sticklebacks parasitized by *G. anomala* noticeably avoided prey items close to the predatory cichlid (figure 4.3). They stayed even farther away from the predator than did the healthy fish (figure 4.1). It is therefore tempting to conclude that either *S. solidus* or *G. anomala* or even both species directly manipulate the stickleback's behaviour to their own ends (Milinski, 1985). The anti-predator behaviour of fish infected with *S. solidus* and of those infected with *G. anomala* as compared with the behaviour of uninfected fish is changed in opposite directions. This apparently corresponds to whether the parasite needs to be transmitted to another host or not.

Since the stickleback is the only host for *G. anomala*, it is surprising that the conspicuous white cysts look as though they were made for attracting predators. Perhaps this is the case however. Parasites which are able to kill local populations of their host – and *G. anomala* infections have the potential to do this – need a vector to transport them to the next uninfected host population, as suggested by Ewald (1983). There may be a further advantage for *G. anomala* when its stickleback host is eaten by a predator: under favourable conditions an infected stickleback can tolerate the parasite well (Canning and Lom, 1986) and therefore the numerous parasite spores, which the fish isolates in a cyst with a tough membrane, would have to wait for a long time until the fish dies and they are released. (I have kept sticklebacks with ripe cysts of *G. anomala* for more than a year in the laboratory and

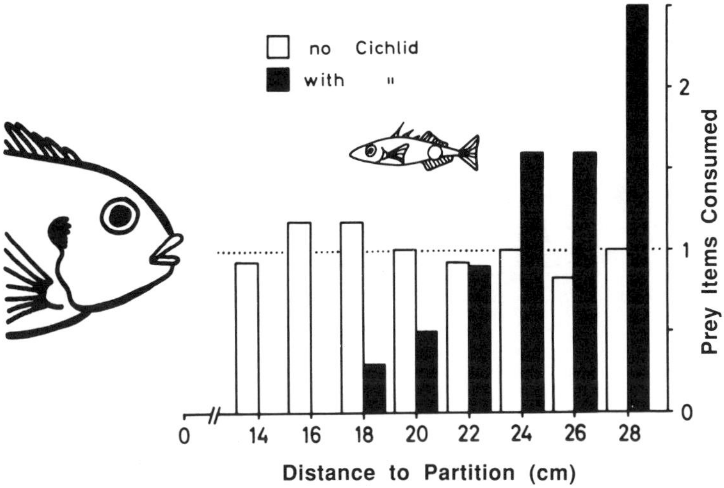

Figure 4.3. Mean number of *Tubifex* worms that sticklebacks, parasitized with *G. anomala*, consumed at different distances from a watchful fish predator behind a glass partition. Each of 11 fish was allowed to eat eight worms. After Milinski (1985).

the spores did not burst out.) If the fish is eaten by a predator, however, the spores may be able to start their next generation straight away. So a mutant of *G. anomala* increasing the stickleback's risk of being preyed upon would therefore be at a selective advantage.

When the cysts are ripe, they seem to utilize much less energy than *S. solidus* (as yet there is no quantitative data on the energy budget). Therefore, the stickleback's decision-making based only on its state of hunger probably would not produce fearless behaviour similar to that of fish infected by *S. solidus*. This agrees with what we observe. However, contrary to this observation these fish should not be less hungry and thus more fearful than uninfected ones. If the conspicuousness of *G. anomala* is in order to increase the chance of its stickleback host being caught by a predator, then the fearful behaviour of these fish is in the interest of the fish and not in that of the parasite. When parasitized by *G. anomala* the sticklebacks could be aware that they are conspicuous and compensate for this conspicuousness by being more cautious, but this needs to be tested. This would be a specific counter-strategy which could have been specifically evolved because sticklebacks with *G. anomala* are able to reproduce (personal observation).

How can we determine whether *S. solidus* manipulates its host's behaviour directly or whether it decreases the stickleback's predator avoidance (figure 4.4) simply by consuming its energy? Orr and Hopkins (1969) found that the plerocercoid becomes infective to the final host after about 10 weeks in the fish. Prior to this stage the *S. solidus* plerocercoid may want its stickleback host to avoid being eaten, although this could be at the expense of collecting food for the parasite. Manipulation is only required if the best compromise between feeding and predator avoidance differs between fish and parasite. If there is direct

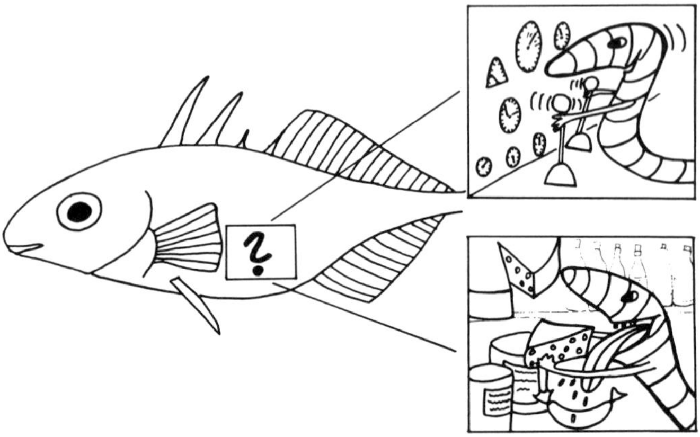

Figure 4.4. The parasite can either manipulate its host's behaviour directly or it can influence its host's decision-making by imposing a constraint, e.g. by causing a drain of energy.

manipulation of behaviour, we would expect a rapid decrease in predator avoidance when the plerocercoid is about 10 weeks old. No such step was found by Giles (1987a) when he considered the decrease of predator avoidance with increasing parasite index, defined as the weight of the parasite as a percentage of the host–parasite complex. One drawback is that the parasite index can be misleading: a high value can be obtained for a small fish harbouring one small worm, or a large fish harbouring either several small worms or one large infective individual. The comparison of the behaviour of equal-sized fish harbouring either several small plerocercoids or the equivalent weight of one mature infected worm would be invaluable in my opinion. Perhaps with this comparison it will be possible to decide whether direct manipulation is required in order to understand altered behaviour of hosts. However, it will still not be possible to rule out direct manipulation.

PARASITIZED OXYGEN AND HOST DECISION-MAKING

Another constraint placed by *S. solidus* on its stickleback host is an increased oxygen demand (Walkey and Meakins, 1970; Lester, 1971; Meakins and Walkey, 1975). For this reason authors have claimed repeatedly that parasitized fish swim closer to the surface where the dissolved oxygen concentration is higher, and where the bird predator, the definitive host, is likely to be waiting. McPhail and Peacock (1983), however, suggested that observations which seemed to confirm this behavioural change could simply be an artefact of the greater conspicuousness of parasitized individuals to the observer (figure 4.5(a)). A similar effect could be brought about by parasites that affect vision as does *Diplostomum spathaceum*. Loss of vision causes melanophores to stay expanded and fish to become darker, (figure 4.5(b)) and therefore more conspicuous (J.M. Behnke, personal communication). In dace (*Leuciscus leuciscus*), for example, it has been demonstrated that the greater the intensity of *D. spathaceum* infection in the eyes of parasitized fish the greater the time spent in the upper 10 cm of the water column (Crowden and Broom, 1980).

Recently Giles (1987a) and Smith and Kramer (1987) working with three-spined sticklebacks and nine-spined sticklebacks (*Pungitius pungitius*) respectively, published papers testing McPhail and Peacock's hypothesis. Both investigations showed that sticklebacks infected with *Schistocephalus* (*P. pungitius* was probably parasitized by another *Schistocephalus* species) moved to the surface earlier than uninfected ones, when the level of dissolved oxygen was experimentally reduced. Giles also tested gravid females which had an egg load comparable to the parasite load of the infected fish. Both the gravid females and fish with *Schistocephalus* exhibited a comparable reduction in the threshold concentration of dissolved oxygen which resulted in movement to the surface. This suggests that increased oxygen demand only and not direct manipulation causes parasitized fish to move to

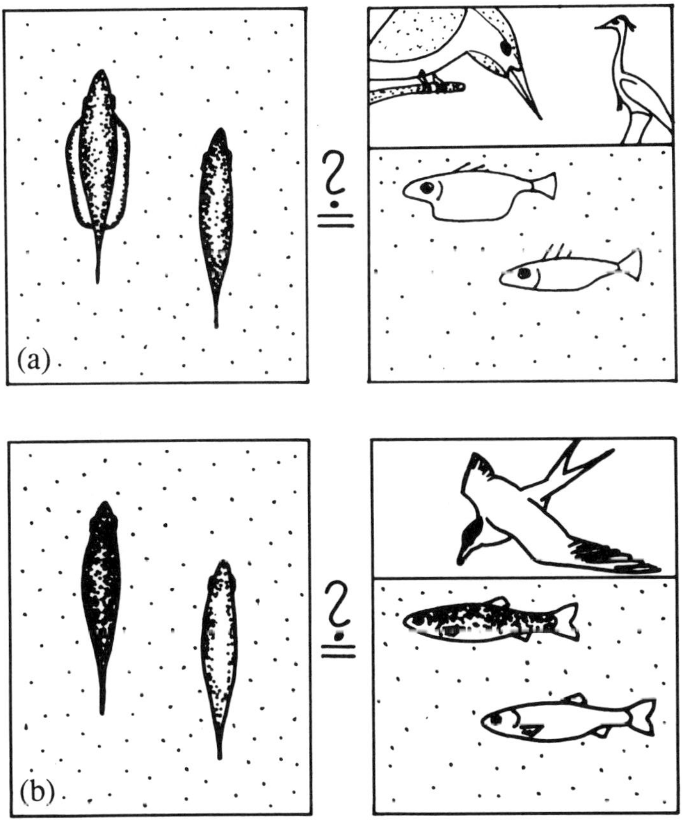

Figure 4.5. (a) Do sticklebacks which are parasitized with *S. solidus* appear to swim closer to the water surface because they are more conspicuous or do they actually swim higher in the water because they need more oxygen? (b) Do dace which are parasitized with *Diplostomum spathaceum* appear to swim closer to the water surface because they are darker or do they actually swim higher?

the surface. Smith and Kramer used a bird model to frighten the fish during surfacing. In response to the predator stimulus unparasitized sticklebacks made longer dives than the parasitized ones when the level of dissolved oxygen was reduced. However, as is indicated in both publications, dissolved oxygen levels below the threshold which initiated moving to the surface are not common in natural conditions.

Nevertheless, the problem is not yet completely resolved since both studies did not include heavily parasitized fish, that is those in which the parasite represented more than 40% of the total host–parasite complex, although such fish are frequently found in natural populations (Arme and Owen, 1967; Walkey and Meakins, 1970). These fish are more likely to harbour a plerocercoid which is infective to the final host. Interestingly there is some suggestive evidence at least, that such heavily parasitized sticklebacks decide to move towards the surface even under normal oxygen concentrations (Giles, 1987a).

To complicate matters further, the mere presence of a parasite species in the habitat may alter host decision-making. For example, the ectoparasite *Argulus canadensis* was observed to change the preferred water depth of three-spined (*G. aculeatus*) and blackspotted (*G. wheatlandi*) sticklebacks in natural conditions. Fish of both species swam near to the bottom when the parasites were absent from the habitat, but swam near the surface in the presence of the parasites which sit near the bottom and wait for potential fish hosts (Poulin and FitzGerald, 1989). In this case, the sticklebacks did decide to move to the surface, possibly to reduce the risk of parasitism.

DECISIONS BASED ON VISUAL ACUITY WITH A FLUKE IN THE EYE

The presence of metacercariae of the digenean *Diplostomum spathaceum* in the eye of a stickleback may interfere with its vision, and so make the fish more susceptible to predation by the birds which are the definitive hosts of this parasite (Wootton, 1976). Furthermore, the presence of the eyefluke should affect the results of all behavioural actions which are based on visual acuity, e.g. finding food. An intensive study by Crowden and Broom (1980) on dace, *Leuciscus leuciscus*, parasitized with *D. spathaceum* has examined this problem.

Although the dace were infected with up to 170 *D. spathaceum*, their feeding was usually vigorous, and there was no indication that any of the fish were suffering from any disability except that of reduced visual acuity. The parasitized fish obviously had difficulty in distinguishing between prey items consisting of live *Gammarus pulex* and apparently inedible objects floating in the water or lying on the bottom of the tank. The more heavily infected individuals made ten times as many mistaken responses as compared to those with fewest parasites. Even if the prey was identified correctly, parasites still seemed to handicap their hosts. There was a strong negative correlation between the number of parasites and the maximum distance from which each fish was seen to approach its prey (figure

4.6(a)). Heavily infected fish also missed more often, usually because the snap was completed before the prey was within range. The percentage of feeding responses resulting in food capture declined with increasing numbers of parasites (figure 4.6(b)). Thus, the dace foraged less efficiently the more the eyeflukes obstructed their view. Surprisingly, however, the number of *Gammarus* eaten by individual fish was not correlated with the numbers of parasites present (figure 4.6(c)). Crowden and Broom found a simple explanation for this result. The parasitized dace compensated for their reduced efficiency by increasing the amount of time spent feeding (figure 4.6(d)). Since the heavily parasitized fish did not consume more food, the parasites were apparently not imposing energetic costs and this is possibly because these metacercariae were 'waiting' stages. It seems obvious that a host spends more time looking for food to satisfy hunger when impaired eyesight causes reduced efficiency in feeding. The fish may use the same decision rule when its feeding efficiency is reduced by a low availability of prey. Therefore, in this case, the host does not need a parasite-specific counter-adaptation to compensate for this parasitic constraint. If there is surplus time, dace do not suffer from their *D. spathaceum* infection. This works quite well in a tank with plenty of prey items in a laboratory lacking hungry gulls and uninfected competitors.

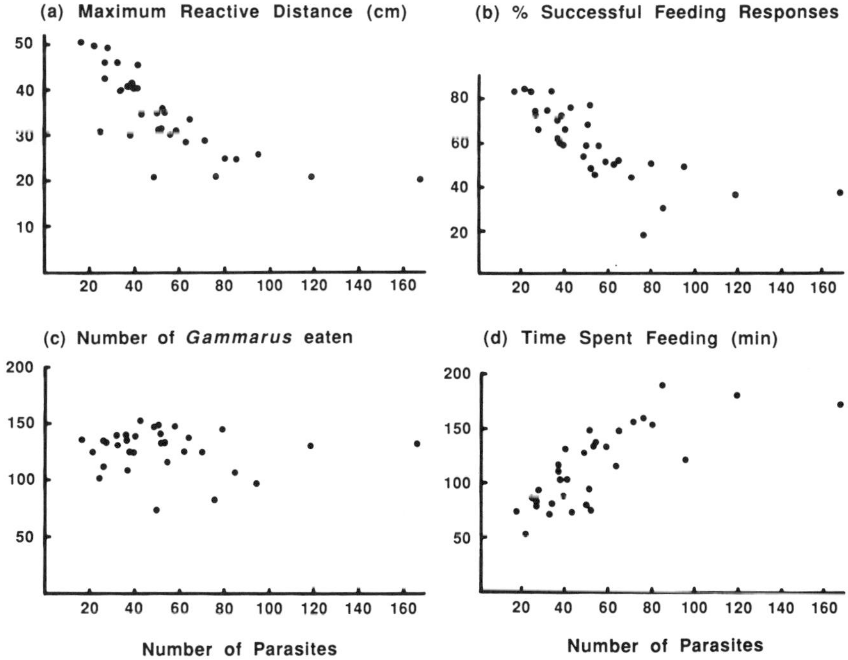

Figure 4.6. Foraging behaviour of dace dependent on the number of *D. spathaceum* parasites in the fishes' eyes. (a) Relationship between maximum reactive distance to prey and numbers of parasites. (b) Relationship between percentage of feeding responses resulting in prey capture and numbers of parasites. (c) Relationship between numbers of prey items eaten and numbers of parasites. (d) Relationship between time spent feeding and numbers of parasites. After Crowden and Broom (1980).

As soon as there is some predation pressure, prolonging the time spent foraging may increase the risk of becoming a victim: the animal is more conspicuous by its inevitable movements while foraging and it has to concentrate on finding food at the expense of vigilance for predators (Milinski, 1986). Furthermore, since under natural conditions food is often a limited resource, parasitized dace would end up hungrier especially when in competition with uninfected fish of superior competitive ability. Is it possible that host decision-making can also compensate for a reduced competitive ability to a certain extent?

POOR COMPETITORS HAVE TO BE CLEVER

Intraspecific competition for food in natural populations of sticklebacks increases in the autumn when prey becomes scarce and immense numbers of young fish from summer broods need more food. At that time of the year the following situation may arise: a good competitor and a poor competitor detect a large prey item and a small prey item at about the same time and distance. If both fish tried to catch the more profitable large item first, the poor competitor would arrive too late and the quicker fish would then catch the small item as well. In such a situation it should pay the poor competitor to prefer the less profitable prey, although it ought to ignore it in the absence of competition. When two randomly chosen sticklebacks of similar size but without apparent parasites were presented with this situation repeatedly, the fish previously determined as the poorer competitor gradually changed its preference for large prey to that of more often preferring the small

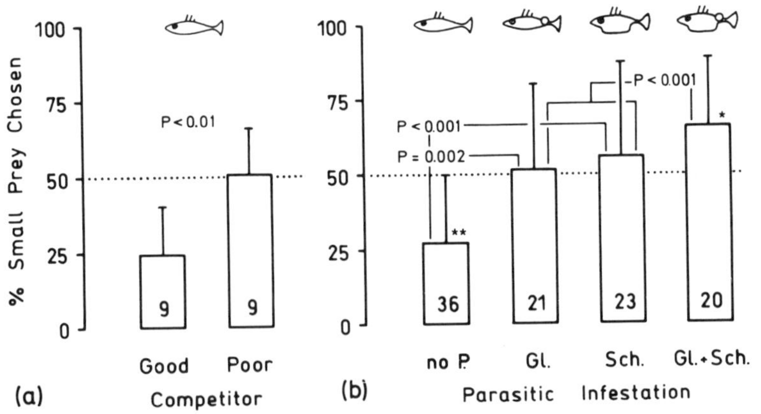

Figure 4.7. Mean percentage (+ 1 SD) of small prey chosen by sticklebacks (a) which were unparasitized and had competed as pairs for large and small prey in the laboratory, (b) which were either unparasitized (no P.), infected with *G. anomala* (Gl.), infected with *S. solidus* (Sch.) or infected with *G. anomala* plus *S. solidus* (Gl. + Sch.) and were tested directly after they had been caught from a pond which had a low food availability. *P* after two-tailed Mann–Whitney *U* test. Asterisks denote significant differences between attacks on large and small prey (*$P<0{\cdot}03$, ** $P<0{\cdot}001$, two-tailed Wilcoxon test). Numbers of fish are given in the columns. (a) After Milinski (1982), (b) after Milinski (1984).

items. In this way the poorer competitors actually increased their overall food intake (Milinski, 1982). The poor competitor actively chose which prey to attack and did not simply consume what was left by the good competitor. When alone the fish continued to choose almost the same diet for a short time as it had under competition (figure 4.7(a)).

Since parasites may reduce their host's competitive ability by utilizing its energy and oxygen, by mechanically impeding its movements, by damaging its organs (including sense organs), and/or by burdening their host with their excretions, we would expect that under natural conditions parasitized sticklebacks would switch to less profitable smaller prey items which comprise a profitable diet for poor competitors. Sticklebacks of four classes, those which had no apparent parasite, those which had either *G. anomala* or *S. solidus* only and those which had both parasites, were caught from a pond which had a high stickleback density and low food availability. Without being fed in the laboratory, all fish were tested singly using the same procedure as had been used at the end of the previous experiments (Milinski, 1984). The results showed that fish which were parasitized by either *G. anomala* or *S. solidus* chose a higher percentage of small prey items than unparasitized fish. Sticklebacks which were infected by two parasite species chose an even higher percentage of small prey than fish which were infected by only one parasite species (figure 4.7(b)). Those fish which had both parasites significantly preferred small waterfleas as opposed to large ones. The parasitized fish must have developed their preference under natural conditions and chose a diet which proved to be a profitable one for poor competitors.

Since *S. solidus* and *G. anomala* seem to affect the competitive ability of their stickleback host in a similar way, at least by reducing maximum swimming speed and manoeuvrability, fish infected by either species gain from using the same conditional strategy, that is by preferring more frequently small, less profitable prey. Fish which are parasitized in their eyes with the digenean *Diplostomum spathaceum* may have a lowered competitive ability because of impaired vision although their swimming performance is not necessarily reduced. These fish would not gain from switching to small less profitable and less visible prey when under competition. Perhaps they would prefer large unprofitable prey when available instead. Further research is required in this area.

COMPETITION AND RISK OF PREDATION

Although parasitized animals may have a reduced competitive ability, they are possibly competitively superior to healthy individuals in the presence of a predator if the parasite makes the host less fearful. This hypothesis can be tested by allowing pairs, consisting of a parasitized fish and a healthy fish to compete for *Tubifex* worms in the presence of a predatory fish (Milinski, 1985). When one stickleback was infected with *S. solidus*, it clearly outcompeted the uninfected fish (figure 4.8(a)). For example, excepting the prey most distant from the predator, the parasitized fish consumed four times as many worms as the healthy fish. It seemed as if the parasitized fish was avoiding the predator in this case, contrary to the experiment in which only one stickleback had been used at a time (figure 4.2). However, when the healthy fish was handling a *Tubifex* worm, usually at a great distance from the predator, the parasitized fish was noticeably attracted by its companion's prey and approached it. Although the parasitized fish could rarely steal the prey from the other fish it usually continued to search in the neighbourhood of the other fish. Thus, the parasitized fish did not really avoid the predator.

When one competitor was parasitized with cysts of *G. anomala*, the healthy stickleback was again significantly outcompeted (figure 4.8(b)), although in this case the parasitized fish preferred feeding at a greater distance from the predator, similar to a previous experiment with single fish with *G. anomala* (figure 4.3). An increased energy requirement of the parasitized fish may explain why it usually started to feed before the uninfected one. When the healthy fish began foraging, the parasitized fish had already consumed some of the worms most distant from the predator. Therefore, the healthy fish did not have the same choice as in the first experiment (figure 4.1). This would explain why it did not prefer the worms farthest away from the predator as in the first experiment. However, there is an additional effect here: two healthy sticklebacks often perform simultaneously so-called predator inspection visits (Pitcher, 1986) towards the cichlid (Milinski, 1987). Pairs consisting of a healthy fish and a stickleback parasitized with *G. anomala* sometimes did simultaneous inspection visits which the healthy fish usually initiated and led. On the way back both fish preyed upon worms close to the predator. I did not observe such cooperative behaviour in trials in which one of the fish had been parasitized with *S. solidus*.

Both parasite species cause their stickleback hosts to be superior competitors as compared with healthy fish in the presence of a predator. However, if both healthy and parasitized fish are starved for a longer period, the situation should be different and the healthy fish should decide to make use of its superior competitive ability.

Normally, animals do not feed as fast as they are able to. Sibly and McFarland (1976) explained this phenomenon functionally assuming that a small deviation from satiation will induce only a small loss of fitness. However, a similar increase of hunger at a lower level of reserve will have a more marked effect. This can be clarified with a human analogy. When a car is burning the last litre of fuel, the driver will be more anxious to find a petrol station than when he had used up the first litre of a full tank. Hence, with increasing hunger an animal should give greater priority to feeding at the expense of other activities and therefore the feeding rate should increase exponentially with period of starvation (figure 4.9(a)). After further starvation the increase in feeding effort is less pronounced, because there are costs and limits associated with performing activities at high rates. When deficits have weakened the animal, its feeding effort will continue to decrease. However, with increased risk of predation the animal will decide to reduce its feeding effort and to delay the increase of feeding rate (see dotted line in figure 4.9(a)). The optimal balance between feeding and avoiding predation depends on the actual value of each need.

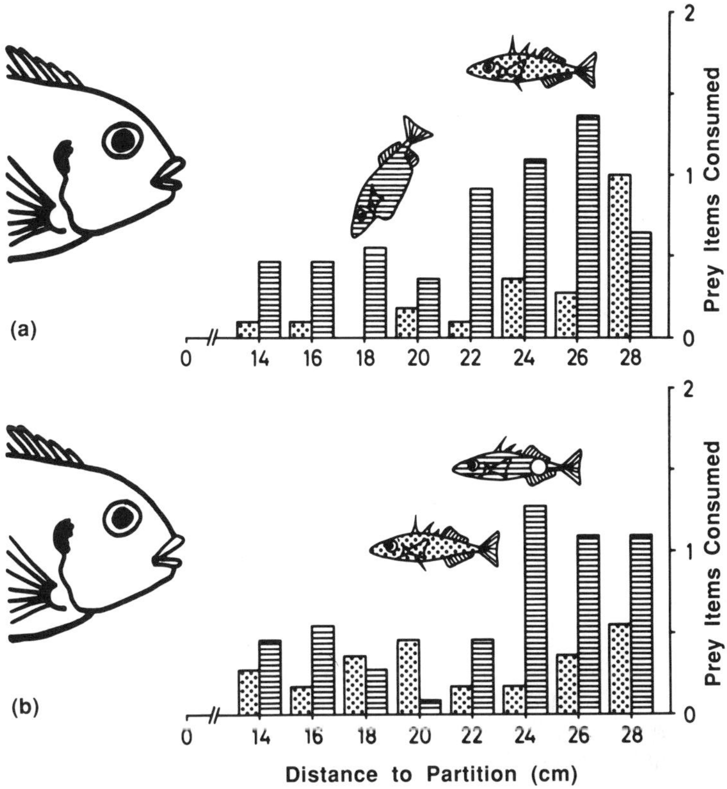

Figure 4.8. Mean number of *Tubifex* worms consumed at different distances from a watchful fish predator by (a) 12 pairs, each consisting of an uninfected (dotted) stickleback and one (hatched) infected by *S.solidus*; (b) 11 pairs, each consisting of an uninfected (dotted) stickleback and one (hatched) infected by *G. anomala*. Data consist of the first eight worms each pair of fish consumed. After Milinski (1985).

Sticklebacks parasitized with *S. solidus* have a lower maximum feeding rate than uninfected fish because of their poorer swimming ability. However, because parasitized fish have a greater need for food, they should increase their feeding effort after a shorter period of starvation than healthy ones. When starving, parasitized sticklebacks are weakened sooner and die earlier than uninfected individuals (Walkey and Meakins, 1970; Pascoe and Mattey, 1977). Increased risk of predation should retard the increase of feeding effort only for a short time in parasitized fish, because their need for food quickly outstrips their ability to increase their feeding rate (figure 4.9(a)). When feeding without a companion, parasitized and unparasitized fish consumed the first eight worms at a similar rate when the cichlid was absent (Milinski, 1985). This could be the case after a starvation time *x* (figure 4.9(a)). After the same period of starvation with additional risk of predation, however, the parasitized fish should have a similar feeding rate but the healthy fish should feed at a very reduced rate (figure 4.9(a)). Therefore it is

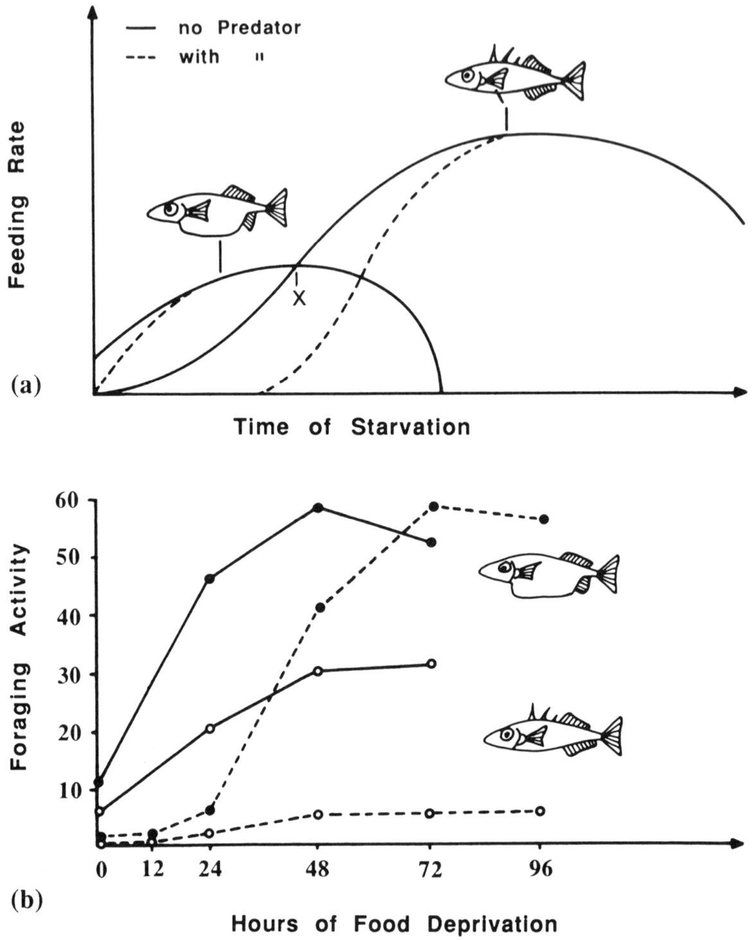

Figure 4.9. Relationship between feeding rate (activity) and the time of starvation of unparasitized sticklebacks and fish parasitized by *S. solidus*. —, No predator present: – – –, with increased predation risk. (a) Relationship predicted theoretically. After Milinski (1985). (b) Median foraging activity scores for groups of 12 parasitized (•) and healthy (○) juvenile sticklebacks after 0–96 hours of food deprivation. Modified after Giles (1987b).

possible that parasitized fish outcompete healthy ones in the presence of a predator.

This graphical model (figure 4.9 (a)) was recently tested experimentally by Giles (1987b). The aim of his study was to test whether the graphs shown in figure 4.9 (a) could be generated by measuring the foraging activity of parasitized and unparasitized sticklebacks after a range of time periods of food deprivation followed (or not followed) by a predator stimulus. Giles used size-matched sticklebacks, which were either parasitized with *S. solidus* or uninfected. With both groups two series of experiments were carried out. The first series involved suddenly dipnetting a fish from its home tank and transferring it to the test tank where its subsequent foraging behaviour was recorded for 10 minutes. By suddenly dipnetting a fish a predator stimulus was simulated and therefore these tests were called 'predator stimulus tests'. In the second series of tests the sticklebacks were left undisturbed in their home tanks and their foraging behaviour was recorded as in the predator stimulus tests. In both series, fish were deprived of food for periods of 0, 12, 24, 48, 72 and 96 hours except 12- or 96-hour periods were omitted in the home tank tests.

Although the general form of the curves which were generated experimentally correspond with the predicted feeding activity, there is a discrepancy between the results and predictions (figure 4.9(b)). The parasitized fish always had a higher foraging activity during the 10-minute tests. This difference was very pronounced in the predator stimulus tests, where the unparasitized fish showed almost no foraging activity even after 96 hours (= 4 days) of starvation. The healthy fish in both experiments had a much lower feeding rate than expected. Giles provided two explanations for this discrepancy: (1) When undisturbed the parasitized fish fed at a higher rate than healthy individuals even after 4 days of starvation because the fish had a lower parasite load and therefore had a lower foraging ability constraint than had been assumed in the model; and (2) When disturbed, the healthy fish did not resume normal feeding activity even after 4 days of starvation because dipnetting is likely to be a more frightening stimulus than a tame cichlid.

If *S. solidus* influences host decision-making only by causing an energy drain, healthy fish should make similar decisions when they have a similar need for energy caused by starvation. It is therefore surprising that even after 4 days of starvation the frightened healthy fish did not decide to shift the compromise between foraging and predator avoidance towards feeding. Does this result provide evidence for direct manipulation of host behaviour by *S. solidus* in the parasitized fish? In my opinion it does not. Since there is no reason to assume that the debilitated, parasitized fish are able to achieve a higher maximum feeding rate than the healthy fish, the feeding activity of the healthy fish should increase further with subsequent starvation (longer than 96 hours) until it is higher than that of the parasitized ones. We would therefore predict that after a much longer food-deprivation period the feeding activity of both disturbed and undisturbed healthy fish should be higher than that of the parasitized fish. Is there any evidence for this prediction? When we know what period of starvation results in the death of parasitized and unparasitized sticklebacks, we would know how far from this point both groups of fish had been in Giles' experiments.

Walkey and Meakins (1970) starved two groups consisting of either 12 parasitized or 12 unparasitized three-spined sticklebacks until all the fish parasitized with *S. solidus* had died. Of this group 50% died after 15 days and none survived more than 31 days (figure 4.10). Uninfected fish, by comparison, showed a slight initial mortality comparable with that of infected fish but a considerably reduced mortality thereafter. Even after 5 weeks of starvation, more than half of the unparasitized fish were alive and active. Thus, the risk of dying of starvation for healthy fish after 4 days without food was fairly low in Giles' experiments. We

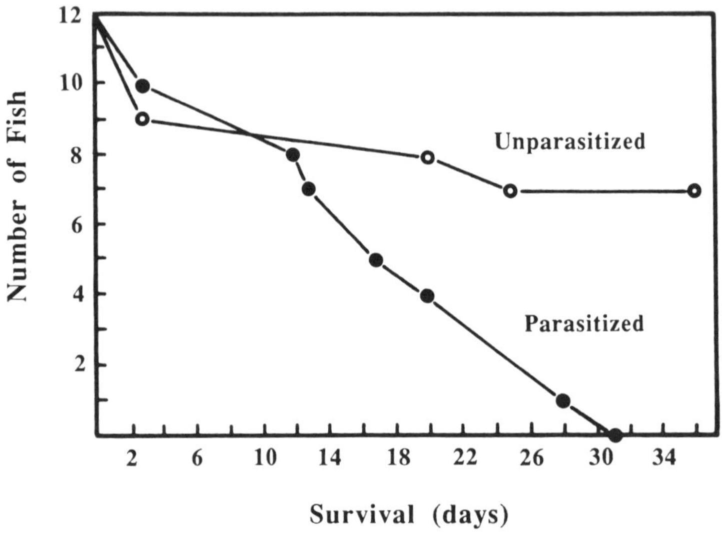

Figure 4.10. Survival of starved sticklebacks. Groups of 12 unparasitized fish and 12 fish parasitized with *S. solidus* were starved to death. After Walkey and Meakins (1970).

would expect that the feeding activity of healthy fish over-rides that of parasitized fish after about 15 days (or 360 hours) of food deprivation. The time-scale in the model (figure 4.9(a)) should read weeks rather than days.

Therefore, after a few days of starvation parasitized sticklebacks should be, and are indeed, competitively superior to healthy fish, but they probably experience an increased risk of predation. In a field study Jakobsen *et al.* (1988) demonstrated that the increased predation risk almost outweighed the benefits of feeding in exposed habitats with high food density in which sticklebacks parasitized with *S. solidus* had decided to forage. Most of these fish did not survive feeding under the increased risk of predation, which shows that they did not achieve more by clever decision-making than by making the best of a bad situation!

However, in the light of the infections considered in this chapter caution needs to be exercised in making generalizations about other parasitic infections, because animals (including humans) do not always increase their feeding activity when infected with parasites that cause an energy drain (Crompton, 1984).

PARENTAL CARE AND HOST DECISION-MAKING

Since a 'well-balanced' host–parasite association is not necessarily one in which the parasite does little harm to its host (Anderson and May, 1982), growing parasites may increasingly reduce their host's ability to reproduce (see Chapter 1). If additionally, the parasite is able to castrate its host or to prevent it reproducing in other ways (Edwards and Barnard, 1987), it may gain the advantages derived from increased host survivorship, increased host growth and/or increased energy availability (Baudoin, 1975; Smith Trail, 1980). For example, whereas female sticklebacks parasitized with *S. solidus* suffer from a delay and a reduction in oocyte maturation (Arme and Owen, 1967; Pennycuick, 1971; Meakins, 1974; McPhail and Peacock, 1983), rudd (*Scardinius erythrophthalmus*) infected with plerocercoids of *Ligula intestinalis*, a cestode closely related to *S. solidus*, rapidly become castrated and subsequently do not reproduce (Orr, 1966).

Although once castrated by a parasite the host may be unable to affect its individual reproductive fitness, it can, under certain conditions, affect its inclusive fitness by 'committing suicide' (Smith Trail, 1980). The host will increase its inclusive fitness, if by its death (and that of its parasite) the level of subsequent parasitism in its kin is reduced relative to that in non-kin. Pea aphids (*Acyrthosiphon pisum*) parasitized by the braconid wasp *Aphidius ervi* exhibit this apparent suicidal behaviour. When parasitized, the stages up to and including the third instar (the preferred stages for parasitism by *A. ervi*) are certain to die before being able to produce offspring. The aphid parasites develop very quickly and there is a high probability that they will infect the young of their host's siblings (McAllister and Roitberg, 1987). Thus, a pea aphid will probably maximize its inclusive fitness by deciding to 'commit suicide' soon after it has become parasitized. When slightly disturbed, parasitized pea aphids commit suicide by dropping from their host plant to the ground where they run a high risk of death due to heat stress and desiccation (McAllister and Roitberg, 1987). Committing suicide is necessarily a coevolved counter-adaptation of the host. Although the host is functionally castrated by the parasite, it is able to reproduce its genes by increasing the reproductive success of its close relatives.

If, however, reproduction is possible only during a short time after the host has become parasitized, the host can make the best of the situation by deciding to alter its life-history strategy. Life-history theory predicts that animals should allocate effort to each reproductive episode in such a way that their overall lifetime reproductive success is maximized (e.g. Williams, 1966; Stearns, 1976). An iteroparous animal may decide not to reproduce maximally during its first reproductive cycle in order to save energy for producing further offspring in later cycles. This trade-off between present and future reproduction will be changed dramatically when an animal 'knows' that its present reproductive cycle will be the last one in its life because it has just become infected by a 'deleterious' parasite. In this case, the best strategy would be to allocate as much effort as possible to the present reproductive cycle.

McPhail and Peacock (1983) found that a lower percentage of gravid than of nongravid female three-spined sticklebacks was parasitized by *S. solidus* in Fuller Lake, Canada. This observation suggests that either gravid females are less prone to infection or that infected females rarely become gravid. Since the parasitized gravid females had a parasite index as low as about 10%, female sticklebacks may not be able to become gravid when the parasite has grown larger than that. If this is true, female sticklebacks could maximize their life-time reproductive success by investing maximally in present reproduction soon after they have become parasitized by *S. solidus*. Surprisingly, although the parasitized gravid females must have suffered from a significant energy drain caused by the plerocercoid, they had slightly more eggs than uninfected ones. Therefore, they must have decided to alter their life-history strategy by concentrating reproductive effort on present reproduction.

A spectacular case of host life-history variation in response to parasitism has been demonstrated in *Biomphalaria glabrata* snails exposed to the trematode parasite *Schistosoma mansoni* (Minchella and Loverde, 1981) (see also Chapter 3). All snails exposed to these parasites appear to compensate for expected future loss in reproductive success by an increase in egg laying soon after exposure. There is a high probability that a schistosome infection will result in the 'reproductive death' of a snail and the eggs produced via fecundity compensation during pre-patency may become a significant proportion of an infected snail's total reproductive output (Minchella, 1985). The response takes advantage of an apparent delay in the parasite's inhibition of host reproduction. Interestingly, the snails which were exposed to infective miracidia but were not parasitized (for unknown reasons) also increased their early reproductive output in the same way as their parasitized counterparts, demonstrating that this behavioural change is the result of host decision-making (figure 4.11). However fecundity compensation has a definite cost in terms of reduced future reproductive output. Snails that are exposed but not parasitized do not show an increase in the rate of mortality, but they lay fewer eggs over the rest of their lifetime than unexposed snails (Minchella and Loverde, 1981).

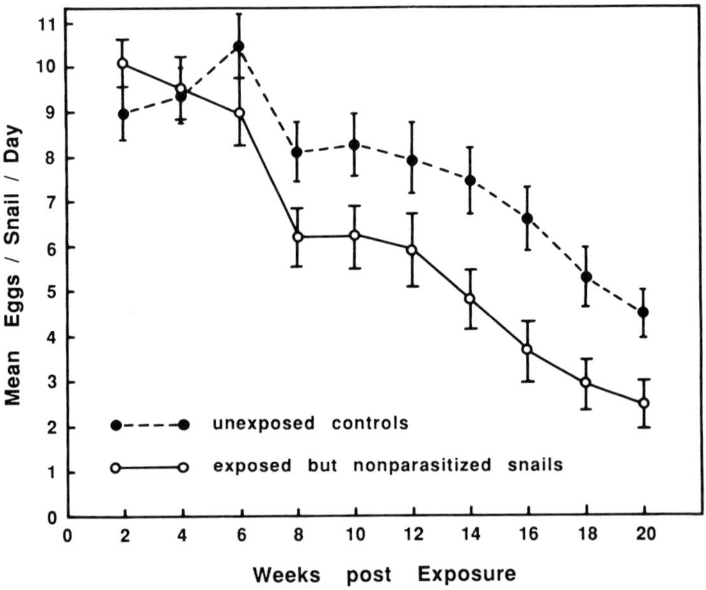

Figure 4.11. A long-term study of the effect of *Schistosoma mansoni* on the fecundity of exposed but unparasitized *Biomphalaria glabrata* snails compared with unexposed controls measured in mean eggs per snail per day ± SE. ○, Exposed but nonparasitized snails; •, unexposed controls. After Minchella and Loverde (1981).

CONCLUSIONS

When we observe that the behaviour of parasitized animals differs from that of unparasitized individuals, the parasite must have altered the host's decision-making. The parasite can either manipulate its host's decision-making directly so that the host behaves to its own disadvantage but in the interest of the parasite or it can change the host's behaviour indirectly by placing constraints on it, such as utilizing its energy and oxygen or reducing its foraging and competitive ability, and so forth. In the second case the variables which are changed are those on which the host's decision-making is based. Therefore the host simply takes into account that costs and benefits of different behavioural options have been changed and makes the best of a poor situation by trying to compensate for the constraints placed on it by the parasite. Since many of the constraints imposed by parasites do not differ qualitatively from those imposed by everyday problems (e.g. an energy drain can be caused by a parasite and by unfavourable weather) the host does not have to have evolved specific counter-strategies to behave 'adaptively' when parasitized. The host's compromise behaviour may be the best solution for both animals given that the parasitic constraint is inevitable. The host maximizes its survival and thereby its value as a resource for the parasite. Although this outcome seems to be cooperative, no coevolution needs to have occurred. We expect this kind of 'adaptive' response to parasites even in castrated hosts which have no opportunity to increase their inclusive fitness.

If the best compromise of the host's decision-making differs between the host and its parasite (e.g. because the parasite needs its present host to be eaten by its definitive host), both host and parasite compete for the host's behaviour. If the host is able to reproduce or to increase its inclusive fitness through alternative strategies when parasitized, coevolution between host and parasite is possible and an evolutionary arms race can occur. Under these conditions we would expect to find specific counter-adaptations in host decision-making such as increased wariness towards definitive hosts or committing suicide (if this prevents close kin becoming infected).

When the host has just become infected (or has some indication of infection) with a parasite which will reduce or stop the host's reproduction in the near future, the host should decide to alter its life-history strategy and allocate as much effort as possible to the present reproductive cycle.

ACKNOWLEDGEMENTS

The editors made very helpful comments on the manuscript and Olivia Lassière kindly improved my English.

REFERENCES

Anderson, R. M. and May, R. M., 1982, *Parasitology*, **85**, 411–426.
Arme, C. and Owen, R. W., 1967, *Parasitology*, **57**, 301–314.
Baudoin, M., 1975, *Evolution*, **29**, 335–352.
Canning, E. U. and Lom, J., 1986, *The Microsporidia of Vertebrates* (London: Academic Press).
Crompton, D. W. T., 1984, *Federation Proceedings*, **43**, 239–245.
Crowden, A. E. and Broom, D. M, 1980, *Animal Behaviour*, **28**, 287–294.
Edwards, J. C. and Barnard, C. J., 1987, *Animal Behaviour*, **35**, 533–540.
Ewald, P. W., 1983, *Annual Reviews in Ecology and Systematics*, **14**, 465–485.
Giles, N., 1983, *Animal Behaviour*, **31**, 1192–1194.
Giles, N., 1987a, *Journal of Fish Biology*, **30**, 631–638.
Giles, N., 1987b, *Journal of Fish Biology* **31**, 37–44.
Holmes, J. C. and Bethel. W. M., 1972, in *Behavioural Aspects of Parasite Transmission* edited by E. U. Canning and C. A. Wright, *Zoological Journal of the Linnean Society*, **51**, 123–149.
Houston, A. I., Clark, C., McNamara, J. and Mangel, M., 1988, *Nature, London.*, **332**, 29–34.
Jakobsen, P. J., Johnsen, G. H, and Larsson, P., 1988, *Canadian Journal Fisheries and Aquatic Sciences*, **45**, 426–431.
Krebs, J. R. and Davies, N. B., 1987, *An Introduction to Behavioural Ecology*, (Oxford: Blackwell Scientific Publications).
Lester, R. J. G., 1971, *Canadian Journal of Zoology*, **49**, 361–366.
McAllister, M. K. and Roitberg, B. D., 1987, *Nature, London.*, **328**, 797–799.
McFarland, D., 1985, *Animal Behaviour: Psychology, Ethology and Evolution* (Oxford: Pitman Ltd.).
McPhail, J. D. and Peacock, S. D., 1983, *Canadian Journal of Zoology*, **61**, 901–914.
Meakins, R. H., 1974, *Zeitschrift für Parasitenkunde*, **44**, 73–79.
Meakins, R. H. and Walkey, M., 1975, *Journal of Fish Biology*, **7**, 817–824.
Milinski, M., 1982, *Behavioral Ecology and Scoiobiology*, **11**, 109–115.
Milinski, M., 1984, *Behavioral Ecology and Sociobiology*, **15**, 35–37.
Milinski, M., 1985, *Behaviour*, **93**, 203–216.
Milinski, M., 1986, In *The Behaviour of Teleost Fishes* edited by T. J. Pitcher, (London: Croom Helm), pp. 236–252.
Milinski, M., 1987, *Nature, London.*, **325**, 433–435.
Milinski, M. and Heller, R., 1978, *Nature, London.*, **275**, 642–644.
Minchella, D. J., 1985, *Parasitology*, **90**, 205–216.
Minchella, D. J. and Loverde, P. T., 1981, *The American Naturalist*, **118**, 876–881.
Moore, J., 1984, *Scientific American*, **250 (5)**, 82–89.
Orr, T. S. C., 1966, *Nature, London.*, **212**, 736.
Orr, T. S. C. and Hopkins, C. A., 1969, *Journal of Fisheries Research Board of Canada*, **26**, 741–752.
Pascoe, D. and Mattey, D., 1977, *Zeitschrift für Parasitenkunde*, **51**, 179–186.
Pennycuick, L., 1971, *Journal of Zoology, London.*, **165**, 143–162.
Pitcher, T. J., 1986, in *The Behaviour of Teleost Fishes* edited by T. J. Pitcher (London: Croom Helm), pp. 294–337.
Poulin, R. and FitzGerald, G. J., 1989, *Canadian Journal of Zoology*, **67**, 14–18.
Sibly, R. and McFarland, D., 1976, *The American Naturalist*, **110**, 601–617.
Smith, R. S. and Kramer, D. L., 1987, *Canadian Journal of Zoology*, **65**, 1862–1865.
Smith Trail, D. R., 1980, *The American Naturalist*, **116**, 77–91.
Smyth, J. D., 1985, *Introduction to Animal Parasitology* (London: Hodder & Stoughton).
Stearns, S. C., 1976, *The Quarterly Review of Biology*, **51**, 3–47.
Walkey, M. and Meakins, R. H., 1970, *Journal of Fish Biology*, **2**, 361–372.
Williams, G. C., 1966, *The American Naturalist*, **100**, 687–690.
Wootton, R. J., 1976, *The Biology of the Sticklebacks* (London: Academic Press).

5. Parasites and the evolution of host sexual behaviour

A. F. Read

INTRODUCTION

Contemporary Western culture aside, men in most societies apparently prefer plump women (Symons, 1979). Human sociobiologists have argued that plumpness can be an indicator of status and health and therefore that using body fat as a cue in mate choice might be adaptive (Rosenblatt, 1974; Symons, 1979). The idea that parasites are likely to have an impact on host sexual behaviour is not a new one (eg. Giard, 1886; Potts, 1906; Huxley, 1938; Haldane, 1949) but has, until recently, received little attention in the literature. In the last decade, interest has increased for two reasons. First, theoretical developments in host–parasite coevolutionary theory show that parasites can provide potent long-term selection on host populations (Clarke, 1976; Anderson and May, 1979; May and Anderson, 1979, 1983). Second, parasites have been invoked as causal agents in the maintenance of sex (Jaenike, 1978; Bremermann, 1980; Hamilton, 1980) and the evolution of elaborate ornaments and courtship rituals used in mate choice (Freeland, 1976; Hamilton and Zuk, 1982). In this chapter, I focus primarily on adaptive modifications by the host of its own sexual behaviour in the face of parasite-imposed selection. Parasites are defined broadly to include viruses, bacteria and those animals found in standard parasitology text books. Probably because the ideas are new, little consensus has emerged about whether particular behaviours are parasite-induced host adaptations. In most cases the empirical evidence is equivocal. This necessarily leads to a somewhat speculative review but, at the very least, I hope to identify those areas which need more theoretical and empirical exploration.

Other things being equal, asexual females pass on twice as much of their genetic material as do sexual females. Asexual reproduction should therefore replace sexual reproduction in wild populations, but clearly it has not usually done so. Most attempts to solve this paradox have attempted to find an advantage for the creation of new genotypes, which is the most obvious consequence of sex (Maynard Smith,

1978). Several authors (Jaenike, 1978; Bremermann, 1980; Hamilton, 1980, 1982) have proposed that one of the few (perhaps only) selective forces which fluctuates rapidly enough and which is sufficiently widespread to account for the ubiquity of sex is that imposed by parasites. Many parasites have much shorter generation times than do their hosts, and therefore have the potential for much greater adaptability. Because sex creates new gene combinations, it is seen as a host adaptation to counter this adaptability. The idea is controversial (May and Anderson, 1983) and alternative hypotheses abound (Michod and Levin, 1987). Bremermann (1987), Seger and Hamilton (1987) and Hamilton *et al.* (in press) discuss the theoretical issues involved, and some empirical support is provided by Lively (1987), Burt and Bell (1987) and Nee (1988).

Whether or not sex is itself an adaptation to counter parasites, hosts might modify their sexual behaviour adaptively in response to selection imposed by parasites (Freeland, 1976). In addition, parasites might modify host sexual behaviour non-adaptively through the pathological consequences of infection, or they might adaptively manipulate host sexual behaviour to their own advantage. Elsewhere in this volume, Holmes and Zohar (Chapter 2) deal with the pathological effects of parasitism on behaviour and Moore and Gotelli (Chapter 7) discuss parasitic manipulation of host behaviour (see also Dawkins (1982), Dobson (1988), Keymer and Read, (in press) Read, in press), so I do not deal with these here. However, distinguishing between host adaptation and parasite manipulation becomes especially problematic in analyses of host sexual behaviour, because there are many reasons to expect that parasites should interfere with sexual behaviour. I therefore discuss briefly why parasites might manipulate sexual behaviour and how this might be distinguished from adaptive responses of the host to parasite infection.

PARASITE MANIPULATION OF HOST REPRODUCTIVE EFFORT

Non-sexually transmitted parasites may benefit from interfering with or preventing reproduction for several reasons (Baudoin, 1975; Obrebski, 1975; Dawkins, 1982):

1. Sexual behaviour may involve activities which decrease transmission rate. For example, where foraging exposes hosts to a greater risk of predation, parasites whose transmission depends on predation might prefer their hosts to feed rather than engage in sexual behaviour.
2. Where the number of parasite propagules released from a host is positively related to host lifespan, prolonged host survivorship might be favoured. The parasite could achieve this by utilizing resources associated with host reproduction, either because they are not part of the host's life support system and can therefore be destroyed without decreasing host life expectancy, or because lower host reproductive effort may decrease the costs of reproduction (Williams, 1966; Partridge and Harvey, 1988) paid by the host in terms of longevity.
3. The parasite might benefit from resources diverted from reproductive effort if this leads to greater resource ingestion by the host (where these resources

become available to the parasite), and increased size of the host, where parasite fecundity is related to the host size.

Given these advantages, we might expect to find parasite adaptations which interfere with host reproductive effort. The most extreme example of this is host castration, a phenomenon often associated with parasites of invertebrate hosts. Baudoin (1975) argues that castration is best viewed as an adaptation by parasites to decrease host reproductive effort, because it can increase parasite fitness (for any of the reasons outlined above). Furthermore, castration is frequently achieved by specialized mechanisms which selectively destroy host gonads or endocrine organs controlling host sexual development, but which have no other apparent function (Reinhard, 1956; Wülker, 1964; see also Chapters 3 and 4).

Similar arguments can be applied to modifications of host sexual behaviour which decrease host reproductive effort. For example, several invertebrate species are attacked by parasites which secrete specialized hormones to suppress the development of traits involved in male–male competition for mating opportunities (Reinhard, 1956; Baudoin, 1975). Parasites could also interfere with host attractiveness or responsiveness. Clearly, the situation is somewhat different for sexually transmitted diseases. Dawkins (1982), for example, suggests that such parasites may actually be selected to increase host copulation rate. Whether such manipulation is possible will depend in part on the extent to which hosts are limiting their copulation rates in the absence of parasites.

Distinguishing host adaptation from parasite manipulation is easy enough when the interests of the host and parasite conflict (complete host castration is unlikely to be an adaptation by the host). However, in many cases the best strategy for both parasites and at least some of the hosts will be very similar. For example, male laboratory mice (*Mus musculus*) mount nematode-infected females less often than expected (Edwards and Barnard, 1987). This could be due to adaptive manipulation by the parasites of host attractiveness or responsiveness (see Chapter 1), which, by diverting resources away from host reproductive effort, could, as argued above, increase parasite fitness: in the wild, parasites of pregnant females are less likely to be transmitted to predators because pregnant females have a lower predation risk (Edwards and Barnard, 1987). But it is also possible that one criterion used in host mate choice is resistance to infection (Hamilton and Zuk, 1982; see Mate Choice below), in which case active host choice based on signs of health might explain the observed mating patterns. Parasite adaptation, host adaptation, or both, could be involved. There is also a third scenario: the pathological consequences of infection alone may result in modifications of sexual behaviour; an explanation couched in terms of host or parasite adaptations need not be required. How can we distinguish between these possibilities?

Careful behavioural studies may allow some adaptive explanations to be eliminated. Edwards and Barnard (1987) found that the avoidance of infected mice was not a consequence of active mate choice of unparasitized individuals, but rather that infected females were more likely to avoid or respond aggressively to mounting attempts by males. Whether decreased female receptivity is due to parasite manipulation and/or whether it is an adaptive response of infected, weakened females to avoid the costs of pregnancy remains an open question. The infection-related disinclination to mate could also be a consequence of the pathology of infection. This latter explanation is more parsimonious than those invoking adaptation, and should perhaps act as a null hypothesis, especially given that infection is often likely to have debilitating side-effects. Identification of a fitness benefit to behavioural modification is important but insufficient evidence for adaptation. Eliminating hypotheses invoking the pathological consequences of infection from those invoking adaptation, and distinguishing host adaptation from parasite adaptation may be possible if the proximate causal mechanism of the behavioural modification can be identified. The most compelling support for adaptive hypotheses is evidence of specialized features possessed by host or parasite which are involved only in producing the modification. In many cases this evidence may be difficult or impossible to obtain.

With this caveat in mind, I now turn to the area of host sexual behaviour which has attracted the most attention from those interested in parasites as selective agents.

MATE CHOICE

Parasite-imposed selection could lead to adaptive modifications of host mate choice behaviour in three ways (Freeland, 1976; Hamilton and Zuk, 1982). (Here, I follow convention and discuss female choice primarily but many of the arguments apply equally well to species in which male choice might be found, such as those in which males provide parental care, or in which sperm production is limited.) First, animals might want to avoid mating with individuals infected with contact-transmitted parasites, such as lice and various sexually transmitted diseases. Second, in species in which males provide parental care, females should choose healthy and vigorous mates. In both instances, the choice is for healthy phenotypes, which need not have a genetic basis. Third, where parasite resistance is heritable, females might want to mate with resistant males so as to obtain resistance genes for their offspring. This latter possibility is an example of a 'good genes' model of sexual selection (Trivers, 1972), where female choice is said to be adaptive because choosy females produce more viable offspring. (Following Maynard Smith (1987) I use the term viability to mean all components of fitness other than mating success. This then includes parental care, as well as resistance.) One of the major criticisms of good genes models comes from population genetics theory, which suggests that, for the most part, heritable variations in fitness in randomly mating populations at equilibrium will be zero (Falconer, 1980; Charlesworth, 1987). If this is true, then choosing mates to obtain good genes will become ineffective in the long-term (Maynard Smith, 1978). Hamilton and Zuk (1982) have suggested that this objection can be overcome if host–parasite coevolution imposes fluctuating selection at loci controlling resistance. The idea is that continuous host–parasite coevolution prevents resistance alleles from becoming fixed in a population. Thus, females who choose males which are resistant to the pathogens afflicting the population at the time can obtain those resistance genes for their offspring. Charlesworth (1987) discusses some other ways heritable variation can be maintained.

Does host–parasite coevolution maintain heritable variation in parasite resistance? There is abundant evidence of genetic differences in resistance within and between populations, and parasite resistance can often be readily increased by artificial selection (reviewed by Read, 1988; Wakelin and Blackwell 188). For example, Hillgarth (in press) used experimentally manipulated parasite exposure to select for resistance in farm populations of ring necked pheasants (*Phasianus colchicus*). But obtaining unequivocal evidence from wild populations is practically impossible. Host–parasite systems in manipulated natural populations can always be said to be experiencing novel selection regimes imposed by the experimenter, which reveal heritable variation that would be removed by long-term selection. For the rest of this discussion I assume that heritable variation in resistance is maintained, at least in some populations.

The presence of sexually-transmitted parasites and/or heritable variation in resistance should lead to mate choice for healthy and/or resistant individuals. There are several possible pathogen-correlated cues which might be used in mate choice: intraspecific competition, chemosensory cues, direct external signs of infection and secondary sexual characters.

Intraspecific competition

In many species, dominance is of prime importance in gaining access to mates, with females mating with the most dominant male (for a recent review see Rees and Harvey, in press). Reasoning that health is likely to affect the outcome of agonistic encounters, Freeland (1976) suggested that females ought to induce male–male competition so as to reveal the disease states of potential mates and then choose dominant males as mates, either to avoid sexually transmitted disease or to acquire good genes.

Several experimental studies have shown that parasites can influence the outcome of intraspecific competition and the establishment and maintenance of dominance in mice (Freeland, 1981; Rau 1983, 1984; Figure 5.1(a)). The situation in the field is less clear-cut. In agreement with intuition and the laboratory experiments, fence lizards (*Scelopous occidentalis*) free from malaria are socially dominant to infected lizards and are better than infected individuals at competing for females (Schall and Dearing, 1987). However, dominance status is not associated with worm burdens in feral horses (*Equus ferus*; figure 5.1(b)), even though parasitic egg outputs are negatively correlated with subsequent body condition (Rubenstein and Hohmann, 1989). Dominant baboons (*Papio cyanocephalus*; figure 5.1(c)) and reindeer (*Rangifer tarandus*), on the other hand, have higher parasite burdens (Hausfater and Watson, 1976; Halversen, 1986). Correlational field evidence from different study systems is notoriously contradictory (see for example, the literature on costs of reproduction; Partridge and Harvey, 1988), so perhaps we should not be surprised by the conflicting evidence here. Differences in genetic quality may be associated with dominance and therefore confound any correlations (e.g. dominant individuals may be able better to tolerate a higher worm burden); equally, dominance may induce stress with consequent immunosuppression and increased vulnerability to infection. Furthermore, dominance may affect the quality of resources to which an individual has access and consequently the risk of exposure to directly transmitted parasites. The direction of any correlation between dominance and exposure risk is therefore likely to depend on the particular host–parasite system under study.

Even given that parasites can potentially affect competitive ability, distinguishing the view that the outcome of male–male competition is a cue in adaptive mate choice from the null hypothesis of no choice is probably impossible. Avoidance of sexually transmitted diseases or any good genes advantages obtained may be an incidental consequence of mating passively with the dominant male.

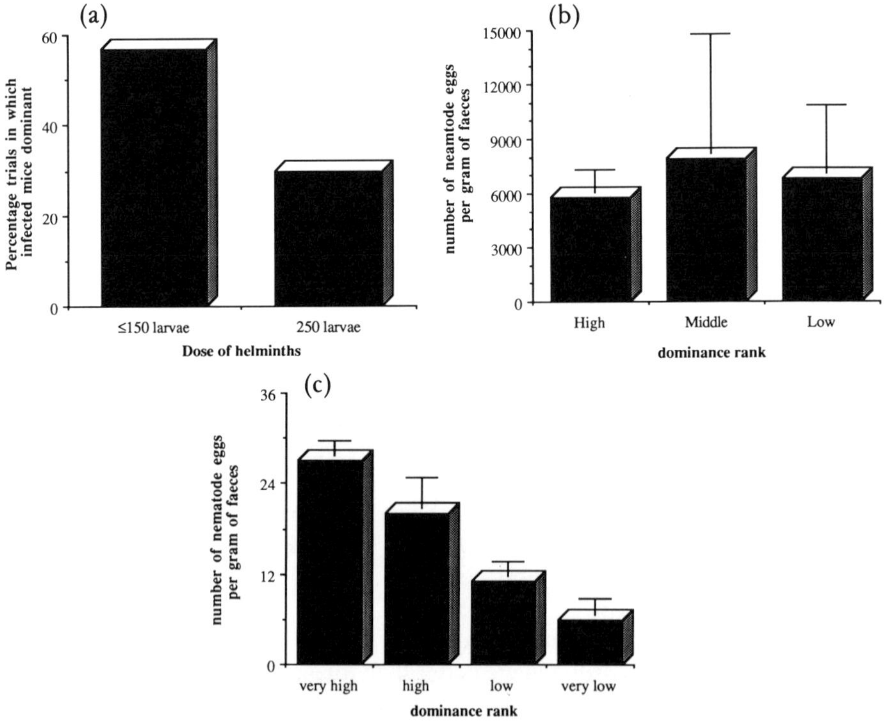

Figure 5.1. Examples of associations between parasites and male dominance. (a) Probability of laboratory mice experimentally infected with the helminth *Heligmosomoides polygyrus* becoming dominant when housed with uninfected mice. Infection at these doses is sublethal and does not significantly affect weight. Mice infected with 150 larvae or less (n = 51 replicates) did no better or worse than the uninfected mice; those receiving 250 larvae (n = 20 replicates) did significantly worse than expected by chance alone. Drawn from data reported by Freeland (1981).
(b) Eggs released from nematodes of wild male horses of different dominance rank (from a sample of 61 adult horses). Redrawn from Rubenstein and Hohmann (1989). (c) Ova emissions from free-living male yellow baboons of different dominance ranks in a troop of 29 adults and subadults. Redrawn from Hausfater and Watson (1976).

Furthermore, there are many other factors associated with dominance that might make such choice rules adaptive (e.g. in some species, dominant males are likely to control better resources).

Chemosensory cues

The major histocompatibility complex (MHC) in mammals is that group of genes coding for molecules that provide the context for the recognition of foreign antigens by T lymphocytes (Klein, 1986, p.21). The MHC is therefore involved in recognizing non-self. Several authors have suggested that there ought to be mechanisms which promote heterozygosity at the MHC loci to provide a diversity of markers used to recognize self and thus protection against a wide range of parasites, often as a corollary of sex (Bremermann, 1980, 1987; Beauchamp *et al.*, 1985; Seger and Hamilton, 1987). One way for this to occur would be for individuals to choose mates with different MHC alleles. Experiments with congenic strains of laboratory mice have demonstrated marked mating preferences by males of females differing only at the MHC (Yamazaki *et al.*, 1976; Beauchamp *et al.*, 1985). Preferences for females of different MHC haplotypes were more common, although this varied somewhat, depending on the male doing the choosing, his haplotype, and the haplotypes of females he was offered. The discrimination is based on chemosensory signals from the urine (Yamazaki *et al.*, 1982), which are apparently fragments of MHC antigens (Singh *et al.*, 1987). Mate preferences for different haplotypes have not been demonstrated in any outbred or natural populations and the finding that some mice preferred individuals with the same haplotype is somewhat problematical. In any case, the use of MHC in mate choice need not be a consequence of selection imposed by parasites for outcrossing; any other selection pressure which favours outcrossing (such as inbreeding depression, for which there is now good evidence (Harvey and Read, 1988)) could explain it (Jones and Partridge, 1983)

Direct external signs

If parasites produce visible signs of infection then females could use these to avoid infected males. The opening of feathers, mutual preening, display and inspection of genital regions and even cloacal pecking in dunnocks (*Prunella modularis*) might be ways of checking potential mates for sexually transmitted ectoparasites (Hamilton, in press; Davies (1983) provides an alternative explanation of cloacal pecking). Courtship behaviours which involve inspection of possible nest sites might have arisen because choice of nest site can affect the risk of exposure to parasites. Both cliff (*Hirundo pyrrhonota*) and barn swallows (*H. rustica*) avoid old nests containing ectoparasites, and in at least some cases this could result in alternate-year nest site or even colony site usage (Burgerjon, 1964; Brown and Brown, 1986; Møller, in press, a).

Female bluntnose minnows (*Pimephales notatus*) apparently prefer males with fewer trematode-induced black spots (Pot, 1986). These black spots are caused by encysted metacercaria in the body wall, and adult males can have up to 150 of them. Males guard the nest after egg laying and females prefer to mate with bigger males, probably because the offspring guarded by larger males are more likely to survive. When size is controlled for statistically, females prefer to mate with males with fewer spots (figure 5.2). Because the parasites are not transmitted directly, selection on females to avoid either themselves or their offspring becoming infected cannot be involved. The following year the number of trematode cysts was not associated with female preferences, possibly because there were lower levels of infection that year. Pot (1986) suggests that females might be using spot number as an indicator of a potential mates' genetic resistance, so as to obtain good genes for their offspring. Pot discounts female choice for healthy phenotypes with high quality guarding behaviour because he was able to detect effects on competitive ability and survivorship of adult males only when the number of cysts exceeded 200. At lower intensities, only juvenile growth and survivorship were affected.

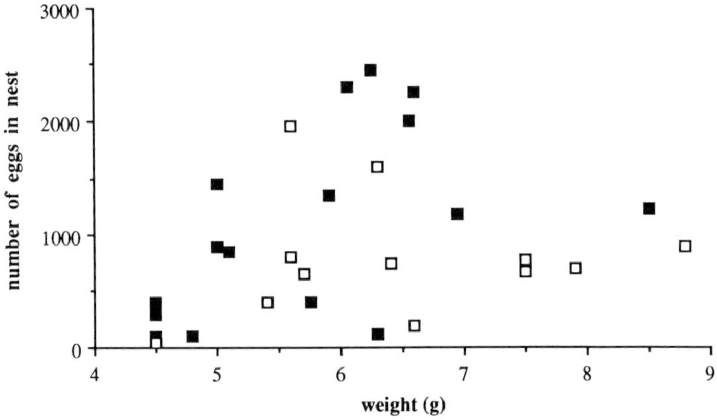

Figure 5.2. Number of eggs laid in the nests of wild minnows (a measure of female preference) as a function of male body size during a single season. ■, fewer than average encysted trematodes in the body wall; □, more than average encysted trematodes. Redrawn from Pot (1986).

Male sage grouse (*Centrocercus urophasianus*) with lice are less likely to mate than are louse-free males (Johnson and Boyce, in press). Lousy males usually have haematomas on their air sacs, and these are readily seen during the male courtship display. In choice experiments with captive grouse, Spurrier *et al.* (in press) mimicked the haematomas by applying artificial red splotches to the air sacs. Prior to application, females were equally likely to choose treatment and control birds, whereas after the splotches were applied, females tended to avoid the treated males. Unfortunately, the experiment lacked adequate controls (females might, for example, dislike the smell of treated birds). But avoidance of lousy males makes sense: by being choosy, females are likely to avoid contracting the infection themselves. They may additionally obtain resistance genes for their offspring.

This experiment suggests that females are choosing as mates those males with fewer external signs of infection. Could the air sacs have evolved *because* they enable females to better assess a male's disease state? Such a view is embodied in the Hamilton–Zuk theory for the evolution of secondary sexual characters.

Secondary sexual characters: the Hamilton–Zuk hypothesis

Hamilton (1982) and Hamilton and Zuk (1982) proposed that secondary sexual characters, such as bright colours and the elaborate plumes and vocalizations of many animals, evolved because they allow females to assess whether potential mates are resistant to the pathogens currently infecting the population. The hypothesis rests on three premises. First, that some host species are continuously co-evolving with parasites in such a way as to maintain heritable variation in parasite resistance. Second, that some of these parasites have appreciable effects on viability. Third, that the full expression of secondary sexual characters depends on the health and vigour of the individual possessing them. It follows from these premises (all other things being equal), that female choice based on the condition of male secondary sexual characters will increase offspring fitness.

This hypothesis is explicitly a good genes model of sexual selection, and is especially contentious because a variety of theorists have suggested that elaborate secondary sexual characters may serve no function other than to attract mates (Fisher, 1958; Lande, 1981; Kirkpatrick, 1982). The debate between proponents of the good genes and 'aesthetics' schools is reviewed elsewhere (Kirkpatrick, 1987a; Bradbury and Andersson, 1987), so here I consider only the theoretical and empirical issues pertaining to the Hamilton and Zuk hypothesis.

Theoretical considerations

Sexual selection is awash with theory and very short on empiricism (Bradbury and Andersson, 1987). Several genetic models of the Hamilton–Zuk hypothesis have recently been developed (Kirkpatrick, 1986; Andersson, 1986; Pomiankowski, 1987a,b, 1988; Tomlinson, 1988), but rather than compare them with each other or with other models of other sexual selection hypotheses (see Bradbury and Andersson, 1987; Pomiankowski, 1988), I discuss below the most detailed of these (Pomiankowski, 1987a) to illustrate the theoretical basis for the Hamilton–Zuk hypothesis and also because it provides a way of thinking about several theoretical and empirical problems which arise when more realistic parasitological assumptions are included.

Traits which signal parasite resistance

An implicit prediction of the Hamilton–Zuk hypothesis is that females should be using as cues in mate choice, male traits which honestly indicate genetic resistance. Clearly there are no viability advantages to be obtained by choosing males who are able to disguise their disease states. There are two classes of characters which could be involved.

The first are handicaps (Zahavi, 1975), which are traits that reduce a male's survival chances but act as a signal indicating viability. Ornate plumage, bright colours and long tails might be viewed in this way. The handicaps must amplify viability differences between males so that they are apparent to females (Maynard Smith, 1985, 1987; Pomiankowski, 1988), and honest signalling will be maintained only if the production and maintenance of secondary sexual characters is costly (Zahavi, 1975; Kodric-Brown and Brown, 1984), that is, energetically expensive. How this cost affects future viability or subsequent ability to attract mates depends on the viability of the male. For example, a low viability male widow bird (*Euplectes progne*) could, by diverting resources from body musculature, presumably produce a tail as long as those males chosen by females (Andersson, 1982), but because of this he may be unable to fly as well with it and will therefore forage or display less ably. Sufficiently high costs, then, will prevent low viability males from cheating.

Maynard Smith (1985) identified three types of handicap: *Zahavi handicaps*, which are produced by all males, but only high viability individuals can survive with them; *condition-dependent handicaps*, which are developed only by high viability males; and *revealing handicaps*, where the condition (i.e. size, brightness, showiness, display rate) of the male's ornament is highly correlated with a male's viability. Hamilton and Zuk's hypothesis is usually thought of as treating secondary sexual characters as revealing handicaps (Maynard Smith, 1985, 1987), and here I consider them as such unless I state otherwise. However, the parasite hypothesis has also been considered to involve condition-dependent handicaps (Andersson, 1986). In practice the two are likely to grade into each other: arriving on a display ground on any particular day might be considered a condition-dependent handicap, whereas the proportion of days attendance on a display ground during the breeding season might be correlated with viability and is therefore a revealing handicap. The qualitative conclusion from Pomiankowski's models (see below) are the same for both types of handicap.

A second type of trait which could be used as a reliable indicator of genetic resistance is direct signs of parasitic infection, such as those discussed under 'Direct external signs' (p. 123). Scar tissue, externally visible encysted parasites, haematomas and in some cases the parasites themselves, provide cheat-proof indications of susceptibility to infection when there is no way that the signs could be removed even if the animal had access to unlimited resources. These sorts of characters ('direct indicator traits') are distinct from other windows on health, such as the colour of exposed flesh (which may reveal anaemia) and fine tail feathers around the cloaca (which might reveal signs of diahorrea), where honesty would have to be maintained by the costs of hiding signs of infection (wattles could be reddened by pigments or production of more erythocytes; or cloacal feathers could be constantly preened) and which would therefore be handicaps.

The distinction, then, between handicaps and direct indicator traits is based on the means by which honesty is maintained: costliness or the impossibility of hiding the traces of parasitism. In principle, some traits which show direct signs of infection might become elaborated or enlarged if this amplifies the previous differences between the males, allowing more accurate assessment of male resistance (somewhat akin to amplifier traits; Hasson, in press). These traits may consequently become elaborated to a point where they are costly to produce and maintain, so that honesty is enforced by both the costs of the trait and the impossibility of hiding the signs of infection. The air sacs of sage grouse discussed above may be one such example, and the long tail feathers of barn swallows, which show visible signs of lice which eat holes in the feathers (Møller, in press, b), another.

The secondary sexual characters which have attracted most attention are the elaborate characters, such as ornate plumes, vigorous displays and complex songs, and in the context of the Hamilton–Zuk theory these must be considered handicaps. In what follows I therefore concentrate primarily on female choice of handicaps, before turning briefly to choice based on direct signs of infection.

MODELS

Pomiankowski (1987a) considers an infinite haploid population in which the males are polygynous and contribute only gametes to their mates. There are three unlinked diallelic loci. The *T* locus codes for a sex-limited secondary sexual character which is a revealing handicap. Males possessing the ornament (T_1) have a reduced chance of surviving to adulthood, whereas those without (T_2) have no viability cost. The *P* locus codes for female mating preference. Females are assumed to encounter males randomly and if a male is rejected, females encounter another and the process is repeated until all females are mated. P_1 females are equally likely to accept both dull (T_2) and ornamented (T_1) males, whereas P_2 females are choosy and prefer ornamented males. This is a fixed relative preference rule (Seger, 1985), but similar conclusions are reached under a best of *N* model (Seger, 1985), where females choose from among a subset of males in the population (Pomiankowski, 1987a). Choosy females incur a cost of mate choice, imposed by greater predation risk or loss of time and energy, which reduces their viability.

The *F* locus codes for parasitic resistance, with the two alleles conferring opposite degrees of immunity against different strains of a single parasite species. Following the verbal model of Hamilton and Zuk (1982), hosts and parasites are assumed to be coevolving in a cyclic, frequency-dependent manner. The genotype of the common parasite imposes selection for the allele resistant to it which therefore increases in frequency. When this in turn becomes common, it imposes selection in favour of the alternative parasite genotype which subsequently spreads and leads to selection for the alternative *F* allele and so on. A variety of models have suggested that such cycles are one possible (though not necessarily very common) outcome of host–parasite coevolution (Clarke, 1976; Hamilton, 1980, 1982; May and Anderson, 1983; Eshel and Hamilton, 1984). Such coevolution means that a single allele at the *F* locus confers resistance for only a few generations, and that a positive parent–offspring correlation in resistance is maintained in the long-term. So long as positive parent–offspring correlations are maintained it actually matters little how fluctuating selection on resistance alleles arises: constantly evolving or invading novel genotypes, or chaotic dynamical behaviour could also be involved (Seger and Hamilton, 1987). In the model of Pomiankowski (1987a) the relative viability of the two host genotypes through time is approximated by a sine curve.

Pomiankowski's simulations show that the introduction of heritable viability advantages increase the likelihood that a revealing handicap will become established in the population through a Fisherian runaway process. This runaway process, common to many sexual selection models arises because males with the preferred trait get more matings and if the trait is heritable, the sons of choosy females will possess the character and consequently get more matings (Fisher, 1958; O'Donald, 1980; Lande, 1981; Kirkpatrick, 1982). Once elaborated, the male trait can only be maintained in the population if females continue to prefer males with it, which will occur if the benefits of choice outweigh the costs of choice. In Pomiankowski's model, the viability benefits obtained by females who mate with males possessing the revealing handicap vary through time and choosy females will only be maintained in the population when the benfits of choice outweigh the costs, summed over the whole coevolutionary cycle. That is, when:

$$\left| \sum_{n=1}^{G} \frac{D_{FP}}{V_F} [\beta^F(m) + \beta^F(f)] \right| > \left| \sum_{n=1}^{G} \beta^P \right| \qquad \text{equation 5.1}$$

where G is the number of generations in a full host–parasite cycle, β^F is a measure of the strength of selection at the resistance locus for males (m) and females (f), β^P is a measure of the strength of selection at the preference locus, V_F is the genetic variance at the resistance locus and D_{FP} is the linkage disequilibrium between the resistance and the preference loci. (Linkage disequilibrium is a measure of the extent to which alleles at two loci are found in the same gamete more often than expected following random segregation at meiosis.) Traits revealing heritable variation in parasite resistance can spread and be maintained when the increased viability of offspring obtained by choosy females (which is a function of the benefits of being resistant to the parasites invading the host population at the time, and on the correlation between the preference and the resistance genes (left hand side of equation 5.1)) outweigh the costs of that choice (right hand side of equation 5.1).

As with many models in which runaway elaboration is involved (Arnold, 1985), some additional mechanism is required to get the frequency of choosy females in the population above the threshold required to initiate runaway (when the preference is rare, the mating advantage gained by possessing the character is too small to overcome natural selection against the trait). Alternative processes, such as pleiotropy or genetic drift, may be needed to get the process started (Kirkpatrick, 1986; Pomiankowski, 1987a).

Other models have led to similar conclusions. Kirkpatrick (1986) studied a three locus haploid model similar to Pomiankowski's, but used a step function to describe the host–parasite coevolution, where one resistance allele is favoured for G generations, then the other is favoured and so on. Tomlinson's (1988) three locus diploid models produced qualitatively similar results to those found in the haploid models. Using a three locus haploid model of a condition-dependent handicap, Andersson (1986) demonstrated that given a continuous supply of alleles with slight viability advantages (such as might arise through fluctuations in selection imposed by host–parasite coevolution), ornamental traits can spread even in a monogamous population where there can be no Fisherian mating advantage, because the male trait and the preference hitch-hike to high frequencies with high viability alleles.

PARASITOLOGICAL ISSUES

The Hamilton–Zuk hypothesis assumes that secondary sexual characters are handicaps which amplify viability differences between males (Maynard Smith, 1985, 1987; Pomiankowski, 1988). In the case of revealing handicaps, the degree of expression of the ornamental character must be highly correlated with heritable genetic variation in viability which, according to the Hamilton–Zuk view is primarily determined by parasite resistance. The weaker this correlation, the less likely it is that choosy females will mate with males with good resistance genes. This will weaken the linkage disequilibrium between the resistance and preference alleles (D_{FP} of equation 5.1), decreasing benefits of being choosy. If this linkage disequilibrium is too low, the costs of choice will outweigh the benfits of being choosy, and consequently choice and secondary sexual characters will be eliminated from the population.

So the crucial question is how strong is the correlation between the degree of expression of secondary sexual characters and heritable genetic variation in resistance?

Environmental variation in fitness has to influence the expression of handicaps which reveal general health (Eshel and Hamilton, 1984; Charlesworth, 1987). For example, nutritional differences resulting from differences in parental territory quality must also affect the expression of secondary sexual characters if they are indicators of health. Consideration of host–parasite ecology suggests several pathogen-related factors which are also likely to weaken the correlation between the degree of expression of the secondary sexual character and heritable variation in resistance:

1. *Genetic resistance for which there is no consistent positive parent–offspring correlation.* Female choice for good genes will be worthless unless there is a consistent positive parent–offspring correlation in parasite resistance. If the correlation is not positive, females who mate with resistant males will not obtain resistance alleles for their offspring. Only a subset of the parasites infecting a host population can be coevolving in such a way as to maintain a positive correlation (Eshel and Hamilton, 1984). For example, where parasite genotypes respond very rapidly to changes in host resistance, present resistance in adult males may be uncorrelated (or even negatively correlated) with future resistance (Hamilton and Zuk, 1982). To date, the types of parasites likely to maintain this positive correlation have received relatively little theoretical examination (for a review see Read, 1988), and the issue requires attention. For example, are the relevant parasites likely to have direct or indirect life cycles, be host generalists or host specialists, micro- or macroparasites, have long or short prepatent periods, and be over-dispersed or under-dispersed? Nevertheless, there are characteristics of some parasites which make them untenable as selective agents promoting

female choice for good genes. Many of the most virulent pathogens, such as some viruses and bacteria with their rapidly evolving genotypes and generation times much shorter than those of their hosts, are unlikely to maintain positive parent–offspring correlations in the long-term (Hamilton, 1982; Eshel and Hamilton, 1984; Seger and Hamilton, 1987). Yet if the expression of a handicap is tightly correlated with health, as it must be to maintain honest advertisement, infection with all diseases will be revealed. Where the parasites which do maintain positive parent–offspring correlations are a small component of overall exposure, variation in male traits may have only a weak relationship to alleles conferring resistance in the next generation.

There is an interesting conflict here between the parasite theory of sex and the parasite theory of secondary sexual characters: to maintain sex, parasites must be widespread and virulent, with sufficiently rapid genotype fluctuations to make sex a worthwhile proposition for host females (Hamilton, 1980, 1982; May and Anderson, 1983; Seger and Hamilton, 1987). Yet such parasites cannot maintain positive parent–offspring heritabilities; indeed that is precisely why sex is supposedly beneficial. If they are sufficiently ubiquitous to maintain sex, then surely they should have major effects on revealing handicaps?

2. *Phenotypic variation in exposure to parasites.* Secondary sexual characters may reveal as much or more about exposure to parasites than they do about genetic resistance. The male with the most ornate trait in the population might be the most resistant male in the population, or he might be the luckiest: he may have not been exposed to as many parasites as were other individuals. This problem has not been considered explicitly in any of the models of the Hamilton–Zuk hypothesis, yet it seems highly unlikely that all individuals will be equally exposed to the parasites currently infecting the population. Measuring the proportion of exposed individuals in a population is difficult because it need not be equal to the prevalence of infection. Serological surveys probably come closest to giving estimates of the proportion of a population who have been exposed to infection; these generally show that there are individuals in a population who have not been exposed (e.g. Grenfell and Anderson, 1985). This poses a problem for the Hamilton–Zuk hypothesis, which is perhaps most clearly seen when exposure is age-dependent. Revealing handicaps will enable old healthy (resistant) males to be distinguished from old sick males, but not from young lucky individuals.

What are the consequences of these uninfected individuals for the maintenance of female preferences for revealing handicaps? The relationship between the costs of parasitism for the hosts (as reflected in revealing handicaps) and exposure level differs for micro- and macroparasites. Microparasites (e.g. protozoans) proliferate within the host, so that exposure to a single parasite propagule can lead to parasitaemias similar to those resulting from exposure to many propagules. Intensity of macroparasitic infection (e.g. helminths), on the other hand, is a function of the number of propagules an individual has been exposed to because almost all macroparasites grow but do not propagate within a single host. All other things being equal, revealing handicaps will, according to the Hamilton–Zuk hypothesis, reveal the presence or absence of microparasitic exposure, and the extent of macroparasitic exposure. To illustrate the qualitative effects of exposure levels on the maintenance of female preferences and showy male traits in a population, I consider the simpler case of microparasitic infection.

Imagine a haploid host population, where males have tails which are costly to develop and maintain. For a given male viability, the cost is directly proportional to the length of the tail and therefore length is perfectly correlated with health, and acts as a revealing handicap. Females use a fixed relative preference rule (following the models of the parasite hypothesis by Kirkpatrick (1986) and Pomiankowski (1987a,b)). Assume that a parasite species is coevolving with this host population as described by Kirkpatrick (1986): there are two parasite strains which alternately infect the host population every G generations. Host resistance is controlled by two alleles at a single locus which confer opposite degrees of immunity to each parasite genotype and are thus favoured alternately every G generations. At any given time, a proportion r of the hosts are resistant to the parasite genotype currently infecting the host population. Kirkpatrick assumed that all hosts were exposed to the parasite genotype, but imagine that a proportion $(1 - e)$ of the potential hosts are not exposed, and that neither of the resistance alleles affects the likelihood of exposure. There are thus four types of individual in the population, with the following frequencies at any point during the host–parasite cycle:

	exposed	unexposed
resistant	re	$r(1 - e)$
susceptible	$e(1 - r)$	$(1 - r)(1 - e)$

Of the four types, only exposed susceptible males will have shorter tails, because it is only these individuals whose health prevented the full elaboration of the tail. (Throughout this discussion I use susceptible to mean that an individual is not genetically resistant and will become infected if exposed.) All other males have equally long tails (asuming that all other things are equal, in particular that there are no costs to resistance; see below) and are therefore equally likely to be preferred by choosy females. By mating with unexposed, susceptible males, females will be getting it wrong: they will not be obtaining good genes for their offspring. Following any particular period, the linkage disequilibrium between the resistance and preference alleles (D_{FP}) of equation 5.1 will therefore be proportional to that fraction of matings by choosy females which are *not* with unexposed susceptible males (J*). That is:

$$J^* = r/[e(r-1)+1] \qquad \text{equation 5.2}$$

In figure 5.3, I have plotted the value of J* as a function of the proportion of resistant hosts in a population for various values of e. The proportion of individuals exposed (e) has little affect on the linkage disequilibrium (J* is large) when a high proportion is resistant. But when the proportion of resistant males is low (and thus where choice for good genes should be most important), even small proportions of unexposed males can cause substantial decreases in the linkage disequilibrium between the resistance alleles and the female preference, and therefore in the benefits of being choosy (left hand side of equation 5.1). J* is likely to fluctuate throughout the host–parasite cycle as r and e change, so that its effects on the linkage disequilibrium will be summed over the entire cycle (equation 5.1). J* will have the greatest effect in D_{FP} when, on average during a host–parasite cycle, a relatively small proportion of the host population is exposed to the parasites and a relatively small proportion of males is resistant. Such host–parasite systems are thus less likely to be involved in sexual selection for good (resistance) genes. (From equation 5.1 it can be seen that increased virulence by the parasite – greater β^F – may offset this effect, but very virulent pathogens are unlikely to maintain heritable variation in resistance in the long-term (Hamilton and Zuk, 1982; Eshel and Hamilton, 1984; Seger and Hamilton, 1987).) By definition, the value of the term (D_{FP}/V_F) in equation 5.1 will tend to zero as the proportion of the population which is resistant tends to zero or one because the variance at the resistance locus will tend to zero. Therefore, at extreme values of r, the value of J* will have little effect on the likelihood of the preference being maintained. Of course, even in the absence of unexposed individuals, if there are on average very few resistant individuals in the population, the search costs associated with finding them may prohibit the maintenance of the preference for them.

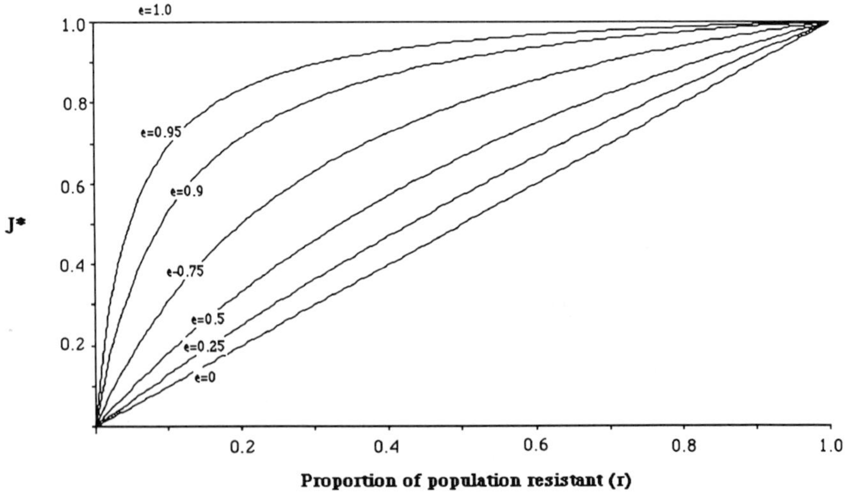

Figure 5.3. J^*, the fraction of the linkage disequilibrium between resistance and preference loci (D_{FP} in equation 5.1) remaining during any one breeding cycle, as a function (equation 5.2) of the proportion of resistant hosts in a population for different proportions of hosts exposed (e). The smaller J^*, the lower the viability advantages gained through female choice for good (resistance) genes because of females choosing unexposed susceptible males as mates.

The viability advantages obtained by choosing mates on the basis of handicaps which reveal genetic resistance to macroparasites will be similarly affected by variation in exposure levels, but in a more complex way. The pathogenicity of macroparasitic infection is an increasing function of the number of parasites found within the host (Anderson, 1978, 1979). If secondary sexual characters are handicaps honestly revealing health, then the degree of elaboration of the ornamental trait will be a function of both a male's genetic resistance to infection and of the number of parasites to which he has been exposed (all other things being equal). The basis for macroparasitic dispersion patterns in host populations is contentious, but it is generally agreed that genetic differences between hosts is only one of many causes (Anderson and Gordon, 1982). Other phenotypic causes of variance in macroparasite burdens, such as differences in exposure due to local environmental conditions or temporal fluctuations in parasite population dynamics, could be involved (Anderson *et al.*, 1978; Keymer and Anderson, 1979). Similar factors are also likely to influence exposure to microparasites.

Exposure levels are likely to be particularly important when heritable variation in resistance confers differing abilities to develop *acquired* immunity to parasites currently invading the host population. Common in many host–parasite systems, acquired immunity can only be developed after an initial infection, yet the initial infection may adversely affect health and therefore the expression of secondary sexual characters (all other things being equal). In the case of microparasitic infection, the acquired immunity itself is only tested after a second exposure, and therefore will not be honestly revealed by a handicap before that; for macroparasitic infection, acquired immunity is a function of the intensity of the initial infection. Thus females using revealing handicaps as an indication of genetic resistance will not only be mating with males who have never been exposed, they will also be avoiding males capable of developing acquired immunity, but who are sick from initial infection.

3. *Costs of resistance.* Structures which contribute to passive immunity, such as thickened cuticles and shells, as well as the infrastructure necessary to produce a facultative immune response, may be expensive to produce. Similarly, mounting an immune response and perhaps maintaining immunity against a disease may be energetically expensive (see also Chapter 1). Direct evidence of these costs is difficult to come by, and the only unequivocal evidence involves bacterial resistance to phage attack (Levin and Lenski, 1985; Keymer and Read (in press), discuss other possible examples). The observations that inadequate nutrition is often associated with infectious disease in vertebrate populations (e.g. Bundy and Golden, 1987), and that susceptible individuals are often present in populations subject to high disease pressure are suggestive of energetic costs associated with resistance, although other interpretations are possible (Anderson, 1986). If resistance is costly, resistant individuals should have less well-developed revealing handicaps than susceptible individuals in the absence of parasites.

The essential point, then, is that revealing handicaps will reveal not only heritable variation in resistance, but also components of health which cannot be involved in female choice for good genes. The wrong sort of genetic resistance, variation in exposure to parasites, and the costs of resistance must all affect handicaps. Insofar as these factors obscure the effect of heritable variation in resistance on the handicaps, assessment of a potential mate's good genes will be less accurate. Linkage disequilibrium between resistance and preferences alleles will therefore be weakened and the genetic benefits of choice based on revealing handicaps lowered (equation 5.1). Given that the heritability of fitness-related traits tends to be low anyway (Mosseau and Roff, 1988), and that other environmentally-determined variation in revealing handicaps (Charlesworth, 1987) further decreases the viability advantages of choice alleles (Eshel and Hamilton, 1984), the likelihood that female choice and secondary sexual characters can be maintained by the Hamilton–Zuk hypothesis depends critically on the magnitude of the costs of mate choice, of which we have few good estimates but which are unlikely to be negligible, at least in some cases (Pomiankowski, 1987a; Slagsvold *et al.*, 1988).

The problems introduced by revealing handicaps reflecting components of health which cannot be obtained by choice for good genes, might be limited by certain rules and cues in choice. If males who have not been exposed to the relevant parasites are common in a population, then open-ended, psychophysical (Lande, 1981) preferences by females, where the males with the most extreme traits in the population are preferred, are unlikely to be maintained. If the male trait really is a revealing handicap, males with the most ornate characters may be those who have not been exposed, or who have not paid the costs of resistance. Preferences for tails over certain fixed length, for example, would at least enable females to distinguish between males paying the costs of resistance and exposed susceptible males.

Mate choice might also be expected to incorporate rules which allow females to exclude the subset of males in a population who have not been exposed, then to use revealing handicaps to decide amongst the remainder. Perhaps females should make their choice amongst males who show cheat-proof signs of parasite exposure, such as scars or encysted parasites, or who live and display in areas where they are liable to be exposed. Older males might also be safer (Halliday, 1983; Weatherhead, 1984). The evidence for this, however, is mixed. Female collared flycatchers (*Ficedula albicollis*) apparently do not prefer to mate with older males (Alatalo *et al.*, 1986), and although older red-winged blackbird (*Ageliaus phoeniceus*) males obtain more mates, this is probably due to the high quality territories they control (Searcy and Yasukawa, 1983). Experiments with black-billed magpies (*Pica pica*) show that choice based on age can occur but this could be choice for experienced fathers rather than good genes (Komers and Dhindsa, 1989). In any case, most of these rules would not enable females to distinguish variation in male traits due to heritable resistance from those due to pathogens for which positive parent–offspring correlations do not exist.

Choice based on male traits which are direct indicators of infection, rather than revealing handicaps (where the honesty of the signal is maintained not by the costs of the trait but by the impossibility of hiding the signs of infection; see p. 126) would overcome several of the problems discussed above in connection with handicaps. Signs of infection which are caused exclusively by parasites for which there is a positive parent–offspring correlation in resistance might be possible. Costs of resistance will not affect the direct signs of infection. However, the problem of variation in exposure levels between individuals remains: is a male bluntnose minnow (Pot, 1986; and pp. 123–124) with no encysted trematode metacercariae in the body wall resistant to infection, or has he not been exposed? Furthermore, direct signs of infection are potentially less informative cues on which to base female choice than are handicaps, because they will be influenced by fewer parasite taxa.

Empirical evidence

INTERSPECIFIC TESTS

In their original paper, Hamilton and Zuk (1982) predicted that if secondary sexual characters have evolved to allow females to assess a potential mate's ability to resist parasites, then showy and elaborate characters should be found in those species which are more subject to attack by parasites. That is, there should be a positive correlation across species between the parasite burden experienced by individuals in a species, and degree of elaboration of secondary sexual characters. In support of this prediction, they reported a correlation between haematozoa prevalence and male brightness, female brightness and song complexity across 109 species of North American passerine birds.

It is now, however, widely appreciated that across-species correlations can arise as a consequence of confounding variables associated with phylogeny (Ridley, 1983, see Chapter 7), or with ecological correlates which vary independently of phylogeny. A variety of statistical methods now exist to deal with these problems (Pagel and Harvey, 1988). Phylogenetic effects are, at least in principle, the easiest to deal with because the number of potentially confounding, non-phylogenetically associated, third variables infinite. For tests of the parasite–showiness prediction, there may be important ecological correlates of both parasitism and sexual selection which need to be controlled for. For example, open nesting birds tend to be less brightly coloured than hole nesters, possibly because bright birds incubating in the open have a higher predation risk (Baker and Parker, 1979). If blood parasites transmitted by vectors which overwinter in nests are more common in hole nesting populations (because they are likely to reuse the same nest), a correlation across species between levels of parasitic infection and brightness might be produced. Whether or not these assumptions are true, the general point is clear, other variables may be independently associated with both brightness and parasites and produce the associations predicted by the Hamilton–Zuk hypothesis. Of course, if these assumptions were true and a relationship between parasite burdens and brightness became non-significant when nest type was controlled for, it could be argued from the Hamilton–Zuk hypothesis that hole nesting species are brighter precisely *because* they have higher levels of parasitic infections. Nevertheless, the identification of confounding variables opens the way to alternative interpretations

Table 5.1. Interspecific tests of the Hamilton–Zuk hypothesis.

Data set	Presumed secondary sexual character	Support across species?	Support when controlling for: Phylogeny	Support when controlling for: Ecological variables	Notes
North American passerine birds (n = 114)	Male brightness	Yes[1,2], Equivocal[3]	Yes[2], No[3]	—	Conclusions depend on brightness scores used[a]
	Female brightness	Yes[1,2]	No[2]	—	
	Male song complexity	Yes[1,4], Equivocal[5]	No[4]	—	
European passerine (n = 113)	Male brightness	Yes[2], Equivocal[3]	Yes[2], Equivocal[3]	Yes[2]	Depends on number of birds per species samples for parasites (n*)[b]
	Female brightness	No[2]	No[2]	—	
Passerine (n = 131)	Song repertoire size	No[4]	No[4]	—	
	Syllable repertoire size	No[4]	No[4]	—	
	Song duration	Opposite direction[4]	No[4]	—	
	Song continuity	No[4]	No[4]	—	
	Song versatility	No[4]	No[4]	—	
	Song rate	No[4]	No[4]	—	
Neotropical birds (n = 526)	Male brightness	Yes[6]	Yes[6]	—	Found across resident but not migrant species[c]
	Female brightness	Yes[6]	—	—	
	Sexual dichromaticism	No[6]	—	—	

Birds of Paradise ($n = 10$)	Male brightness	Yes[7]	—	Yes[7]	Correlation with haematozoa intensity but not prevalence
	Female brightness	No[7]	—	No[7]	Relationship depends on n*[d]
	Size dimorphism	Yes[7]	—	—	Relationship depends on n*[e]
British and Irish freshwater fish ($n = 24$)	Sexual dichromaticism	Yes[8]	Yes[8], No[9]	Yes[8]	
North American freshwater fish ($n = 108$)	Sexual dichromaticism	No[9]	No[9]	No[9]	

Yes, positive support for the hypothesis; Opposite direction, significant negative relationship between parasite burden and the degree of elaboration of secondary sexual characters; No, no evidence of the predictal relationship; n, number of species in the sample.

[a] Correlations found using subjective colour scores obtained from independent assessors are weaker than those found using Hamilton and Zuk's original colour scores (References 1, 2), and the correlations with the former cannot be distinguished from phylogenetic associations (Reference 3).

[b] Relationship exists across species for which few individuals have been sampled, but is not significant across species where more than ten individuals have been sampled (Reference 3). This effect is independent of phylogenetic associations.

[c] Zuk (Reference 6) argues that this is because host–parasite coevolutionary cycles are more likely to be found where hosts are attacked by one set of parasites continuously.

[d] Significant only across species for which more than ten individuals have been sampled.

[e] Not significant across species for which more than ten individuals have been sampled.

References: 1 = Hamilton and Zuk, 1982; 2 = Read, 1987; 3 = Read and Harvey, 1989; 4 = Read and Weary, 1990; 5 = Harvey *et al.*, in press; 6 = Zuk, in press; 7 = Pruett-Jones *et al.*, in press; 8 = Ward, 1988; 9 = Cabana and Chandler, in press.

of any showiness–parasite correlations (which, in the hypothetical case of nest type as a confounding variable, would need to provide other explanations for the link between brightness and hole nesting).

Given that it is impossible to identify (much less test) all potential confounding variables, the best interspecific support for the hypothesis would be provided by positive results from a diverse range of host and parasite taxa, because the likelihood that any one confounding variable is responsibile for producing the same association in many different host–parasite systems (and for many different secondary sexual characters) is small. To date, we are still well short of such a range of tests, much less the positive support (table 5.1). None of the studies which has controlled for both phylogenetic associations and known ecological correlates of the secondary sexual character under study provides clear support for the hypothesis.

Perhaps most worrying for future comparative tests of the hypothesis is the problem faced by Zuk (in press) in her study of over 500 species of neotropical birds: too little is known about the ecology of the species involved even to begin to consider controlling for confounding variables which are not associated with phylogeny. Zuk found positive associations between male brightness and parasites within more taxa than expected by chance alone, but any confounding variable that varies within taxa could also produce such an effect. I mentioned nest type as a possibility above; likewise, any other ecological correlate of bird colour (see Baker and Parker, 1979) might be independently associated with parasite prevalence. In principle, it is easy to control for these variables (see Read, 1987), but detailed knowledge of the ecology of the species involved is required. Such data sets are rare, and are likely to remain so. Alternative is analyses of a diverse range of taxa (lizards, tropical fish, invertebrates, other bird taxa) which differ in ecology, parasite taxa and secondary sexual traits might overcome this problem.

As well as confounding variables, there are two other possible explanations for any associations found between the elaboration of secondary sexual characters and parasite load. The first is that the causality is opposite to that proposed by Hamilton and Zuk. Perhaps species with secondary sexual characters are more susceptible to parasitic infestations because they have to produce bright plumage, long tails or other costly ornaments and therefore have fewer resources to devote to resisting parasitic invasions. If this hypothesis were correct, then within a species, preferred males should have more parasites, a prediction opposite to that of the Hamilton–Zuk hypothesis, and not supported by recent intraspecific studies (see below and table 5.2). A second possible explanation for supporting comparative evidence is that in species in which males provide parental care, females could choose males on the basis of their secondary sexual characters, not to obtain good genes as Hamilton and Zuk suggested, but rather as indicators of current vigour and therefore of the quality of parental care a male is able to provide. This idea predicts that there should be a relationship across species which do provide parental care, but not across species which do not. Unfortunately, conclusions of the comparative tests published to date are too equivocal to assess this prediction (table 5.1). Of course, this alternative hypothesis of choice for healthy male phenotypes is unable to explain the most extreme cases of elaboration of secondary sexual characters found in species in which males do not provide any parental care (Darwin, 1871).

On the other hand, there are a disturbing number of highly plausible *post hoc* rationalizations which can be invoked to explain why the predicted association might not have been found and to save the hypothesis:

1. The crucial parasite taxa were not included in the analysis (perhaps the parasites which were assayed were not coevolving so as to maintain a positive parent–offspring correlation in resistance).
2. Differences in study sites and techniques of parasite sampling between studies, together with sampling error, especially in species for which only a few individuals have been assayed for parasites, caused a type II error (false acceptance of the null hypothesis).
3. The measurement of species susceptibility to parasitism was not accurate enough (parasite prevalence and number of parasite species found in each host species are the only comparative measures used to date, and both of these are, at best, correlated with the infection level, not the proportion of individuals exposed to parasites or the level of exposure).
4. Subjective scoring of 'brightness', 'showiness' or 'song complexity' introduces noise and increases the chance of a type II error.

Of these, perhaps only the last can be dealt with adequately. Quantitative measures can to some extent remove error imposed by subjectivity. However, there is the broader issue of how adequately to quantify the elaboration of a secondary sexual character. Often the evidence that a single character (like plumage brightness) is used as a cue in female choice is available for only a few of the species in any particular comparative data set, and what evidence there is, is usually correlational. The male trait might be involved in mate choice in some species but not others. There is compelling evidence that song repertoire size, for example, is an important cue in female choice in some species, but there is equally compelling evidence that it is not in others (Read and Weary, 1990). There is the problem of how to compare a secondary sexual character in one species with that in another: is a red tail more elaborate than a yellow tail? Does a bird species with a large repertoire of songs consisting of only a few syllables have a more elaborate song than a species with a small song repertoire made up of a few long complex songs? This comparability problem (Krebs and Kroodsma, 1980), well known in the song literature, applies to all comparative studies of sexually selected traits, and can always be invoked as an explanation for failure to find a relationship.

Several authors (see table 5.1) have endeavoured to examine whether sexual dimorphism is greater in species more liable to attack by parasites. There are, however, many reasons why no relationship might be found, even if the Hamilton–Zuk hypothesis were correct. Dimorphism is the result of the joint pressures of sexual and natural selection acting independently on each sex (Clutton-Brock, 1983; Slatkin, 1984). Differential selection on the two sexes, such as might be imposed by predators in species where only the female incubates (Baker and Parker, 1979), could therefore give rise to dimorphism, irrespective of any parasite-

Table 5.2. Intraspecific tests of the Hamilton–Zuk hypothesis.

			Relationship between intensity of infection and host viability		Heritable variation in parasite resistance	Expression of secondary sexual character varies with parasite burden		Females choose males with fewer parasites	
Reference	Host	Parasite	Correlational[a]	Experimental[a]		Correlational[a]	Experimental[a]	Correlational[a]	Experimental[a]
1.	Field crickets	Gut protozoan	—	Y/X[b]	—	X	—	X[c]	—
2.	*Drosophila melanogaster*	Nematode	Y	—	—	Y?[d]	—	Y?[e]	—
3.	Guppies	Gut nematode, ectoparasitic digenean	Y	—	Y[f]	Y (figure 5.4)	Y	Y	—
4.	Fence lizards	Malaria	Y	—	—	Y[g]	—	—	—
5.	Satin bowerbirds	Head louse	—	—	—	Y/X[h]	—	Y	—
6.	Sage grouse	Lice, malaria[i]	Y	—	—	Y	—	Y	Y/N[j]
7.	Barn swallows	Haematophagous mite, lice	Y	Y	—	Y	—	Y[k]	—
8.	Rock doves	Lice	—	—	—	—	Y[l]	—	Y (figure 5.5)
9.	Jungle fowl	Nematode	Y	Y	—	—	Y	—	Y
10.	Birds of Paradise	Haematozoa	—	—	—	Y	—	N	—
11.	Pheasants	Various coccidia, nematodes and protozoa	Y	Y	Y[m]	—	N	Y	N

Y, demonstrated; X, negative evidence; —, not tested or known; X/Y, evidence mixed.

[a] Type of evidence, correlational or experimental, where experimental evidence is defined as those studies that experimentally manipulated the key independent variable.

[b] Negative effects demonstrated only in one of two species, and only for individuals on suboptimal diets.

[c] Parasite load had no effect on male attractiveness but females spent longer mating with uninfected individuals, who therefore have time to provide more spermatophores.

[d] Parasitized males have distended abdomens, lighter than the usual dark orange. Obvious to a human, but it is not clear that these features are secondary sexual characters.

[e] Male–male competition cannot be ruled out.

[f] Laboratory experiments (Madhavi and Anderson, 1985).

[g] Malarious males look brighter and less female-like to a human than uninfected individuals but there is no data on whether females prefer bright or dull males (similar results are reported by Schall (1986) on another lizard host–parasite system, although for an apparently avirulent parasite).

[h] Bower holders had fewer parasites but other features of bowers uncorrelated.

[i] Also some unknown antibiotic-sensitive parasites.

[j] Females prefer males treated with antibiotics over controls, but are indifferent to deloused males.

[k] As both mates and for extra-pair copulations.

[l] Display rate lower in lousy males but no treatment effect on plumage.

[m] Selection experiment on captive birds.

References: 1 = Zuk, 1987a–c; 1988; 2 = Jaenike, 1988; 3 = Kennedy *et al.*, 1987; 4 = Ressel and Schall, 1989; 5 = Borgia, 1986; Borgia and Collis, 1989; 6 = Spurrier *et al.*, in press; Johnson and Boyce, in press; 7 = Møller, in press a,b; 8 = Clayton, in press; 9 = Zuk *et al.*, in press; 10 = Pruett-Jones *et al.*, in press; 11 = Hillgarth, in press.

Scientific names of species 1 = *Gryllus veletis* and *G. pennsylvannicus*; 5 = *Ptilonorhynchus violaceis*; 10 = *Parotia lawseii*; 11 = *Phasianus colchicus*.

imposed selection. Similarly, species differences in the relative importance of male choice of females will have an important effect. Untangling these factors statistically should be possible, at least in principle, but will require considerable information about selection pressures on secondary sexual characters in each species.

There is one other reason why a relationship between showiness and parasites might not be found. Genetic models of the Hamilton–Zuk hypothesis (discussed above) suggest that female choice for resistant males could lead to runaway elaboration of secondary sexual characters, and although parasites may be important in maintaining the elaborate trait in the population (Pomiankowski, 1987a,b), the point at which the elaboration is halted may be uncorrelated with the initial parasite load: natural selection against further elaboration may be the most important determinant of the extent of elaboration. If so the degree of elaboration need not be correlated with parasite load.

So if a null result cannot falsify the prediction, what of a negative association between sexual showiness and parasites? In their original paper Hamilton and Zuk (1982) made an explicitly unidirectional prediction: the antagonistic coevolution between parasite and host, and invasions by new parasite genotypes, are assumed to be never-ending, so that in the long-term, *species*' vulnerability to parasitism is likely to be unaffected by female choice for parasite resistance. However, as Borgia (1986) has pointed out, a negative relationship might not be problematical for the hypothesis: perhaps the evolution of strong female preferences in a population will increase the proportion of females mating with resistant males, leading to an increase in the number of resistant offspring in the population, and subsequently a net decrease in the vulnerability of the population to parasitic attack (i.e. fewer parasites). The issue hinges on whether female choice will increase population fitness or leave it unaltered (that is, whether parasite numbers drop or remain at similar levels). Of the theoretical studies published to date, only that of Kirkpatrick (1986) examines this issue. He concludes that mean population fitness does not rise but also that this conclusion is inherent in the assumptions of his model. Further theoretical investigation is needed, but until then the interspecific prediction is probably best considered a bidirectional prediction, since minor modifications of the hypothesis can probably incorporate a negative correlation between showiness and parasites. (This argument raises another possible reason for failing to find supporting comparative evidence: vulnerability to parasitic infection might be unchanged by the evolution of strong female preferences in some species, but decreased in others.)

Thus, the interspecific prediction of Hamilton–Zuk hypothesis is easier to support than to reject, although only positive results from well controlled studies on a wide range of host–parasite systems involving a diversity of secondary sexual characters can provide compelling support. Null results, on their own, are unable to lead to rejection of the hypothesis because there are many other reasons for finding no association. Indeed the prediction is unfalsifiable. Only an accumulation of null results is likely to lead to acceptance of the null hypothesis of no association: it is therefore crucial that non-significant associations be reported. It is tempting to wonder how many comparative analyses which could be added to table 5.1 are languishing in files labelled 'unpublishable'.

INTRASPECIFIC TESTS

There are four major premises of the Hamilton–Zuk hypothesis to be tested intra-specifically:

1. That there is heritable variation in parasite resistance;
2. That there is a negative relationship between parasite load and host viability;
3. That the expression of secondary sexual characters is negatively correlated with variation in parasite burden over the range of infection intensities which affect host fitness and which are usually found in natural populations;
4. That preferred males should be those males which are more resistant to parasites infecting the population at the time.

There has been a recent proliferation of studies involving a wide range of hosts and parasites where data has been gathered on one or more of these premises with the explicit aim of testing the Hamilton–Zuk hypothesis (table 5.2; figures 5.4 and 5.5). No clear picture emerges. Evidence which supports and contradicts the predictions have been reported, and the only study which has attempted to experimentally test all the predictions (Hillgarth, in press) found that experimentally increasing parasite burdens did not affect the outcome of female choice, nor any of the characters presumed to secondary sexual traits.

If parasites are important in the evolution of secondary sexual characters, then they must affect the expression of these traits. Many studies on invertebrates have

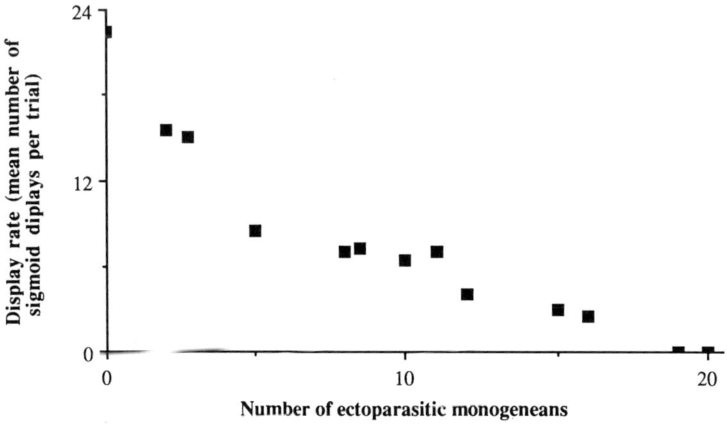

Figure 5.4. Rate of display by male guppies (*Poecilia reticulata*) to females following experimental exposure to *Gyrodactylus*, an ectoparasitic mongenean. All males were given the same level of exposure. More heavily infected males display less ($r = 0{\cdot}94$, $P < 0{\cdot}001$), and experimentally parasitized males also performed significantly fewer displays than unparasitized controls. Redrawn from Kennedy *et al.* (1987).

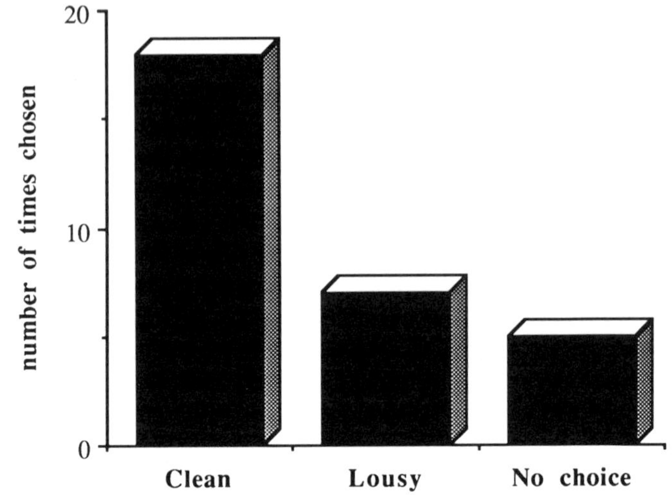

Figure 5.5. Preferences of rock dove females offered a choice of males with experimentally increased louse loads and unmanipulated control males. Female choice scored as the male with whom a female was in close proximity most of the time during a trial. Redrawn from Clayton (in press).

found that this does occur, often in association with host castration (see reviews by Reinhard, 1956; Wülker, 1964; Baudoin, 1975). But a clear corollary of the Hamilton–Zuk hypothesis is that secondary sexual characters should be more revealing of parasite resistance than are other morphological features. Otherwise female choice for resistant males could be based on any health-related cue, such as general body condition, and need not give rise to the elaboration of ornate secondary sexual characters. To date, only two studies have produced experimental data related to this. Zuk *et al.* (in press) reared male jungle fowl (*Gallus gallus*) chicks in sterile conditions, and infected half the chicks with intestinal round worms at similar intensities to those found in free-ranging zoo populations. Once the chicks matured, choice experiments showed that unparasitized males were preferred by females. Control and treatment males could not be separated statistically using measurements of non-ornamented characters (body weight, length of tarsus, bill and back feathers and bill width) but could be separated using measurements of purported secondary sexual characters (combs and tail feathers were smaller in parasitized males, combs, back and tail feathers were duller, and parasitized males lost their breeding plumage earlier). Clayton (in press) experimentally manipulated intensities of lice infection on adult male rock doves (*Columbia livia*). Louse load was manipulated by inserting metal 'bits' between birds' mandibles which prevent efficient preening but not feeding. Females avoided lousy males in choice experiments (figure 5.5) even though lice were not visible directly. There was no difference in weight between control and treatment groups and plumage was unaffected. Human observers were unable to distinguish the two groups, and the only correlate of female choice Clayton could identify was display rate. These experiments provide perhaps the strongest supporting evidence reported to date (table 5.2), although in neither case is there experimental (e.g. Andersson, 1982; Møller, 1988) or field evidence showing that the purported secondary sexual characters are indeed cues in female choice.

That females will mate with males resistant to contemporary parasite strains is a critical prediction of the Hamilton–Zuk hypothesis. Following Hamilton and Zuk (1982), this is usually taken to mean that females will prefer males with fewer parasites. This is likely to be the correct prediction most of the time, but the problem of varying exposure levels (see p. 130–133) complicates the issue. Females would do best by choosing males who have been exposed to levels of parasitism which can affect viability because the resistance of these males has been tested. It is difficult to imagine how females might achieve this, although, choosing among males who show signs of exposure to parasites or whose habitat, behaviour or age increases the chances of exposure are possible mechanisms (see pp. 134–135). If females can do this, it seems plausible, at least in principle, that they will avoid highly parasitized males as well as those that are uninfected (and in the case of macroparasites, also those with subclinical infection intensities), since the latter may be lucky rather than resistant individuals. In the absence of information about variation in exposure levels, comparisons of the reponses of preferred and avoided males to experimental challenges with natural pathogens are needed to test properly the premise that preferred males are more resistant to parasites currently invading the population.

But imagine that, as Hamilton and Zuk predict, good experimental evidence from a single host–parasite system becomes available which shows that parasites which affect host viability at natural intensities and for which there is heritable resistance in the host population, affect secondary sexual characters more than non-ornamental characters, and that preferred males are more resistant. This would provide strong support for the hypothesis, and it is difficult to imagine stronger support from a single system. But such evidence is also open to alternative interpretations which may be impossible to distinguish empirically from the Hamilton–Zuk hypothesis.

First, if the evidence is from a host species in which males provide parental care, females might be assessing a male's phenotypic vigour, with any heritable resistance being an incidental consequence. Even if the evidence was from a species in which males only provide sperm there is a second possibility. Secondary sexual characters may have evolved for other reasons, such as non-adaptive runaway (Kirkpatrick, 1987a) and been elaborated into structures or displays which are energetically expensive. If so, then disease-free males are more likely to be able to produce the most ornate characters and vigorous displays. Female choice of resistant males may thus be an incidental consequence of female preferences, not the reason they evolved or are maintained (Kirkpatrick, 1987a,b; Ryan, 1988). This is a particularly problematical alternative interpretation, because there are good reasons to believe that parasites will affect energetically expensive secondary sexual characters, either as an effective means of manipulating host reproductive effort or as a pathological side-effect of infection (see pp. 118–120). Indeed, the pathological consequences of parasitism are precisely the reason secondary sexual characters are supposed to reveal resistance (Hamilton and Zuk, 1982).

Thus any alternative model of sexual selection through female choice (which can explain the maintenance of secondary sexual characters despite the costs of choice; Pomiankowski, 1987a) potentially provides an alternative interpretation of even the best possible experimental support for the Hamilton–Zuk hypothesis. So far as I can see, such alternatives can only be rejected in favour of the Hamilton–Zuk hypothesis if compelling comparative support for the hypothesis becomes available. And not just any interspecific association with parasites will do. Hamilton and Zuk (1982, p.385) expect secondary sexual characters to be well developed in those species subject to high parasite pressure (at least before female choice and male traits evolve). Thus, either positive or negative correlations between the expression of secondary sexual characters and parasites might be found across species, depending on the impact of female choice on the mean level of population resistance (see Interspecific tests, p. 142). Models of sexual selection by female choice which do not invoke parasites as causal agents could explain the negative association by claiming that choice incidentally results in an increase in the proportion of females mating with resistant males, thus decreasing the number of hosts infected with parasites. But these models make no claim that secondary sexual characters are more likely to evolve in species subject to initially high parasite pressure, and hence cannot accommodate a *positive* association between parasites and male showiness. Thus, the prediction of a positive interspecific association is the only prediction of the Hamilton–Zuk hypothesis which cannot be explained by the alternative view that obtaining resistance genes is an incidental side-effect of choice which has evolved for some other reason. Unfortunately, the available comparative evidence is sufficiently equivocal to be accommodated by both models (table 2.1).

So experimental studies cannot provide exclusive support for the Hamilton–Zuk hypothesis. But perhaps even more disturbing, the Hamilton–Zuk hypothesis is unlikely to be falsified with intraspecific data. Hamilton and Zuk (1982,p.385) state that 'Our hypothesis is contradicted if within a species preferred mates have most parasites . . . '. Yet by any view, this is unlikely to be the case. If the theory is actually wrong, then a far more likely outcome is a failure to reject the null hypothesis of no association between mate preference and infection levels. The problem here is that, as with the comparative prediction, there are many entirely plausible *ad hoc* explanations which can be invoked to explain this outcome and save the hypothesis:

1. The wrong parasite was studied: it is not coevolving so as to maintain positive parent–offspring correlations in resistance;
2. Not all the relevant parasite taxa were assayed;
3. Both the male trait and the female preference are maintained by periodic epidemics and the study was too short to pick these up. This is a possibility raised by Pomiankowski (1987a,b). Selection imposed by the costs of female choice will not remove preference alleles (and consequently male traits) immediately. Several, perhaps tens of host generations (depending on the

magnitude of the costs of female choice) will be required so that, in the short-term, ornamental traits can be maintained in the absence of any viability advantage conferred by obtaining resistance genes. If relatively few individuals are exposed or exposure levels are low in a particular breeding season, choice might not correlate with parasite load even though it may be maintained by parasites in the long-term. Correlations between choice and parasites in some but not all breeding seasons have been reported in two studies (Pot, 1986; Borgia and Collis, 1989).

Each of these possibilities is plausible but make the hypothesis essentially unfalsifiable. As with tests of the comparative prediction, only the accumulation of results from carefully conducted studies which fail to reject the null hypothesis is likely to lead to the hypothesis being abandoned.

MATING SYSTEMS

As is so dramatically illustrated by the AIDS epidemic, parasitism is a potentially potent selective force in the evolution of host mating systems. Avoidance of sexually-transmitted diseases might, for example, impose selection for either long-term pair bonding or polygyny with healthy males (Hamilton, in press). Where there is heritable variation in resistance, non-sexually transmitted diseases might also have effects on the evolution of the mating systems of their hosts, perhaps promoting polyandry, polygyny and extra-pair copulations.

Polyandry

Consider the situation where it is impossible for individuals to assess the genetic contribution of potential mates to offspring viability in advance of mating (perhaps individuals that have recovered from infection are indistinguishable from those that were never susceptible or from those that were never exposed). In such cases, multiple mating by females might increase the genetic diversity of the progeny and therefore the chance that at least some of the offspring will be resistant to parasite genotypes infecting offspring viability. Such a possibility has been discussed in the past in connection with multiple mating and sperm competition, but usually founders because it is difficult to imagine that the advantages of any extra diversity obtained from multiple mating over that resulting from outcrossing with a single mate will be greater than the costs of obtaining extra matings (Parker, 1984; Drummond, 1984). This is especially unlikely in those species with small brood sizes. However, avoidance of disease has been proposed as the factor promoting multiple mating in the social hymenoptera (Hamilton, 1987; Sherman *et al.*, 1988). Polyandry in the social hymenoptera is a rare but well documented phenomena (Queller *et al.*, 1988) which challenges the traditional kin selection explanation of eusociality, since multiple mating has the effect of lowering the average genetic relatedness within colonies. Hamilton (1987) and Sherman *et al.* (1988) porposed that polyandry increases genetic diversity within the colony and thus reduces the likelihood that the colony will be exterminated by a pathogen, to which a less diverse, 'pure stand' (Hamilton, 1987) would be susceptible. According to this hypothesis, the haplodiploid genetic system of the social hymenoptera makes such susceptibility particularly important. There is little evidence supporting the idea and there are many alternative hypotheses purporting to explain the phenomenon (reviewed in Sherman *et al.*, 1988). However, the hypothesis makes the readily testable prediction of a correlation between the number of matings with different males by queens and the prevalence or diversity of parasites infecting a species or population. Evidence from 14 ant species suggests that multiple mating is commoner in larger colonies, and this is consistent with the hypothesis if parasite load increases with colony size (Sherman *et al.*, 1988).

Extra-pair copulations

In species in which both males and females cooperate to raise the brood extra-pair copulations (EPC) are easily understood as an alternative male reproductive tactic (Trivers, 1972). In many species, extra-pair copulations appear forced and thus apparently involve no female choice (McKinney *et al.*, 1984). However, where female cooperation in copulation is essential, or where females actively solicit EPCs, EPCs can be interpreted as an alternative reproductive tactic for females as well (see Halliday and Arnold (1987) for a different interpretation). The benefits females obtain through EPCs are not easily explained because EPCs can be risky. Females might, for example, risk mate desertion (Trivers, 1972; Gladstone, 1979) or injury (McKinney *et al.*, 1984). Several hypotheses purport to explain active female involvement in extra-pair copulations (Harvey and Bennett, 1985; Halliday and Arnold, 1987), and there is currently little consensus. One possibility is that where there is heritable variation in parasite resistance, females that have had to mate with unresistant males should attempt to solicit extra-pair copulations with males who are resistant to parasites currently invading the population in an effort to obtain good genes for their offspring (Hamilton, in press). Hamilton considers that in species which pair for more than just copulation, secondary sexual characters might have evolved primarily because of their role in attracting females for EPCs. He argues that courtship is often a lengthy affair in birds, so there are ample opportunities for directly observing stamina and health. He suggests that flashy colours or elaborate songs may be involved in female assessment of males on neighbouring territories for potential extra-pair copulations, since detailed, direct assessments of their health are not possible because of the risk of mate desertion.

This view makes several predictions. First, males obtaining extra-pair copulations should be more resistant than neighbouring males who failed to obtain the EPC. Second, females should solicit copulations from other males more frequently when their own mate is less resistant, as might be the case for females arriving late on the breeding grounds who have to make do with males avoided by early arriving females. (Increased solicitation is crucial here; more EPCs could also be explained by poorer mate guarding by less healthy males.) Third, if secondary sexual characters are handicaps revealing resistance, as Hamilton and Zuk (1982) suggest,

EPCs should be obtained by males with the most well-developed secondary sexual characters. Fourth, in species in which males and females cooperate to raise the brood, interpopulation or interspecific differences in showiness will correlate with the frequency of extra-pair copulations (Hamilton, in press).

EPCs are rarely seen in the field and there are few data on EPC frequency in different species and very little on which males and females within a population are engaging in EPCs. Recent advances in DNA fingerprinting techniques should help resolve this difficulty. In the meantime, there are a few tantalizing suggestions of female choice for good genes. Thirteen successful black-capped chickadee (*Parus atricapillus*) EPCs were observed (in 14 years field work!) where the breeding status and overwinter dominance of the males involved was known (Smith, 1988). In all cases, females had EPCs with males that had higher dominance rank in the previous winter than that of their mates. Furthermore, in nine of these EPCs, females copulated well inside the territory of the neighbouring male, suggesting active solicitation. Møller (1988) found that male swallows with artificially lengthened tails obtained mates earlier, and also had more EPCs even though they did not make more attempts to obtain EPCs after treatment. Furthermore, females participating in EPCs were those mated with control males and males with artificially shortened tails. If female choice in swallows is concerned with obtaining resistance alleles (and Møller's (in press a,b) correlational evidence is consistent with this view; table 5.2), then females soliciting EPCs with males with long tails will be obtaining viability advantages. Møller (in press, b) shows that males who successfully obtain EPCs have lower parasite burdens.

Unfortunately, intraspecific tests of the view that females solicit EPCs to obtain resistance alleles have many of the same problems of alternative interpretations from which intraspecific tests of the Hamilton–Zuk hypothesis suffer (see pp. 143–147). For example, resistant males might still obtain more EPCs as a result of female choice if parasite manipulation or the pathological side-effects of infection affect a male's ability to attract females, and female involvement in EPCs has some other cause.

Polygyny

In accordance with theory (Trivers, 1972), polygyny is the most common mating system found in animals, but it is relatively rare in birds (Lack, 1968). The usual reason given for this discrepancy is that the provision of male parental care in birds has a major impact on the reproductive success of males. The existence of polygyny in those bird species with monogamous close relatives has thus aroused considerable theoretical and empirical interest (for a recent review see Rees and Harvey, in press). If there are viability advantages associated with female preferences for resistant males, as suggested by Hamilton and Zuk (1982), then females might prefer to forego mating monogamously and join an already-mated male, despite the reduced parental care she will receive as a beta female, in order to obtain good genes. Such choice could be based on the pathogen-correlated cues discussed above. There are, of course, other reasons for polygynous mating (see for example,

Lightbody and Weatherhead, 1988) but the possibility that female preferences to obtain resistance alleles could result in polygyny in some populations is worth considering and, all other things being equal, follows from the premises of the Hamilton–Zuk hypothesis (p. 125).

It is straightforward to include polygyny in Pomiankowski's models of the Hamilton–Zuk hypothesis: the costs of mating with an already mated male appear as selection acting on the female preference (β^P) in equation 5.1 above. Preference for male traits that reveal a male's past and present parasite load will be maintained when the costs of the preference (i.e. of mating polygynously and therefore fledging fewer young) are less than the benefits obtained (in terms of offspring viability), summed over the host–parasite cycle. The costs imposed by mating with an already mated male will be additional to those of being choosy, and probably much greater (reviewed in Lightbody and Weatherhead, 1987), making it less likely that the preference will be maintained, especially given that the difficulty of accurately assessing heritable resistance from revealing handicaps tends to decrease the linkage between the preference and resistance alleles (p. 129–135). Nevertheless, if resistance is heritable and honest indicators of past and present parasite burden exist, it is possible, at least in principle, that the Hamilton–Zuk hypothesis may explain polygyny in some populations.

The idea that polygyny may be promoted by mate choice for resistance alleles makes several predictions. For example:

1. Already-mated males with whom additional mates settle will be more resistant to parasites currently invading the population than any unmated potential mates those females rejected (all other things, particularly territory quality, being equal).
2. For a given territory quality, the last female opting to settle with a particular already-mated male should produce offspring of higher viability than females who are the last to subsequently settle with the earlier-rejected males. Viability is probably best measured as survivorship, and particularly the likelihood of being recruited into the breeding population where the important parasites affect juvenile survivorship. Given that secondary females inevitably have lower fledging success than alpha females, it is a necessary corollary of the hypothesis that the reduction in the number of offspring fledged by a beta female is more than compensated for by greater offspring survival to maturity.
3. The frequency of polygyny should be correlated with the extent to which a population or species is liable to attack by parasites.

Intraspecific evidence bearing on these predictions is currently absent. None of the intraspecific tests of the Hamilton–Zuk hypothesis have involved a polygynous species (table 5.1) and I am not aware of any data on which comparisons of the recruitment rates of the offspring of alpha and beta females on territories of similar quality can be made. There are, however, some data on polygyny in human populations supporting the third prediction. Low (1988; in press) has demonstrated a correlation between the degree of polygyny and degree of exposure to severe pathogens across 93 human societies (r_s ranging between 0·20 and 0·32 and the P

value from 0·046 to 0·002). Low rules out group size, maximum harem size and geographical differences as confounding variables, but concludes that 'the causal relationships giving rise to these patterns are not simple . . . non-random mate choice resulting in polygyny is associated with male control of resources independent of pathogen risk, and in many societies, female choice appears to be non-existent or seriously compromized' (Low, 1988, p.123).

As with the Hamilton–Zuk theory of the evolution of secondary sexual characters, the idea that polygyny might be promoted by parasites is difficult to falsify, because there are many *post hoc* explanations for a failure to support it, not the least of which is the quite reasonable claim that there are likely to be many causes of polygyny, any one of which might be involved in any particular case, or which might confound comparative tests. Again, I suspect that only the steady accumulation of studies which fail to reject the null hypothesis could make the hypothesis redundant.

GROUP LIVING AND SOCIALITY

Alexander (1974) and Freeland (1976) suggested that increased risk of parasitic infection might be a major cost of group living because in bigger groups more parasite species are likely to survive stochastic fluctuations of both parasite and host populations, and directly-transmitted parasites may have higher transmission rates. There is now abundant evidence from bird and mammal populations of a correlation between group size and parasite levels (Hoogland and Sherman, 1976; Freeland, 1979; Hoogland, 1979; Brown and Brown, 1986; Møller, 1987; Shields and Crook, 1987; Moore *et al.*, 1988; Rubenstein and Hohmann, 1989; but see Gregory (in press) for a discussion of the difficulties of interpreting such correlations when sampling effort is not controlled for statistically). Severe fitness effects of parasitism have been experimentally demonstrated in several of these (Brown and Brown, 1986; Shields and Crook, 1987; Møller, in press, a), suggesting that parasites can impose selection on group size.

However, this selection may not simply favour living in smaller groups, and more theory and experimental data are needed. For example, parasites transmitted by vectors which move freely between hosts, such as adult biting flies, may affect the costs and benefits of group living in a different way. One study on primates (Freeland, 1979), two on feral horses (Duncan and Vigne, 1979; Rubenstein and Hohmann, 1989) and one experimental study involving reindeer-shaped flytraps giving off reindeer-like CO_2 emissions (Hille and Aspi, 1983) have shown that individual hosts receive fewer bites from potential disease vectors when they are in large groups, presumably as a result of a dilution effect. All three studies produce correlational evidence to show that animals form larger aggregations when biting fly numbers are high (see also Chapter 10). This 'selfish herd' effect (Hamilton, 1971) is thus working in the opposite direction to selection imposed by parasites such as those discussed above. It therefore seems likely that, at least in some host species, parasites have a role in determining the breeding group size, and decisions about whether to breed colonially or which colony to join, but how they do so will likely depend on the specific details of the host–parasite system involved. For example, worm burdens increase with increasing harem size in feral horses, but biting fly numbers per horse drop (Rubenstein and Hohmann, 1989). When flies are abundant, members of a harem aggregate more tightly, but do not move off to join larger harems. Rubenstein and Hohmann (1989) argue that this shows that the benefits of small harems, which may include reduced intrasexual competition as well as reduced exposure to some parasites, outweigh the benefits of decreased fly harassment experienced by bigger harems.

SUMMARY AND CONCLUDING REMARKS

Parasites are likely to affect host sexual behaviour by imposing selection for adaptations which minimize the effects of parasitism. Parasites are also likely to manipulate host reproductive effort to maximize their own reproductive success, and the pathological side-effects of infection can also influence host sexual behaviour (see also Chapters 1, 2 and 3). This chapter deals with the first of these, but the three possibilities may be difficult or even impossible to distinguish empirically.

In the face of parasitic infection, host mate choice may be adaptively modified so that individuals infected with contact-transmitted parasites are avoided. Choice of healthy mates for good phenotypes (parental care) and good genes could also occur. Several cues might facilitate such choice: the outcome of intrasexual competition, chemosensory stimuli, direct external signs of infection and the condition of secondary sexual characters.

The latter possibility, suggested by Hamilton and Zuk (1982) as a 'good genes' theory of sexual selection, has attracted most attention. Genetical models show that the hypothesis can work. If secondary sexual characters are to act as honest indicators of health, they must be costly to produce and will as a consequence reveal components of health which are not associated with good genes, such as exposure levels. These other factors decrease the viability advantages obtained by choosy females, since the assessment of a male's heritable resistance is less accurate. Whether female preferences (and therefore secondary sexual characters) can be maintained by choice of resistance genes despite this, depends critically on the costs of female choice.

Empirical tests of the Hamilton–Zuk hypothesis have proved largely equivocal and are likely to remain so. Evidence in a direction opposite to that predicted by the Hamilton and Zuk hypothesis is unlikely (within a species, females are unlikely to be mating with unhealthy individuals), or could be viewed as supporting a slightly modified version of the hypothesis (across species, greater elaboration of secondary sexual characters could be associated with lower parasite burdens). Plausible *post hoc* rationalizations can be invoked to explain a failure to reject the null hypothesis and save the theory (e.g. the wrong parasites were assayed). Even the best possible supporting experimental evidence (which does not yet exist) can be interpreted as supporting other models of sexual selection. However, positive experimental evidence, together with supporting evidence from carefully controlled comparative analyses from a wide range of host–parasite systems, may provide the best way of supporting the Hamilton–Zuk hypothesis while excluding other hypotheses. Such evidence does not yet exist. It is difficult to see how the hypothesis can ever be falsified by a single test; however, it is likely to be abandoned if many studies are unable to reject the null hypothesis. It is therefore crucial that studies which fail to find supporting evidence be reported.

Parasites could also influence the evolution of mating systems. How they might do so depends on the details of the host–parasite system involved. Promotion of monogamy, polyandry, polygyny and extra-pair copulations have all been suggested, the latter two as logical extensions of the Hamilton–Zuk hypothesis and therefore open to the same difficulties. Furthermore, there are many alternative explanations for the evolution of such mating systems, any one of which can be invoked to explain why parasites might not be involved in particular instances, so that the generality of the ideas are difficult to test. For no one host–parasite system is there experimental evidence that parasites affect the mating system.

Finally, an individual's parasite burden can depend on the size of the group in which it is living, and this can lead to adaptive variation in host group sizes (colonies, harems). How parasites will affect optimum group sizes will depend on the hosts and parasites involved.

This survey has shown that parasites can, in principle, promote adaptive modifications of host sexual behaviour to counter the negative effects of parasitism. In many cases it seems intuitively obvious that they should and, at least in the case of the evolution of secondary sexual characters, genetical models suggest that they could. However, empirical tests of these ideas are difficult because little is usually known about host–parasite ecology and coevolution in any particular system, and there are usually several alternative explanations for any particular outcomes. Distinguishing host adaptations from parasite manipulations and non-adaptive consequences of infection remains a major stumbling block.

Note added in proof

Hamilton and Zuk (1982) predicted positive interspecific associations between showiness and parasite load. But such associations might also be found if species producing costly ornaments (for reasons other than those given by Hamilton and Zuk) are more susceptible to parasitic infection. I argued above (p. 138) that if this is so, then within a species, showier males should have more parasites. Because this is not the case (Table 5.2), I rejected the idea. However, Andrew Balmford (pers. comm.) has pointed out that within species, pattern need not refute this alternative interpretation of the interspecific association. Mean population susceptibility to infection might be increased because all males attempt to produce and maintain secondary sexual characters, but within populations, only relatively healthy males would produce the most elaborate traits.

I now have difficulty envisaging any empirical evidence which could provide exclusive support for the Hamilton and Zuk hypothesis.

ACKNOWLEDGEMENTS

I have benefitted from regular discussions with Richard Gregory, Tim Guilford, Paul Harvey, Anne Keymer, Sean Nee, Mark Pagel and Andrew Pomiankowski, all of whom commented on earlier drafts of this chapter. Nick Barton raised the possibility that the production of secondary sexual characters might increase average susceptibility to infection within a species. G. Borgia, M. Boyce, G. Cabana, D. Clayton, R. Gregory, W. D. Hamilton, B. Low, A. Møller, S. Pruett-Jones, D. Rubenstein, J. Schall and M. Zuk sent preprints. I am supported by a Junior Research Fellowship at Christ Church.

REFERENCES

Alatalo, R. V., Gustafsson, L. and Lundberg, A., 1986, *American Naturalist*, **127**, 241–245.
Alexander, R. D., 1974, *Annual Review of Ecology and Systematics*, **5**, 325–383.
Anderson, R. M., 1978, *Parasitology*, **76**, 119–157.
Anderson, R. M., 1979, In *Population Dynamics* edited by R. M. Anderson, B. D. Turner and L. R. Taylor (Oxford: Blackwell Scientific Publishers), pp.245–281.
Anderson, R. M., 1986, In *Immune Mechanisms in Invertebrate Vectors* edited by A. M. Lackie (Oxford: Clarendon Press), pp.239–274.
Anderson, R. M. and Gordon, D. M., 1982, *Parasitology*, **85**, 373–398.
Anderson, R. M. and May, R. M., 1979, *Nature*, **280**, 361–367.
Anderson, R. M., Whitfield, P. J. and Dobson, A. P., 1978, *Parasitology*, **77**, 189–200.
Andersson, M. B., 1982, *Nature*, **299**, 818–820.
Andersson, M., 1986, *Evolution*, **40**, 804–816.
Arnold, S. J., 1985, *Experientia*, **41**, 1296–1310.
Baker, R. R. and Parker, G. A., 1979, *Philosophical Transactions of the Royal Society (London) B.*, **287**, 63–130.
Baudoin, M., 1975, *Evolution*, **29**, 335–352.
Beauchamp, G. K., Yamazaki, K. and Boyse, E. A., 1985, *Scientific American*, **253**, 66–72.
Borgia, G., 1986, *Behavioural Ecology and Sociobiology*, **19**, 355–358.
Borgia, G. and Collis, K., 1989, *Behavioural Ecology and Sociobiology*, **25**, 445–454.
Bradbury, J. W. and Andersson, M. B., 1987, *Sexual Selection: Testing the Alternatives. Report of the Dahlem workshop on sexual selection* (Chichester: John Wiley and Sons).
Bremermann, H. J., 1980, *Journal of Theoretical Biology*, **87**, 671–702.
Bremermann, H. J., 1987, In *The Evolution of Sex and its Consequences*, edited by S. C. Stearns (Basel: Birkhäuser Verlag), pp.135–161.
Brown, C. R. and Brown, M. B., 1986, *Ecology*, **67**, 1206–1218.
Bundy, D. A. P. and Golden, M. H. N., 1987, *Parasitology*, **95**, 623–637.
Burgerjon, J. J., 1964, *Ostrich*, **35**, 77–85.
Burt, A. and Bell, G., 1987, *Nature*, **326**, 803–805.
Cabana, G. and Chandler, M., in press, *Animal Behaviour*.
Charlesworth, B., 1987, In *Sexual Selection: Testing the Alternatives* edited by J. W. Bradbury and M. B. Andersson (Chichester: John Wiley and Sons), pp.21–40.
Clarke, B., 1976, In *Genetic Aspects of Host-Parasite Relationships* edited by A. E. R. Taylor and R. Muller (London: Blackwell), pp.87–103.
Clayton, D. H., in press, *American Zoologist*.
Clutton-Brock, T. H., 1983, In *Evolution from Molecules to Men* edited by D. S. Bendall (Cambridge: Cambridge University Press), pp.457–481.
Darwin, C., 1871, *The Descent of Man, and Selection in Relation to Sex* (London: John Murray).
Davies, N. B., 1983, *Nature*, **302**, 334–446.
Dawkins, R., 1982, *The Extended Phenotype* (Oxford: Oxford University Press).
Dobson, A. P., 1988, *Quarterly Review of Biology*, **63**, 139–165.

Drummond, B. A., 1984, In *Sperm Competition and the Evolution of Animal Mating Strategies* edited by R. L. Smith (Orlando: Academic Press), pp.291–370.
Duncan, P. and Vigne, N., 1979, *Animal Behaviour*, **27**, 623–625.
Edwards, J. C. and Barnard, C. J., 1987, *Animal Behaviour*, **35**, 533–540.
Eshel, I. and Hamilton, W. D., 1984, *Proceedings of the Royal Society (London) B*, **222**, 1–14.
Falconer, D. S., 1980, *Introduction to Quantitative Genetics*, 2nd ed., (London: Longman).
Fisher, R. A., 1958, *The Genetical Theory of Natural Selection* (Oxford: Clarendon Press).
Freeland, W. J., 1976, *Biotropica*, **8**, 12–24.
Freeland, W. J., 1979, *Ecology*, **60**, 719–728.
Freeland, W. J., 1981, *Science*, **213**, 416–462.
Giard, A., 1886, *Comptes Rendues*, **103**, 84–86 (cited by Reinhard, 1956).
Gladstone, D. E., 1979, *American Naturalist*, **114**, 545–557.
Gregory, R. D., in press, *Functional Ecology*.
Grenfell, B. T. and Anderson, R. M., 1985, *Journal of Hygiene*, **95**, 419–436.
Haldane, J. B. S., 1949, *La Ricerca Scientifica*, **Suppl. 19**, 68–75.
Halliday, T. R., 1983, In *Mate Choice*, edited by P. Bateson (Cambridge: Cambridge University Press), pp.3–32.
Halliday, T. and Arnold, S. J., 1987, *Animal Behaviour*, **35**, 939–941.
Halvorsen, O., 1986, *Oikos*, **47**, 71–74.
Hamilton, W. D., 1971, *Journal of Theoretical Biology*, **31**, 295–311.
Hamilton, W. D., 1980, *Oikos*, **35**, 282–290.
Hamilton, W. D., 1982, In *Population Biology of Infectious Diseases* edited by R. M. Anderson and R. M. May (New York: Springer-Verlag), pp.269–296.
Hamilton, W. D., 1987, In *Animal Societies: Theories and Facts* edited by Y. Ito, J. L. Brown and J. Kikkawa (Tokyo: Japan Science Society Press), pp.81–102.
Hamilton, W. D., in press, *American Zoologist*.
Hamilton, W. D., Axelrod, R. and Tanese, R., in press, *Proceedings of National Academy of Sciences USA*.
Hamilton, W. D. and Zuk, M., 1982, *Science*, **218**, 384–387.
Harvey, P. H. and Bennett, P. M., 1985, In *Human Sexual Dimorphism. Society of the Study of Human Biology. Symposium Vol. 25* edited by J. Ghesquiere, R. D. Martin and F. Newcombe (London: Taylor and Francis), pp.43–59.
Harvey, P. H. and Read, A. F., 1988, *Nature*, **336**, 514–515.
Harvey, P. H., Read, A. F., John, J. L., Gregory, R. and Keymer, A. E., In *Parasitism: Coexistence or Conflict?* edited by A. Aeschlimann and C. Toft (Oxford: Oxford University Press).
Hasson, O., in press, *Proceedings of the Royal Society (London) B*.
Hausfater, G. and Watson, D. F., 1976, *Nature*, **262**, 688–689.
Hille, T. and Aspi, J., 1983, *Acta Zoologica Fennica*, **175**, 129–131.
Hillgarth, N., in press, *American Zoologist*.
Hoogland, J. L., 1979, *Behaviour*, **69**, 1–35.
Hoogland, J. L. and Sherman, P. W., 1976, *Ecological Monographs*, **46**, 33–58.
Huxley, J. S., 1938, *American Naturalist*, **72**, 416–433.
Jaenike, J., 1978, *Evolutionary Theory*, **3**, 191–194.
Jaenike, J., 1988, *American Naturalist*, **131**, 774–780.
Johnson, L. L. and Boyce, M. S., in press, In *Ecology, Behaviour and Evolution of Bird-Parasite Interactions* edited by J. E. Loye, C. Van Riper and M. Zuk (Oxford: Oxford University Press).
Jones, J. S. and Partridge, L., 1983, *Nature*, **304**, 484–485.
Kennedy, C. E. J., Endler, J. A., Poynton, S. L. and McMinn, H., 1987, *Behavioural Ecology and Sociobiology*, **21**, 291–295.
Keymer, A. E. and Anderson, R. M., 1979, *Parasitology*, **79**, 195–207.
Keymer, A. E. and Read, A. F., in press. In *Parasitism: Coexistence or Conflict?* edited by A. Aeschlimann and C. A. Toft (Oxford: Oxford University Press).
Kirkpatrick, M. 1982, *Evolution*, **36**, 1–12.
Kirkpatrick, M., 1986, *Journal of Theoretical Biology*, **119**, 263–271.
Kirkpatrick, M. 1987a, *Annual Review of Ecology and Systematics*, **18**, 43–70.
Kirkpatrick, M., 1987b, In *Sexual Selection: Testing the Alternatives* edited by J. W. Bradbury and M. B. Andersson (Chichester: John Wiley and Sons), pp.67–82.
Klein, J., 1986, *Natural History of the Major Histocompatibility Complex* (New York: John Wiley and Sons).

Kodric-Brown, A. and Brown, J. H., 1984, *American Naturalist*, **124**, 309–323.
Komers, P. E. and Dhindsa, M. S., 1989, *Animal Behaviour*, **37**, 645–655.
Krebs, J. R. and Kroodsma, D. E., 1980, In *Advances in the Study of Animal Behaviour, Vol. 2*, edited by J. S. Rosenblatt, R. A. Hinde, C. Beer and M. C. Busnel (New York: Academic), pp.143–177.
Lack, D., 1968, *Ecological Adaptations for Breeding in Birds*, (London: Methuen).
Lande, R., 1981, *Proceedings of the National Academy of Sciences USA*, **78**, 3721–3762.
Levin, B. R. and Lenski, R. E., 1985, In *Ecology and Genetics of Host-Parasite Interactions* edited by D. Rollinson and R. M. Anderson (London: Academic Press), pp.227–242.
Lightbody, J. P. and Weatherhead, P. J., 1987, *Animal Behaviour*, **35**, 1670–1684.
Lightbody, J. P. and Weatherhead, P. J., 1988, *American Naturalist*, **132**, 20–33.
Lively, C. M., 1987, *Nature*, **328**, 519–521.
Low, B. S., 1988, In *Human Reproductive Behaviour: a Darwinian Perspective* edited by L. Belzig, M. Bergerhoff Mulder and P. Turke (Cambridge: Cambridge University Press), pp.115–127.
Low, B. S., in press, *American Zoologist*.
McKinney, F., Cheng, K. M. and Bruggers, D. J., 1984, In *Sperm Competition and the Evolution of Animal Mating Systems* edited by R. L. Smith (Orlando: Academic Press), pp.523–545.
Madhavi, M. and Anderson, R. M., 1985, *Parasitology*, **91**, 531–544.
May, R. M. and Anderson, R. M., 1979, *Nature*, **280**, 455–461.
May, R. M. and Anderson, R. M., 1983, *Proceedings of the Royal Society (London) B*, **219**, 281–313.
Maynard Smith, J., 1978, *The Evolution of Sex*, (Cambridge: Cambridge University Press).
Maynard Smith, J., 1985, *Journal Theoretical Biology*, **115**, 1–8.
Maynard Smith, J., 1987, In *Sexual Selection: Testing the Alternatives* (Chichester: John Wiley and Sons), pp.9–20.
Michod, R. E. and Levin, B. R., 1987, In *The Evolution of Sex: an Examination of Sex*, (Sunderland: Sinauer Associates).
Møller, A. P., 1987, *Animal Behaviour*, **35**, 819–832.
Møller, A. P., 1988, *Nature*, **332**, 640–642.
Møller, A. P., in press a, *Ecology*.
Møller, A. P., in press b, In *Ecology, Behaviour, and Evolution of Bird-Parasite Interactions* edited by J. E. Loye, C. Van Riper and M. Zuk (Oxford: Oxford University Press).
Moore, J., Simberloff, D. and Freehling, M., 1988, *American Naturalist*, **131**, 22–32.
Mousseau, T. A. and Roff, D. A., 1988, *Heredity*, **59**, 181–197.
Nee, S., 1988, *Trends in Ecology and Evolution*, **3**, 218.
Obrebski, S., 1975, *Science*, **188**, 1314–1316.
O'Donald, P., 1980, *Genetic models of sexual selection* (Cambridge: Cambridge University Press).
Pagel, M. D. and Harvey, P. H., 1988, *Quarterly Review of Biology*, **63**, 413–440.
Parker, G. A., 1984, In *Sperm Competition and the Evolution of Animal Mating Strategies* edited by R. L. Smith (Orlando: Academic Press), pp.2–60.
Partridge, L. and Harvey, P. H., 1988, *Science*, **241**, 1449–1455.
Pomiankowski, A. N., 1987a, *Journal of Theoretical Biology*, **128**, 195–218.
Pomiankowski, A. N., 1987b, *Proceedings of the Royal Society (London) B*, **127**, 123–145.
Pomiankowski, A. N., 1988, *Oxford Surveys in Evolutionary Biology*, **5**, 136–184.
Pot, W., 1986, PhD thesis, University of Guelph, Canada.
Potts, F. A., 1906, *Quarterly Journal of the Microscopic Science*, **50**, 599–621.
Pruett-Jones, S. G., Pruett-Jones, M. A. and Jones, H. I., in press, *American Zoologist*.
Queller, D. C., Strassmann, J. E. and Hughes, C. R., 1988, *Science*, **242**, 1155–1157.
Rau, M. E., 1983, *Parasitology*, **86**, 319–322.
Rau, M. E., 1984, *Parasitology*, **88**, 371–373.
Read, A. F., 1987, *Nature*, **327**, 68–70.
Read, A. F., 1988, *Trends in Ecology and Evolution*, **3**, 97–102.
Read, A. F. and Harvey, P. H., 1989, *Nature*, **339**, 618–620.
Read, A. F. and Weary, D. M., 1990, *Behavioural Ecology and Sociobiology*, **26**, 47–56.
Rees, J. A. and Harvey, P. H., in press, In *Mating and Marriage*, edited by V. Reynolds (Oxford: Oxford University Press).
Reinhard, E. G., 1956, *Experimental Parasitology*, **5**, 79–107.
Ressel, S. and Schall, J. J., 1989, *Oecologia*, **78**, 158–164.
Ridley, M., 1983, *The Explanations of Organic Diversity. The Comparative Method and Adaptations for Mating* (Oxford: Clarendon Press).
Rosenblatt, P. C., 1974, In *Foundations of Interpersonal Attraction*, edited by T. L. Huston (New York: Academic Press), pp.79–95.

Rubenstein, D. I. and Hohmann, M. E., 1989, *Oikos*, **55**, 312–320.
Ryan, M. J., 1988, *American Zoologist*, **28**, 885–898.
Schall, J. J., 1986, *Journal of Herpetology*, **20**, 318–324.
Schall, J. J. and Dearing, M. D., 1987, *Oecologia*, **73**, 389–392.
Searcy, W. A. and Yasukawa, K., 1983, *American Scientist*, **71**, 166–174.
Seger, J., 1985, *Evolution*, **39**, 1185–1193.
Seger, J. and Hamilton, W. D., 1987, In *The Evolution of Sex*, edited by R. E. Michod and B. R. Levin (Sunderland: Sinauer Associates), pp. 176–193.
Sherman, P. W., Seeley, T. D. and Reeve, H. K., 1988, *American Naturalist*, **131**, 602–610.
Shields, W. M. and Crook, J. R., 1987, *Ecology*, **68**, 1373–1386.
Singh, P. B., Brown, R. E. and Roser, B., 1987, *Nature*, **327**, 161–164.
Slagsvold, T., Lifjeld, J. T., Stenmark, G. and Breiehagen, T., 1988, *Animal Behaviour*, **36**, 433–442.
Slatkin, M., 1984, *Evolution*, **38**, 622–630.
Smith, S. M., 1988, *Behaviour*, **107**, 15–23.
Spurrier, M. F., Boyce, M. S. and Manly, B. F., in press, In *Ecology, Behaviour, and Evolution of Bird-Parasite Interactions* edited by J. E. Loye, C. Van Riper and M. Zuk (Oxford: Oxford University Press).
Symons, D., 1979, *The Evolution of Human Sexuality* (Oxford: Oxford University Press).
Tomlinson, I. P. M., 1988, *Heredity*, **60**, 283–293.
Trivers, R. L., 1972, In *Sexual Selection and the Descent of Man* edited by B. Campbell (Chicago: Aldine Publishing Comany), pp. 136–179.
Wakelin, D. and Blackwell, J. M., 1988, *Genetics of Resistance to Bacterial and Parasitic Infection* (London: Taylor and Francis).
Ward, P. I., 1988, *Animal Behaviour*, **36**, 1210–1215.
Weatherhead, P. J., 1984, *American Naturalist*, **123**, 873–875.
Williams, G. C., 1966, *Adaptation and Natural Selection* (Princeton: Princeton University Press).
Wülker, W., 1964, *Experimental Parasitology*, **15**, 561–597.
Yamazaki, K., Beauchamp, G. K., Bard, J., Thomas, L. and Boyse, E. A., 1982, *Proceedings of the National Academy of Science USA*, **79**, 7828–7831.
Yamazaki, K., Boyse, E. A., Miké, V., Thaler, H. T., Mathieson, B. J., Abbott, J., Boyse, J., Zayas, Z. A. and Thomas, L., 1976, *Journal of Experimental Medicine*, **144**, 1324–1335.
Zahavi, A., 1975, *Journal of Theoretical Biology*, **53**, 205–214.
Zuk, M., 1987a, *Behavioural Ecology and Sociobiology*, **21**, 65–72.
Zuk, M., 1987b, *Ecological Entomology*, **12**, 341–348.
Zuk, M., 1987c, *Ecological Entomology*, **12**, 349–354.
Zuk, M., 1988, *Evolution*, **42**, 969–976.
Zuk, M., in press, In *Ecology, Behaviour, and Evolution of Bird-Parasite Interactions* edited by J. E. Loye, C. Van Riper and M. Zuk (Oxford: Oxford University Press).
Zuk, M., Thornhill, R., Ligon, J. D. and Johnson, K., in press, *American Zoologist*.

6. Host behaviour and opportunism in parasite life cycles

R. C. Tinsley

PARASITE TRANSMISSION STRATEGIES

Continuous release of infective stages

Parasite life cycles are traditionally characterized by enormous reproductive output. Each *Fasciola hepatica*, the liver fluke of sheep and cattle, produces around 20 000 eggs per day, *Ascaris lumbricoides*, the intestinal roundworm of man, produces over 200 000 and *Diphyllobothrium latum*, a tapeworm of fish-eating mammals releases around 1 000 000. This prodigious output is sustained day after day and *D. latum* may live for up to 15 years. High reproductive rate is generally considered an essential adaptation to counter the hazards of parasite transfer between hosts, a reflection of the problems of locating favourable patches in an otherwise inhospitable landscape. Indeed, this characteristic is incorporated into modern definitions of parasitism. In particular, the capacity of parasites for a higher reproductive rate than the host is important in distinguishing parasitism from predation where there is a similar antagonistic relationship with the partner, the prey. Predators have a far lower reproductive output than their prey species. However, a very high reproductive rate is developed by some free-living animals too; for the sedentary burrowing clam (*Mya arenaria*) the problems of patch location in a predominantly unfavourable environment are met by a strategy of mass production of about 3 million offspring per year – only a small proportion of which may encounter a suitable settlement area. The problem is largely a statistical one, the probability that a given egg of *Fasciola hepatica* will survive to complete development in the external environment, that the miracidium will encounter and invade the specific snail intermediate host, that this snail will survive for the developmental period of the sporocyst, redia and cercaria stages . . . and so on. For the first of these hurdles, Wilson *et al.* (1982) predicted that even in the most favourable period of the year (spring and summer), only 17 out of every 100 000 eggs laid will survive to hatch.

In some helminth groups, the parasite may counter the low probability of each successive event not only by the very high initial output of eggs but those relatively few individuals which succeed at each stage may boost their numbers by specializations of the life cycle involving further asexual proliferation. Thus, a single invasion of a snail by a digenean miracidium may be followed by prolonged production of very large numbers of cercariae generated by sporocyst and/or redia division, greatly increasing the population of infective stages entering the next step in the life cycle (see Whitfield and Evans, 1983). In *Schistosoma mansoni*, up to 200 000 cercariae may be produced from one miracidium.

The spatial and temporal difficulties of transfer from one host to another are significantly reduced in cases where parasite transmission exploits complex ecological associations between successive hosts. *Raillietiella gigliolii* is a pentastomid parasite infecting the lungs of the worm-lizard, *Amphisbaena alba*. This burrowing reptile is a facultative inquiline of nests of the leaf-cutting ant (*Atta cephalotes*). It locates nests by following the trail pheromones of foraging ants and moves within the underground galleries feeding on arthropod prey, occasionally (and perhaps only accidentally) the ants, but more often the larvae of the rhinoceros beetle (*Coelosis bilboa*) which are themselves inquilines in ant nests. These beetles live in the chambers within the nest in which the ants dump unwanted debris. Pentastomid eggs passed in the faeces of the amphisbaenians are infective to *Coelosis* larvae but they may be deposited throughout the nest and its galleries, out of reach of the refuse dump beetles. However, the ants play a vital role in transmission: they clean all debris from the tunnels, transfer it to the refuse chambers and literally deliver the parasite eggs to the intermediate host (Winch and Riley, 1985; Riley *et al.*, 1986). Strong ecological links between the various components of this life cycle contribute to a marked increase in transmission efficiency, the comparatively low reproductive output of only 100 eggs per female parasite per day is sufficient to maintain a very high prevalence of infection (86%).

In contrast, a major problem with transmission strategies of maximal but relatively non-selective reproduction, characteristic of very many parasites, is that the opportunities for contact between infective stages and potential hosts may be very scattered in space and time. The processes which maximize numbers tend to continue regardless of the short-term probability of success. Thus, the many thousands of eggs which *Fasciola hepatica* produces each day are released into the external environment throughout the year. However, embryo development is halted below 10 °C, and the eggs are killed by freezing. Therefore, eggs passed out onto pasture from October to March (in the UK) will almost invariably die and the total reproductive output of the parasite is entirely wasted for over half the year. Additionally, during this long period the enormous potential of the intramolluscan stages is also unrealized since development within the intermediate host is inhibited by the same low temperatures. Of course, the adult liver fluke, maintained at constant high temperatures in a closely-controlled environment, has no mechanism for detecting that external temperatures are completely inimical to the survival of its mass-produced progeny. Indeed, a strategy for optimizing reproductive investment would actually require a forecast of weather conditions during the weeks or months which follow egg deposition, an unreasonable demand even for a meteorologist!

Even when weather conditions permit transmission of *Fasciola*, all parasite eggs deposited when the host has moved onto relatively higher, drier areas of pasture are unlikely to survive development or to encounter moisture-demanding snails. Only those eggs passed out with the faeces when the host is grazing poorly-drained areas of wet pasture will have a chance of contributing to transmission. The liver fluke, deep within the body of its perambulating host, can have no clue as to the suitability of the habitats outside for its offspring nor, as we have established, how long favourable conditions may persist.

It has been argued (see for instance, Calow, 1983) that this high and apparently profligate reproductive output is not just a reflection of the hazards of parasite life cycles, rather it may be fuelled by a superabundance of nutrients and that mass production of progeny is actually achieved at little cost to the parasite. Alternatively, it could be envisaged that the apparently very inefficient reproductive strategy of *Fasciola hepatica* (when quantified by data from the UK and Northern Europe) actually evolved in warmer climatic regions where there are not such major losses of annual reproductive investment. Despite all these difficulties, the effectiveness of transmission is clearly demonstrated when the normal constraints are relaxed; parasite reproductive investment in just a few months of an unusually warm and wet UK summer provides the potential for major epidemics of fascioliasis. (Although, of course, the severity of a disease such as this is also strongly affected by man's influences, including grazing practices, livestock density etc.)

Discontinuous release of infective stages

Seasonal cycles

In some host–parasite systems, where opportunities for transmission are restricted in time, parasite reproductive effort may be concentrated into the more favourable periods. Seasonal temperature cycles are responsible for part of the annual cyclical variation in parasite infection. In endothermic hosts, internal parasites are buffered from external temperature changes; in ectotherms, parasites may be exposed to temperature fluctuations of both long and short duration, related to natural seasonal and diurnal changes in external conditions and to host behavioural activity – periodic movement between deep, cooler waters and shallow, warmer waters by some fish, basking in the sun by many reptiles, and so on. In these cases the parasite will have access to much more information about the external environment, although the environment will also have a more direct effect on parasite physiology. For monogeneans in ectothermic vertebrates, the functioning of the egg assembly apparatus generally shows close correlation with temperature. Thus, Jackson and Tinsley (1988a) found that the rate of egg production by *Protopolystoma xenopodis* increases by around 10% for each degree centigrade rise in temperature. For many helminths that infect temperate freshwater fish low winter temperatures may inhibit egg production, contributing to a seasonal decline in parasite transmission and infestation levels. This is followed in spring and summer by increased parasite populations attributable to increases in all temperature-limited processes (of egg production, larval development etc.), and to changes in host behaviour, population density etc. Indeed, the combination of all these seasonal effects may also give rise to summer maxima of parasite populations in helminths of endotherms. Thus, as we have seen, in *Fasciola hepatica* the host's homoeostatic control of the adult parasite's environment, resulting in year-round egg production, is completely overridden by the effects of temperature on the life cycle stages in the external environment (and in the ectothermic snail host).

Many of these seasonal cycles of parasite transmission are complicated by other factors (see Chubb, 1977; Kearn, 1986). However, in general, the seasonal concentration of parasite transmission into periods favourable to host invasion are automatic consequences of the effects of temperature on biological processes. For instance, egg production by the gill parasitic monogenean *Diplozoon paradoxum* is largely inhibited during the winter but resumes within 48 hours of transferring infected hosts to a warm aquarium. In these, and many other cases, there is no real adaptation on the part of the parasite to optimize reproductive effort. It is, therefore, of exceptional interest that some parasites show the exact opposite of these features so that their transmission occurs when environmental effects such as temperature would seem to be most unfavourable. Thus, Bauer (cited by Kearn, 1986) reported that the main period of infection of the sturgeon by the monogenean *Diclybothrium armatum* occurs in the winter. This is correlated with the behaviour of the fish which congregate together in depressions in river beds during the winter, providing favourable circumstances for infection despite the low temperatures. Evidence that the helminth egg assembly apparatus can function with high efficiency at very low temperatures is provided by the studies of Combes (1972) on *Polystoma integerrimum*. Annual reproductive output is concentrated into about a week when the common frog (*Rana temporaria*) spawns, and in the Pyrenees both host and parasite lay eggs in ponds which are often still partially ice-covered. Paradoxically, *P. integerrimum* achieves, under these circumstances, the highest rate of egg production recorded for a monogenean: up to 2500 eggs in 24 hours at 4 °C (though this exceptional output is of short duration). This demonstrates how, against all expectations, parasites may adapt physiologically to meet the ecological demands of brief opportunities for transmission.

The strategy adopted by a closely-related polystomatid monogenean illustrates how parasites may adapt their reproduction to meet a repeated opportunity for seasonal infection each year. In Mediterranean regions of France, the toad *Pelobates cultripes* is terrestrial for most of the year, providing no opportunity for transmission involving aquatic infective stages. However, this toad may enter water for two periods of spawning, one in the spring and the other in autumn. The reproductive processes of *Polystoma pelobatis* are exactly correlated with host behaviour, leading to two periods of transmission (Combes, 1968).

Circadian rhythms

There is also good documentation of how parasites can adapt their reproductive output to exploit some daily rhythms of host availability. The monogenean *Diplozoon homoiongracile* infects the gills of *Barbus meridionalis*, a fish which is relatively more active by day and tends to rest in sheltered sites in streams during the night. Macdonald and Jones (1978) recorded that, whilst egg production is a continuous process, a higher proportion of each 24 hour output is manufactured during the night. In consequence, these eggs will be deposited on areas of stream bed where, 5–6 days later, the hatching oncomiracidia are likely to encounter stationary hosts. This reproductive pattern produces a relative concentration of infective stages to coincide with more favourable opportunities for transmission. However, since this also permits some transmission at other times when different groups of fishes may meet, it therefore promotes a higher degree of parasite genetic mixing than would be achieved solely by re-infection within the same host 'dormitory'.

A diurnal concentration of potential infective stages is produced, in the case of *Diplozoon*, by a response of the adult parasite to light, modifying its rate of egg production. The same result may be achieved by a relatively constant rate of egg assembly and a response by the adult parasite to the optimum moment for release of eggs. Thus, Kearn (1986) suggested that in *Oculotrema hippopotami*, a monogenean ectoparasite of the hippopotamus, there may be temporary storage of eggs in the parasite uterus by night (when the host is foraging on land) and that these are deposited by day (when the host is in water and vulnerable to aquatic infective stages).

Cyclical opportunities for transmission involving swimming parasite larvae may also be controlled primarily by environmental factors – the behaviour of the habitat rather than that of the host. Tidal rhythms on rocky shores produce a regular, twice daily, alternation of conditions in littoral pools, the confined static water at low tide provides favourable circumstances for host invasion but this is abruptly replaced by turbulent water with each returning high tide. The life cycle of the digenean *Parorchis acanthus*, a parasite of the bursa of Fabricius and rectum of herring gulls, is adapted to exploit these brief opportunities for infection of the first intermediate host, *Nucella lapillus*. The eggs complete development within the uterus of the adult worm and hatch just before or after release into the host's rectal contents or subsequently in seawater. Visits of infected gulls to newly-exposed rock pools may then result in deposition of immediately-infective miracidia into the water, and these have up to 12 hours to locate and penetrate the stationary dog whelks. It is not clear, in this case, whether eggs are passed out continuously with the host's faeces (and many miracidia may then be deposited far from *Nucella* habitats) or whether oviposition might be regulated in some way by the parent worm.

More clear-cut rhythms which exploit transmission opportunities may result from a well-defined parasite response. Microfilariae, the infective stages of nematodes such as *Wuchereria* and *Loa*, appear in the peripheral blood of man during the period when the insect vectors are more likely to feed. For *Loa loa* this restriction is diurnal, the principal feeding period of anthropophilic *Chrysops* species; for *Wuchereria bancrofti* this is generally nocturnal, coinciding with the night-biting habits of mosquitoes throughout much of the parasite's geographical range. However, in parts of the South Pacific, the mosquito vectors preferentially bite by day, and here the strains of *W. bancrofti* show a corresponding diurnal extension of microfilaraemia.

Irregular opportunities for transmission

In these examples of circadian rhythms, the opportunities favouring transmission occur regularly and it is easy to envisage how natural selection would favour those variants showing gradual shifts in activity: it will be these genotypes which will be more likely to complete their life cycles and contribute disproportionately to further reproduction. However, in some parasite life cycles a more truly opportunistic strategy is evident, where the parasite is able to exploit irregular and often entirely unpredictable possibilities for transmission, where opportunities are brief and are separated by circumstances which are completely unsuitable for transmission. In these cases it is difficult to visualize how parasite adaptation could arise from a progressive and gradual concentration of effort into optimum periods: a hit or miss situation may require a high degree of accuracy for any effective adaptation to become fixed at all. Here, 'opportunism' emerges as an even more remarkable strategy within evolutionary ecology. This principle is illustrated by the life cycle of the Guinea worm (*Dracunculus medinensis*) in the Old World Tropics. Mammalian hosts including man become infected by drinking water containing copepods harbouring larval nematodes. The larvae migrate into the subcutaneous tissues, develop to maturity, and the fertilized females accumulate up to 1 million first stage larvae *in utero*. These must be released into water to be ingested by a copepod and continue the life cycle. To accomplish this the gravid female migrates to an area of the host's body likely to be immersed in water (usually the lower leg), creates an ulcer at the surface, and at the moment when the limb enters water the female protrudes and discharges a fluid containing many thousands of immediately-infective larvae. The life cycle tends to be roughly annual, so that hosts infected during one dry season (when people have to depend on diminishing water supplies) then develop a patent infection which initiates further transmission at the next (Muller, 1971). This strategy ensures release of offspring into the environment where the life cycle can proceed: there is no guarantee of intermediate hosts but we may assume that this is a reasonable certainty in endemic areas.

Host sexual behaviour often provides a restricted but very reliable opportunity for parasite transmission. Whilst the general timing may be seasonal and therefore relatively predictable, the specific window for infection may be extremely narrow. This is true to a greater or lesser extent for all infections which exploit host copulation. Most sexually-transmitted microparasites are transferred passively with host body fluids. However, amongst helminths, one monogenean exploits an exceedingly brief opportunity by an active behavioural response. *Isancistrum subulatae* is an ectoparasite of the squid, *Alloteuthis subulata*, and *in situ* viviparous reproduction can give rise to burdens of several thousand worms per host. However, the behaviour of squid generally precludes host-to-host transfer: squid tend to swim continuously, never rest on the sea-bottom, and rarely come in contact with one another except during mating. Llewellyn (1984) proposed that transmission occurs during copulation (even though this may take only 20 seconds). It was found in experiments that worms were able to transfer when the arms of squid were in contact for less than 5 seconds.

The cuckoo (*Cuculus canorus*) is a brood parasite which practices a carefully-timed response, taking advantage of the brief absence of the prospective host from its nest, rapidly laying a single egg of its own and ejecting one from the existing clutch. The cuckoo synchronizes her egg-laying with the early stages of clutch production by the host. Most birds take 20 minutes or longer to deposit an egg; if the cuckoo were to spend more than a fraction of this time at the nest the chances of detection would be increased and the foster parents would be more likely to reject the cuckoo egg. The female cuckoo deposits her egg remarkably quickly, completing the 'infection process' within 10 seconds (Davies and Brooke, 1988).

There are now very elegant demonstrations of how parasites may exploit or even manipulate aspects of host behaviour in order to facilitate their own transmission (see Chapter 7). In multi-host life cycles, the details of these strategies usually relate to one or a few specific links in a chain of events but behavioural information on the other stages is often incomplete. In direct life cycles, the ecological requirements of transmission are usually easier to identify and the exploitation of the behaviour of the single host species involved may be more clear-cut. Some of the best examples of opportunistic transmission are provided by members of one class of platyhelminths characterized by direct life cycles, the Monogenea, and the adaptations of these show how parasite biology may be intimately enmeshed with host behaviour.

Opportunistic transmission in polystomatid monogeneans

Monogeneans are almost always transmitted by a swimming, ciliated infective stage, the oncomiracidium, and the great majority of species are parasitic on fish. One family of monogeneans, the Polystomatidae, has radiated onto tetrapod vertebrates and occurs on a diverse assemblage of hosts, including lungfish, anuran and urodele amphibians, chelonians and one mammal (the hippopotamus). These share one common feature: they all occur in water at certain times and this is the only opportunity when transmission can take place. In consequence, the life cycles of polystomatids show a very close correlation with the ecology and behaviour of the host.

The hosts of polystomatids vary in the degree of ecological association with water (figure 6.1). The period spent by the host in water provides the only opportunity for the release of parasite infective stages and for invasion of new hosts. Variations in the frequency and duration of entry into water govern the life cycle strategies of the various polystomatid genera. At one extreme *Xenopus*, the African clawed toad, is totally aquatic so transmission of its polystomatid (*Protopolystoma xenopodis*) can

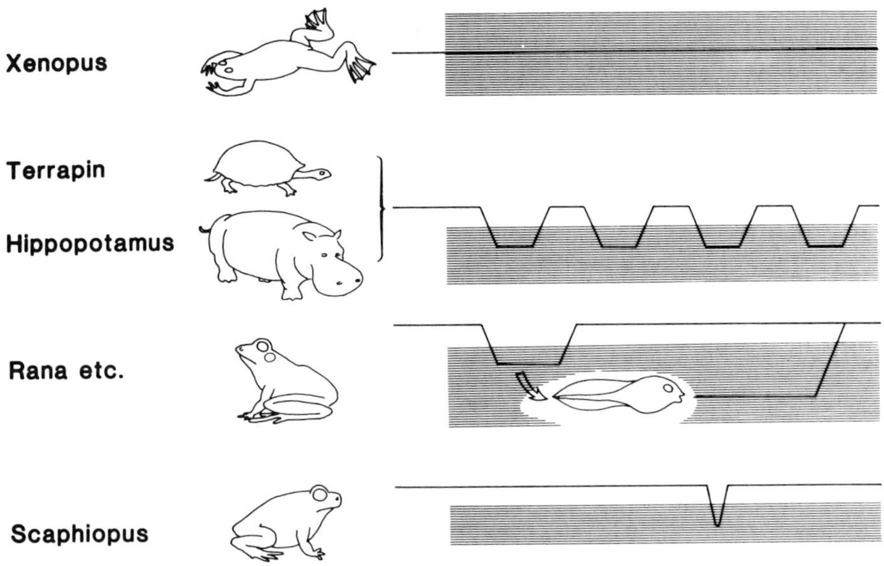

Figure 6.1. *Polystomatid monogeneans*: correlation of parasite transmission with host behaviour.

occur virtually continuously (Tinsley and Owen, 1975). For amphibious hosts, such as many chelonians and the hippopotamus, there is a regular diurnal alternation of aquatic and terrestrial periods (the hippopotamus out of the water at night to browse, turtles out of water by day to bask in the sun), so transmission is likely to be a diurnally periodic process. Amongst anuran amphibians, there is much variation in parasite life cycle patterns. For *Polystoma*, in species of *Rana* etc., transmission is typically vertical with parasites in adult frogs depositing eggs during host spawning but the eventual targets of invasion being the tadpole offspring. In most cases where the ecology of the anuran host is largely mesic, the transmission of its polystomatid parasites is tied to the one period when the host can be guaranteed to enter water, i.e. for spawning. There is enormous diversity in reproductive modes among anurans (Duellman, 1985) and not all groups are infected by polystomatid parasites. Nevertheless, there are frogs which spawn in environments which seem almost impossibly difficult for parasite transmission: in torrential mountain streams, in briefly-flooded tropical forests and in deserts. There are even tree frogs which construct foam nests and never return to water to breed. Examples of each of these are known to be infected by polystomatid monogeneans. There is little ecological information on most, but one case has been examined in detail and a series of remarkable adaptations for opportunistic transmission has been revealed.

Life cycle of Pseudodiplorchis americanus*: a case study*

The ecology of the North American spadefoot toads, *Scaphiopus* species, precisely defines the opportunity for transmission (figure 6.2). The toads live in deserts and escape the most hostile conditions by hibernating for 9–10 months each year, up to a metre below the soil surface. They become active when the annual rains arrive, emerge from the desert soil and enter briefly the newly-formed pools to spawn. They then feed intensively on desert invertebrates during their short activity season before arid conditions force them to return to hibernation. The 1–3 days when spawning occurs is the only period each year when the hosts are available for infection by a swimming oncomiracidium. However, this 'window' is actually even more restricted. The toads are strictly nocturnal, entering water only after darkness (9 p.m.) and they leave water before dawn (around 4 a.m.). If there are several rainstorms, males may enter water on up to 3 nights, giving a total period for transmission still less than 24 hours each year. Females enter water only once and they leave immediately after they have spawned, and their exposure to infection may be less than 4 hours in the entire year. This restriction on annual transmission is probably greater than for any other helminth parasite, but around 50% of *Scaphiopus couchii* in field sites in south east Arizona, USA, are infected by the polystomatid *Pseudodiplorchis americanus* (see Tinsley, 1989, 1990). Invasion of *Scaphiopus couchii* by the aquatic infective stage is restricted to 1–3 days each year, when the toads enter water to spawn, and to a maximum of 7 hours on these days, during nocturnal host activity.

The essential reproductive adaptation of *Pseudodiplorchis* involves ovoviviparity. The adult worm has a long uterus, capable of accommodating several hundred egg capsules, and the offspring develop to an infective stage *in utero* during the long period when the host is hibernating. When the toads enter water to spawn, the encapsulated larvae are discharged *en masse*, pass into the water with the host's urine, hatch and can immediately invade other toads which are congregating in the same water.

The details of the life cycle of *Pseudodiplorchis americanus* have no precedents elsewhere in the Monogenea (figure 6.3). Although the adult parasites occur in the site typical of polystomatids in anurans, the urinary bladder, the infective stages invade via the nostrils. During the week following invasion, the larvae migrate from the nostril sinuses into the mouth and all associated cavities (beneath the tongue, the vocal sacs, the eustachian tubes) and then through the glottis into the lungs. They are the only monogeneans to infect this site (Tinsley and Earle, 1983). Here, they feed on blood and grow rapidly. By day 21, the juveniles begin to return to the oral cavity, and from day 28 postinfection they migrate to the urinary bladder. This involves rapid movement (5–20 minutes) through the length of the alimentary tract, which is also completely without precedent within the Monogenea. With very few exceptions, monogeneans are unable to survive exposure to host digestive enzymes. *Pseudodiplorchis* accomplishes its rapid migration protected by tegumental vesicles which are discharged over its surface membranes during the passage through the gut (Cable, 1989). The activation of this tegumental protection is entirely dependent on a trigger factor which is linked, in some way, with host activity (Tinsley and Jackson, 1986).

Once the juvenile worms migrate to the bladder, they develop rapidly, becoming sexually mature 3–4 weeks postmigration, and begin to accumulate embryos *in*

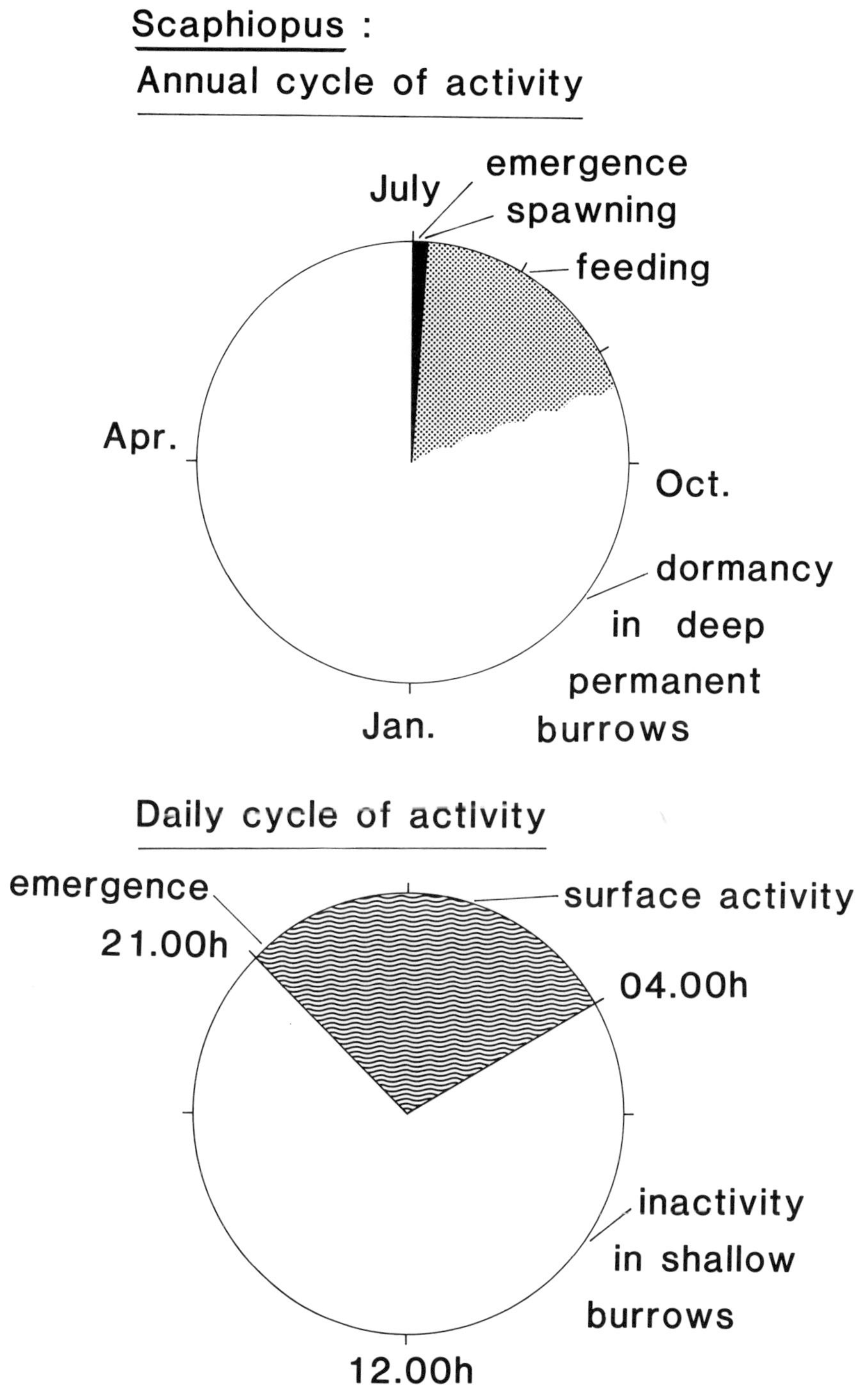

Figure 6.2. *Pseudodiplorchis americanus*: the control of transmission opportunities by host activity.

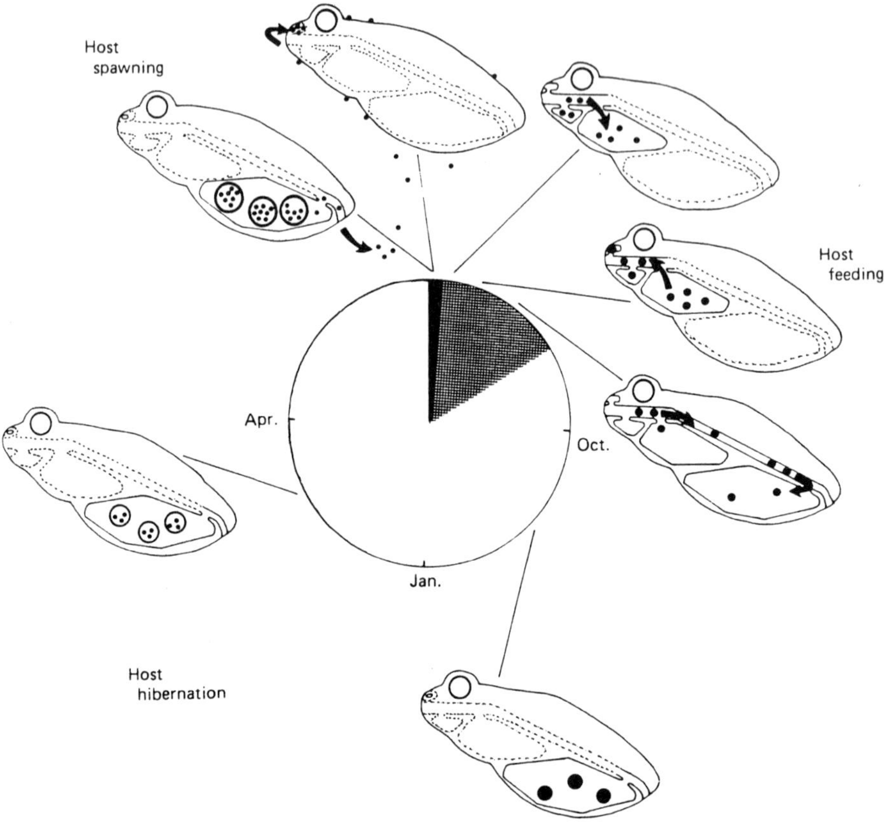

Figure 6.3. The life cycle events of *Pseudodiplorchis americanus* correlated with the annual activity cycle of the host, *Scaphiopus couchii*: transmission during host spawning (▼); juvenile development and internal migration during host feeding (▼); maturation and accumulation of embryos *in utero* during host hibernation (▽). The cycle follows one cohort of parasites, but adult worms (⊛) producing infective stages (•••) in one season's transmission may also survive to reproduce again a year later. Additionally, juveniles failing to migrate may remain in the respiratory tract throughout hibernation and then migrate to the bladder after the resumption of host activity.

utero ready for the next brief opportunity for transmission at the start of the following summer. By then, the uterus may be packed with up to 400 developing embryos, although gravid adults more usually carry around 100 fully-developed offspring, each enclosed within a flexible, membraneous egg capsule (Tocque, 1990; Cable and Tinsley, 1990).

When transmission eventually occurs (generally early in July in Arizona), each episode is limited to a maximum of 7 hours on one night. Field studies in Arizona have documented the effectiveness of this very restricted opportunity (Tinsley and Jackson, 1988). The typical environmental events follow a long period of drought and steadily increasing heat through early summer. Storms often arrive suddenly, sometimes with half the annual rainfall in one night. The spadefoot toads are aroused from hibernation, below ground, by the low frequency vibrations of the rainfall drumming on the surface (Dimmitt and Ruibal, 1980); they emerge after darkness and migrate to the rapidly-forming pools. Assemblies of several hundred toads gather by midnight in pools a few metres in diameter, females are attracted by the chorusing of the males, and spawning is rapid. At dawn, all the toads leave the water and bury themselves in shallow burrows to escape the daytime heat (figure 6.2). This departure abruptly stops all parasite transmission but the toads carry away with them, in their respiratory systems, a record of all parasite invasion they experienced during their brief exposure.

Samples of mating populations of *Scaphiopus couchii* taken at intervals during one night show a more or less exponential increase in infection levels during the 7-hour exposure. In the study described by Tinsley and Jackson (1988), toads which had just emerged from hibernation and which had not experienced any parasite invasion for nearly a year, were infected almost immediately they entered the spawning ponds. By midnight, after a maximum of only 3 hours transmission, over 80% of males carried newly-invaded *Pseudodiplorchis americanus*, with a mean of 2·6 worms per host. Four hours later, at dawn, these infection levels had increased by a factor of nearly ten, so that departing toads carried a mean of 24 worms per host. Unusually, in this study, torrential rain fell again on two further nights during the summer activity season, triggering renewed assemblies of toads in the same site, and providing a total of three pulses of invasion into the male toad population. The result was that every male *Scaphiopus couchii* became infected, with an approximate doubling of mean worm burdens at each exposure, and a cumulative mean parasite load of over 100 worms per host. Frequency distributions of infection levels demonstrate the steady increase in parasite burdens in individual toads. The maximum at midnight on the first night was nine, rising to 95 at dawn after the first exposure, and to 158 after the second exposure. After the third night, half of the male toad population carried burdens exceeding 100 parasites per host. Females generally enter only one spawning (and transmission) site for part of one night, and these acquired a mean of around 40 newly-invaded parasites per host (Tinsley and Jackson, 1988). Few host–parasite systems can demonstrate equivalent speed and efficiency of transmission affecting the entire host population (figure 6.4).

With such a brief exposure period it has proved possible to document the dynamics of host invasion in some detail, and field studies have revealed a very close correlation with aspects of host behaviour (Tinsley and Jackson, 1988; Tinsley, 1989, 1990). The toads migrate varying distances to sites where water has accumulated, so the host population steadily increases during the night. As their parasites discharge stored offspring, the concentration of infective larvae in the water progressively increases at the same time as the density of targets. In consequence, 10% of the night's total invasions occur up to midnight (after a maximum of 3 hours exposure), but more than 30% occur in the final hour before dawn (3–4 a.m.). It follows that differences of only an hour or so in exposure time have a significant effect on worm burdens acquired. However, in addition to exposure duration, invasion levels are strongly determined by the timing of entry of targets into the water. Thus, a female which arrives at the breeding site early, spawns, and then

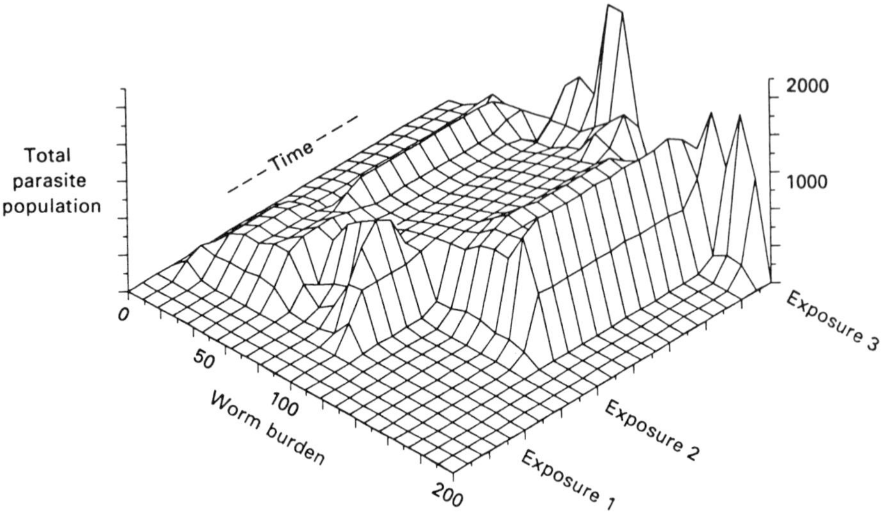

Figure 6.4. Isometric surface to represent changes in the total larval parasite population infecting a population of male *Scaphiopus couchii* during one complete transmission season (3 exposures: 9/10, 10/11 and 21/22 July). Data record total burdens of invading *Pseudodiplorchis americanus* in each infection class (1–10, 11–20, etc. parasites/host) standardized for a cohort of 100 toads. One sample (03.00h, 11 July) omitted because of its small size (n = 6), other samples average n = 17. Time axis (Standard Mountain Time, Arizona) truncated between successive exposures.

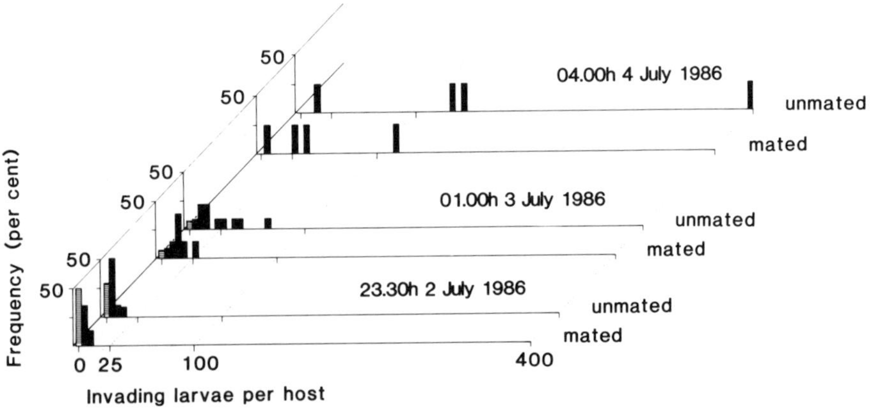

Figure 6.5. Correlation of levels of invasion by *Pseudodiplorchis americanus* larvae with the reproductive success of male *Scaphiopus couchii*. Host samples (mean n = 14) taken at intervals from a mating assembly following the start of the summer rains (Sonoran Desert, Arizona, July 1986). Data from Tinsley (1989).

leaves by midnight will be exposed to much lower concentrations of infective stages than one which arrives at 1 a.m. and leaves at dawn. Males remain in water from the time of their arrival at the mating site until all the assembled toads disperse just before dawn, and so tend to experience heavier invasions than the females. For each male, these heavier burdens acquired at one exposure are then multiplied by the number of mating assemblies which it enters.

Other, unexpected, effects of host sexual behaviour on transmission success have emerged from these field studies (Tinsley, 1989). Males always outnumber females in spawning assemblies of *Scaphiopus couchii*, there is intense male competition, and a major fraction of the male population fails to mate. These unsuccessful males acquire heavier larval invasions than those which succeed in mating with a female (figure 6.5). After less than 3 hours of transmission, the first sample (23.30h) showed low levels of invasion, but the relative density (mean larval burden amongst the total sample) was greater in unmated (2·7/toad) than in mated (1·8/toad) males. Only 1·5 hours later, the mean number of invaded larvae in the host population had increased by a factor of more than 7; recruitment into males which were not amplexed (relative density 20·5/toad) was 50% greater than into amplexed males (relative density 13·5/toad), and the maximum burden in the former group was twice that in the latter. At the end of the second night of spawning (following further rainfall on 3 July), small samples of amplexed and unamplexed males showed a further exaggeration of these differences (means of 46 and 168 larvae/host respectively). (Data from Tinsley, 1989.) An explanation is provided by the different behaviour patterns of males in an assembly: unpaired males are relatively more mobile, swimming to search for a female or to wrestle with other competing males, and will therefore travel through a greater volume of water, encountering more swimming infective stages than a successfully-paired male engaged in the more sedentary process of spawning.

These differences between male and female toads in the dynamics of their exposure result in much heavier oncomiracidial invasion of males, and this is reflected, a year later, in the distribution of adult parasites within the host population (figure 6.6). Overall, the prevalence of infection in male toads is almost twice that in females, and the mean intensity is about 30% greater. So, with each gravid worm containing around 100 infective stages, the parasites in male toads can potentially make a much greater contribution to transmission. Chance plays an important role in the population biology of transmission. The entry of a few heavily-infected toads into a spawning assembly will have a disproportionately greater effect on transmission in comparison with a much larger number of lightly-infected toads (Tinsley, 1989).

Alternative strategies amongst polystomatid monogeneans

The life cycle of European *Polystoma integerrimum* is also geared to meet an extreme temporal restriction of host availability. In this case, the parasite has adopted a strategy of storing gametes and vitelline cells in preparation for rapid assembly of these components during the short period which the frogs spend in the

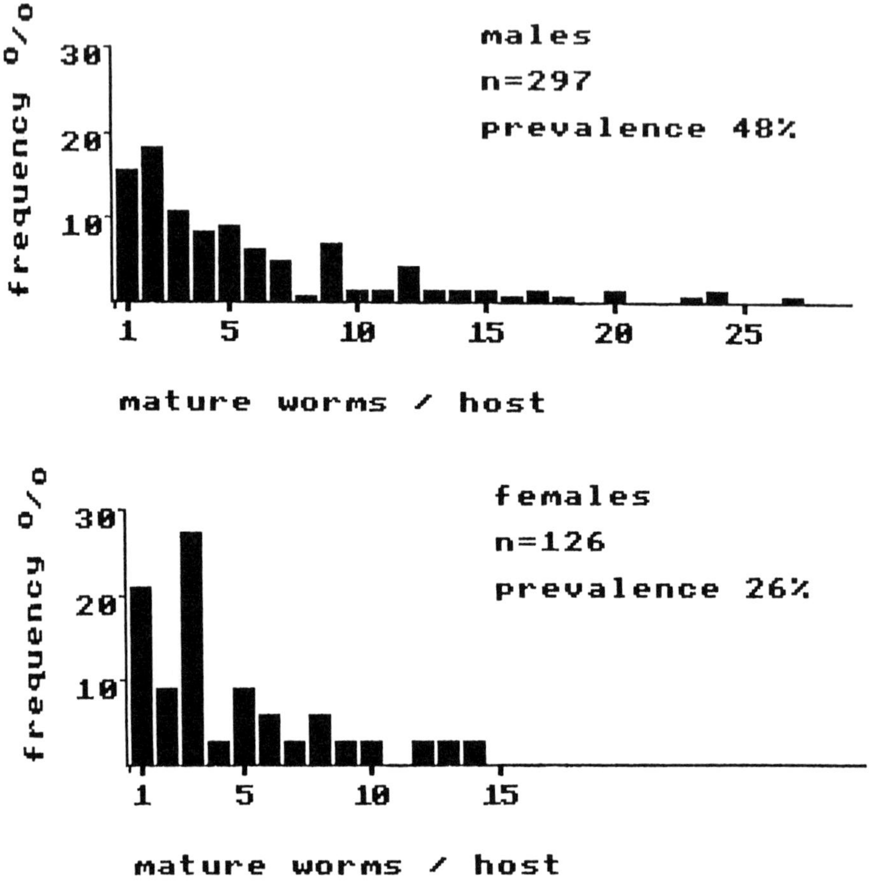

Figure 6.6. Distribution of infection levels of adult *Pseudodiplorchis americanus* (parasites containing infective stages *in utero* and therefore one year or more postinfection). Data from *Scaphiopus couchii* populations in the San Simon valley, Sonoran Desert, Arizona, showing the greater prevalence and mean intensity of parasites in male compared with female hosts (cf. Tinsley, 1989).

breeding ponds. Each *Polystoma* produces and deposits up to 4000 eggs in a few days, a maximum rate of nearly 2 eggs per minute (Combes, 1972). Although this process achieves a concentration of reproductive investment in time (to exploit the brief aquatic phase of the adult host) and in space (to exploit the aggregated tadpole populations, the targets of infection), the resulting life cycle incorporates a series of major risks. The deposition of newly-assembled egg capsules requires embryonic development to occur in the external environment, initially at very low temperatures, and the infective larvae hatch only after 7–11 weeks (Combes, 1968). In addition to the hazards of loss of eggs through burial in sediments, fungal and bacterial attack etc., the strategy of egg deposition relies on the continuing suitability of environmental conditions and on the availability of the host targets when the oncomiracidia eventually hatch. The targeting of invasion on the tadpole produces even greater risks: this is the most vulnerable stage in the host's life and there is, automatically, no opportunity for the parasite to contribute to onward transmission for 3 years, until the host reaches sexual maturity and returns to the spawning sites (figure 6.7). In contrast, the tadpole of *Scaphiopus couchii* is not involved in the life cycle of *Pseudodiplorchis americanus*. Although the spadefoot tadpoles develop very quickly, metamorphosing in 10 days, many die because the pools dry up quickly in the desert heat. Since the life cycle involves 'instant' transmission between adult toads, the parasite avoids the mortality risks associated with the pre- and postmetamorphic stages of the host. Additionally, this strategy provides the first opportunity for further transmission after 1 year when the mature toads enter the next spawning assembly.

A comparison of the bionomics of the life cycles of *Polystoma integerrimum* and *Pseudodiplorchis americanus* demonstrates the supreme effectiveness of transmission involving immediately-infective offspring (discussed in detail by Tinsley, 1983). Although the reproductive strategies of both species are designed to achieve a similar opportunistic response to almost identical constraints imposed by host ecology, the intrauterine development of offspring transforms the potential of opportunistic transmission.

Other polystomatid genera exploit anuran amphibians which are vulnerable only very occasionally to water-borne infection. In these too there is major modification of reproductive morphology to achieve greatly increased storage capacity by the uterus. In species of *Metapolystoma*, *Mesopolystoma*, *Eupolystoma*, *Parapolystoma*, *Diplorchis*, *Pseudodiplorchis* and *Neodiplorchis*, the uterus can accommodate several hundred embryos throughout development. Detailed comparison of internal anatomy suggests that the elongation of the uterus to occupy most of the internal body space has been accomplished by a series of different evolutionary routes in these genera (Tinsley, 1974, 1978a, 1983). The adaptations appear to represent independent solutions to the common problem created by the unpredictable ecology of the hosts. There is little information on the transmission biology of these parasites, apart from *Pseudodiplorchis*, but there are brief references in the literature to infection levels. These suggest that invasion is extremely efficient despite the apparently overwhelming difficulties. *Parapolystoma crooki* is a parasite of *Rana magna* which lives in mountain torrents on forest-covered volcanoes in the Philippines; Vande Vusse (1976) recorded a prevalence of 60%, mean 10·5 (range 1–40) parasites per host. *Mesopolystoma samiriensis* infects the tree frog (*Osteocephalus taurinus* in Amazonian rain forest. This host lives high in the canopy and descends only during torrential rain to spawn in the rapidly-rising floodwaters, when, somehow, transmission must take place. Like both these species, Japanese *Diplorchis ranae* also has an elongated uterus capable of storing immediately-infective offspring (up to 350 per parasite); Ozaki (1935) recorded a prevalence of infection of 80% with up to 38 worms per host. These figures contrast with some of the highest levels recorded for *Polystoma integerrimum* in Europe: 12–24% prevalence, mean intensity about two worms per host (Combes, 1968; Lees, 1962). Murith (1981a) correlated the reproductive biology of three polystomatids in West

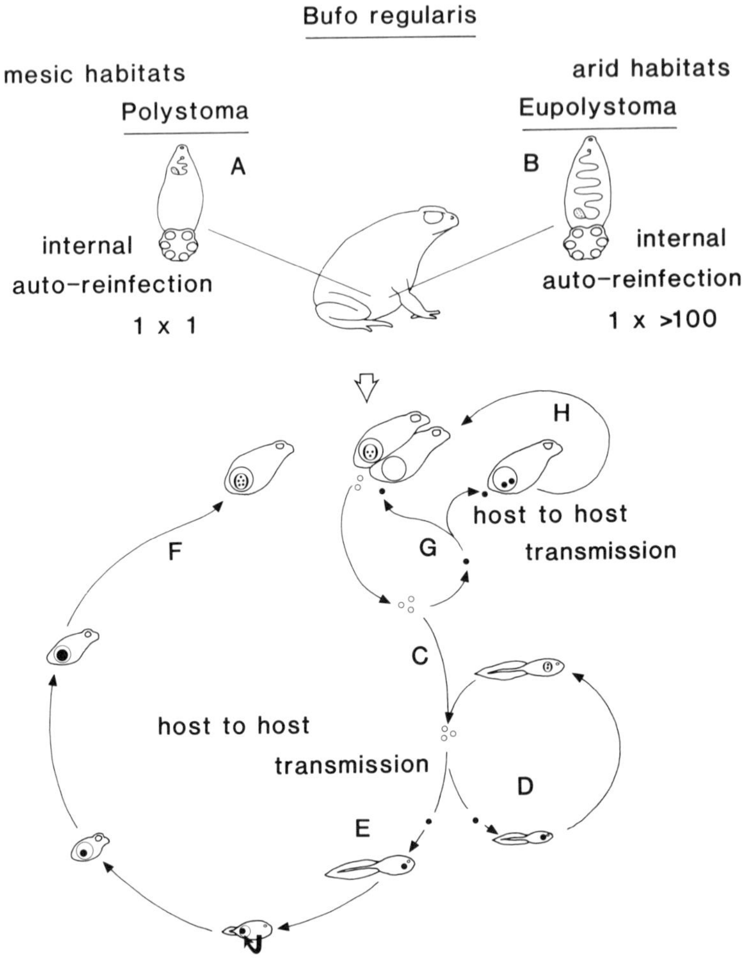

Figure 6.7. Comparison of the life cycle strategy of *Eupolystoma alluaudi*, which infects *Bufo regularis* in arid regions of Africa, with that of *Polystoma africanum* infecting the same host species in mesic habitats. The retention and hatching of one egg in the short uterus of *Polystoma* (after the end of normal egg output) can give rise to internal auto-reinfection, increasing burdens by one larva per adult parasite (A). On the other hand, the elongated uterus of *Eupolystoma* permits accumulation and internal infection by over 100 larvae per adult worm (B). Host to host transmission for both parasites is generally correlated with the spawning of *Bufo regularis*. For *Polystoma*, the eggs develop in the external environment (C). Hatched infective oncomiracidia follow the classical life cycle pattern with alternative pathways according to the stage of tadpole development. On young tadpoles (D), the parasites develop rapidly on the gills, mature precociously, and produce eggs which continue transmission involving other tadpoles – the so-called 'neotenic' cycle. On older tadpoles (E), the parasites follow 'normal' development on the gills, migrate over the ventral skin to the urinary bladder at tadpole metamorphosis, and then reach maturity only when the toad matures (F). The laying of undeveloped eggs by the parasites in breeding toads precludes immediate external reinfection of adult hosts during their stay in the transmission site. In contrast, the elongated uterus of *Eupolystoma* permits deposition of fully-developed oncomiracidia which hatch immediately and bring about 'instant' infection of adult toads assembled in water for breeding (G). The tadpole is not involved and the parasite reaches maturity within a few months (H).

African forest with the niche separation of their respective frog hosts which differ in their activity seasons, spawning habitats and timing of reproduction. The parasite of the most aquatic host has a typical *Polystoma* morphology and life cycle. The most arid-adapted host is infected by a distinct form, *Metapolystoma cachani*, characterized by a long uterus within which around 200 eggs develop. Combes *et al.* (1976) studied the polystomatid infections of the toad (*Bufo regularis*) along a transect from tropical forest to arid savanna in the Ivory Coast. At the boundary of these biomes, there is a change over from infections by *Polystoma africanum* (with transmission involving oviparity directed at the host tadpole) to *Eupolystoma alluaudi* (which accumulates immediately-infective oncomiracidia for instant invasion of briefly-available adult toads) (figure 6.7).

The prolonged storage of eggs within an elongated uterus, seen in these examples, has two major effects on reproductive biology. First, it permits the development of offspring to an infective stage, eliminating the hazards and time delays of development in the external environment. Thus, it enables a virtual synchrony in the two elements of transmission, the release of progeny from one host and their invasion of the next. Second, it permits the accumulation of relatively large numbers of infective stages whose simultaneous release will increase the probability of successful invasion. However, these adaptations facilitating opportunistic transmission make important demands on parasite reproductive biology.

THE REQUIREMENTS AND CONSEQUENCES OF OPPORTUNISM

Factors triggering transmission

Obviously, an opportunistic strategy for host invasion relies above all on timing; a fundamental requirement is for precise cues which signal the occurrence of the favourable conditions vital for transmission.

Most parasites which exploit specific opportunities for infection have life cycles employing active transmission. In contrast, passive transmission generally involves a sit-and-wait strategy for potential hosts to ingest a protected, resistant egg or cyst more-or-less by chance. There are exceptions. For instance, the pinworms (intestinal nematodes of mammals) deposit immediately-infective eggs around the perianal skin, and these stick to the fur and are ingested during grooming. Typically, the eggs of the rodent pinworms (*Syphacia* spp.) are laid around the anus in greatest numbers when the host is asleep and are then readily ingested when the hosts groom, within the nest, after they awake. In humans, for whom the lack of spinal flexibility and more restricted grooming precludes this direct ingestion of recently-deposited eggs, the pinworm (*Enterobius vermicularis*) discharges its eggs (principally at night) onto the perianal skin in a sticky, irritating fluid and the stimulus to scratch promotes the indirect ingestion of eggs via contaminated fingers. In cases such as this, passive infection is facilitated by a specific parasite-induced effect on host behaviour (see Chapter 7).

Dracunculus and the filarial nematodes employ a combination of active and passive transmission strategies: they are transmitted by larvae which are passively ingested by the arthropod intermediate host, but the parasites have specific behavioural adaptations which synchronize transmission. For *Dracunculus*, the response is made by the adult worm which controls the timing of larval release and creates an exit from the enclosed tissue habitat. In filarial nematodes such as *Wuchereria* and *Loa*, release of larvae into the bloodstream is non-selective and the blood-feeding vector provides the means of exit, but the larvae respond to the optimum time for ingestion by appearing in large numbers in the peripheral bloodstream at just the right time to be taken up in the blood meal. The stimulus for *Dracunculus* to discharge uterine larvae may be a local lowering of temperature in an area of the host's body immersed in water, usually the lower limbs. The circadian rhythms of microfilaraemia shown by *Wuchereria* and *Loa* employ cues reflecting physiological changes in the host's diurnal rhythms: differences in oxygen tension and in body temperature. The parasite rhythms can be reversed experimentally, by reversing host activity cycles and by administration of oxygen during sleep. However, there is no satisfactory explanation why the microfilariae concentrate in the pulmonary capillaries at times of day when their insect vectors do not feed, rather than continually circulating in the bloodstream.

For the majority of parasites which exploit unpredictable opportunities for transmission, invasion is accomplished by an active larval stage which hatches from the egg capsule and is responsible for host location. Equivalent principles are important in synchronizing transmission with favourable opportunities: the timing of oviposition (which is recognized by the parent parasite within the host), and the timing of hatching (in which the response is made by the infective larva within the egg capsule, usually some time after it has been deposited into the external environment).

Control of hatching

Straightforward textbook accounts of parasite life cycles often seem deceptive in their simplicity, and a moment's reflection makes the statistical likelihood of successive events seem highly improbable. This is very striking in the case of monogenean parasites of marine fish, when one considers the spatial dimensions of the environment, the effects of water currents and shifting sediments on parasite eggs developing on the sea bed, the mobility of potential hosts, and the speed of the hosts' locomotion in comparison with that of the tiny ciliated swimming stages which must locate and infect them. Whilst random dispersal of infective stages provides a reasonable strategy, statistically, for location of dispersed hosts, it seems incredible that life cycles may operate successfully in these circumstances. The studies of Kearn and co-workers have revealed a series of remarkable adaptations demonstrating how transmission is finely tuned to host behaviour.

RESPONSE TO LIGHT–DARK CYCLES

Entobdella soleae is a skin parasite of the common sole (*Solea solea*) a flatfish which is nocturnally-active and spends the daytime buried in sediment on the sea bed. The parasites release eggs continuously and these take about 30 days to complete development at 13 °C. The egg capsules are tethered to sand particles by sticky appendages and this ballast prevents the eggs from being carried away horizontally to areas of sea bottom not inhabited by their hosts. Having been deposited in localities frequented by the hosts, these eggs are then likely to hatch where the chances of invasion are improved. In a series of elegant experiments, Kearn (1973) showed that hatching exhibits an endogenous circadian rhythm with most of the larvae which hatch on a particular day emerging during the first 4 hours of illumination, and none hatching during darkness (figure 6.8(a)). This focuses host location by the relatively slow-moving oncomiracidia into the period when the target is stationary. Additionally, larvae which hatch soon after dawn will benefit from the maximum period of host inactivity and this will also provide optimum circumstances for detection of chemical cues (Kearn, 1986).

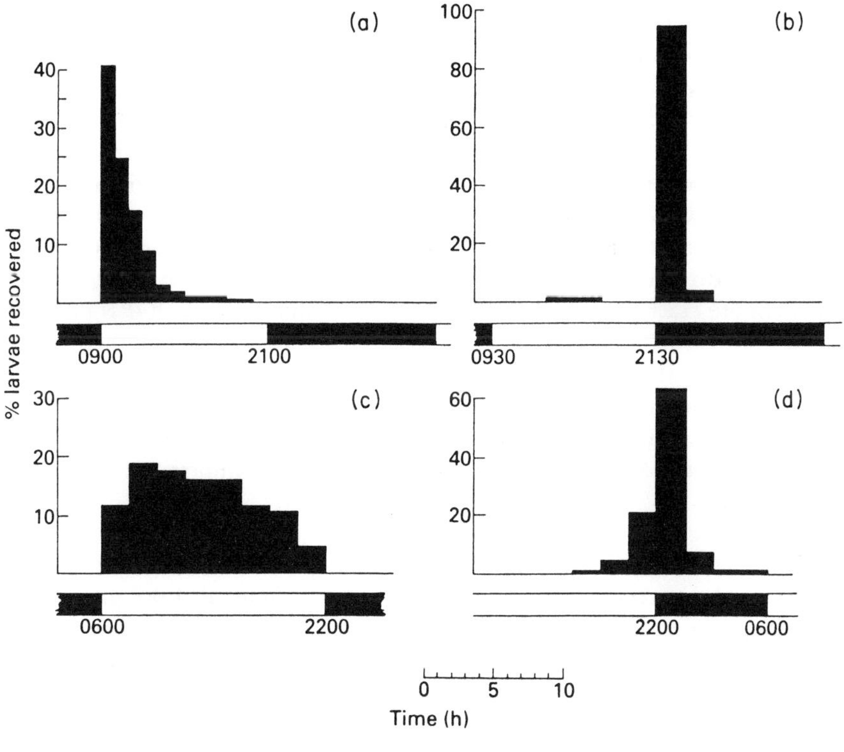

Figure 6.8. The daily patterns of egg hatching in (a) *Entobdella soleae*, (b) *Entobdella hippoglossi*, (c) *Polystoma integerrimum*, (d) *Diplozoon homoion gracile*. Each histogram summarizes observations made over a period of 3 days or more; in each case recoveries of larvae for each of 24 daily 1-hour periods (a) or 12 daily 2-hour periods (b,c,d) have been added together and given as percentage recoveries. (a,b) fluorescent lighting turned on and off abruptly; (c,d) natural illumination. Reproduced with permission from Kearn (1986), *Advances in Parasitology*, **25**, Fig. 48, p. 253.

The related *E. hippoglossi* infect the halibut (*Hippoglossus hippoglossus*), and the hatching of its larvae shows the opposite periodicity with most emerging during the first 2 hours of darkness (figure 6.8(b)). This correlates with what is known about the behaviour of the host, a diurnally-active predator which probably rests on the sea bed at night (Kearn, 1974a).

Circadian rhythms cued by light–dark cycles have also been documented in monogeneans infecting hosts in freshwater habitats. In *Diplozoon homoion gracile*, in addition to the increased rate of egg production which occurs nocturnally (see above), there is a clearly-defined hatching rhythm with the majority of larvae emerging at dusk (figure 6.8(d)). These eggs will have accumulated in areas of river bed where the hosts (southern barbel, *Barbus meridionalis*) occur most commonly at night, and then they are adapted to hatch at the start of the nocturnal period when host targets are again stationary (Macdonald and Jones, 1978). The eggs of *Polystoma integerrimum* also show a hatching rhythm precisely correlated with diurnal movements of the host population, tadpoles of *Rana temporaria*, in Pyrenean lakes. The adult frogs spawn around the margins of lakes and simultaneously their bladder-dwelling polystomatids deposit their eggs. The tadpoles hatch first and congregate by day in these shallow warmer areas, but at night they move into deeper water. The larvae of *Polystoma* hatch over the next few weeks, specifically between the hours of dawn and dusk when they would encounter aggregated tadpoles, and none hatches at night (Macdonald and Combes, 1978) (figure 6.8(c)).

Clearly, these circadian rhythms synchronize hatching only with the general likelihood of host availability, although we must deduce that this probability is sufficiently strong to have resulted in selection and fixation of these specialized adaptations. In some cases, knowledge of the parasite's life cycle strategy is now actually more exact than the details of host behaviour with which it is evidently linked. Thus, some aspects of the activity rhythms of halibut and southern barbel are inferred, in the above examples, from the specific timing of the invasion response. The studies on marine monogeneans may also suggest that the host fish occupy defined areas of the sea bed and return to local sites for resting periods. In these cases, parasitology may provide a stimulus for more comprehensive studies of host behaviour.

Response to shadows

Further insight into the opportunism of some transmission strategies is provided by the finding of Kearn (1982) that sudden extinction of illumination stimulates the rapid hatching of *Entobdella diadema*, a parasite of the stingray, *Dasyatis pastinaca*. The oncomiracidia emerge within 3 seconds of sudden darkness and then swim vertically upwards. Kearn interpreted this as a shadow-induced response to the host, a fish of shallow, well-illuminated waters which is likely to cast a strong shadow on the bottom when it rests or cruises slowly. Following this rapid hatching, the geonegative behaviour of the larvae could lead to contact with the host. This strategy may be effective for a parasite of a large, diurnally-active flatfish with the habit of resting intermittently on the sea bed. However, it presupposes that the host fish is sufficiently common in local areas of the sea bottom to make this non-specific response advantageous. Again, detailed information on host behaviour and distribution lags behind the predictions made from specific parasite adaptations.

RESPONSE TO CHEMICAL STIMULI

Greater precision is possible in synchronizing hatching with brief opportunities for invasion if the cue provides a specific indication of the host, rather than a probability based on its usual habits. Several monogeneans of fishes employ a chemical trigger to stimulate hatching. In *Entobdella soleae*, a proportion of encapsulated larvae do not respond immediately to photoperiodic cues but instead will hatch readily when stimulated by chemical substances from a nearby host (Kearn, 1974b). This response enables the infective stages to take advantage of every opportunity when the common sole might come to rest near them on the sea bed, including the period of darkness when emergence in the absence of a host is inhibited by the circadian hatching rhythm.

However, the hatching factor is not specific to *Solea solea*: skin mucus from a range of fish which are not hosts for *E. soleae* also induce emergence. Kearn (1974b) argued that this may not be as disadvantageous as it may seem if there is ecological segregation of fish species on the sea bed resulting from preferences for specific habitats, bottom deposits etc. Non-host fish may be relatively uncommon in areas inhabited by sole and a non-specific stimulus may still have significant survival value (as in the shadow-induced hatching of *E. diadema*).

Kearn and Macdonald (1976) attempted to define the chemical nature of the host hatching factor and reproduced natural responses with excretory products of fish skin including urea, a small molecule which could diffuse across the egg capsule to stimulate the enclosed larva. However, Kearn (1986) reported that test solutions comprizing the purest chemical grade of urea fail to stimulate hatching of *E. soleae* larvae, suggesting that other unidentified trace substances may provide the essential trigger.

This opportunistic hatching strategy is specialized further in *Acanthocotyle lobianchi*, a skin parasite of rays (*Raja* spp.), to the extent that the infective larvae will only emerge in response to host chemical cues. The encapsulated oncomiracidia are fully developed after 15 days (at 13 °C) and will remain alive but unhatched for nearly 3 months irrespective of photoperiod, shadowing and mechanical stimulation. However, the addition of host skin mucus to these eggs triggers virtually immediate hatching, i.e. within 2–4 seconds (Macdonald, 1974). This response is attributable specifically to urea, and treatment of ray mucus with urease destroys the stimulatory effect (Kearn and Macdonald, 1976; Kearn, 1986). The threshold concentration required to trigger the hatching of *A. lobianchi* larvae (around $0 \cdot 017 \mu$mol/ml) is several times higher than that for *E. soleae* (using impure urea). This undoubtedly reflects the higher levels of urea in elasmobranch mucus and provides a basis for a more specific host stimulus for the ray parasite. *A. lobianchi* is totally committed to a sit-and-wait strategy since the oncomiracidium is unciliated and relies entirely on the ray to settle on the sea bed on top of it. The larva clearly has the capacity to wait for a very considerable period for this chance to arise.

However, an essential requirement for this opportunism is that the encapsulated larva is equipped with substantial energy reserves, and Macdonald (1974) suggested that lipid droplets provide these resources.

In addition to *A. lobianchi*, parasitic on the skin, *Raja naevus* is infected by *Dictyocotyle coeliaca* in the body cavity. In contrast to the total dependence on host hatching factors in *A. lobianchi*, *D. coeliaca* shows no response to photoperiod, host chemicals or other detectable cues. Instead, after a long development ciliated larvae hatch in small numbers continuously throughout the day and night over several weeks (Kearn, 1975). Kearn interpreted these different devices in the two parasites of the ray as alternative solutions for infection of a host lacking any well-defined, rhythmical, daily activity pattern.

However, with this example we return to our starting point: that, in the case of *Dictyocotyle coeliaca*, we must envisage a relatively mobile host releasing parasite eggs which become scattered over a wide area of the sea bed, and from which very low densities of infective stages will emerge to locate unpredictable host targets, a task which daunts one's imagination. On the other hand, perhaps this species responds to as yet unstudied environmental cues, such as tidal rhythms.

Control of release of infective stages by the parent parasite

A different solution to the difficulties of exploiting intermittent opportunities for host invasion is adopted by those parasites in which the parent parasite identifies the moment and responds by releasing infective stages.

There are examples from a range of parasite groups, including the nematode *Dracunculus medinensis* described earlier. The brood parasitic cuckoo employs a complex behavioural repertoire when parasitizing a host, with each part of the procedure designed to defeat host defences. The adaptations involved in this exploitation have been examined in detail by Brooke and Davies (1988), Davies and Brooke (1988, 1989a,b), and summarized by Davies *et al.* (1989). The rapid laying of a small mimetic egg is carried out during one afternoon early in the host's laying period. Accurate synchronization requires that the female cuckoo will watch the intended victim, often for several days. This process also requires initial identification of the appropriate species of foster parent. The cuckoo (*Cuculus canorus*) is divided into genetic races or gentes, each specialized on a particular host species, and each female lays eggs that generally match those of the future foster parent of its young.

The glossiphoniid leech, *Marsupiobdella africana* is uniquely specialized for brooding its offspring: up to 50 young are retained within an internal chamber and they are carried, protected by the parent, throughout their development. When a prospective host, the African clawed toad (*Xenopus laevis*), comes within range the leech responds rapidly, moves towards it and at the moment of contact the parent discharges the infective stages more or less explosively onto the host's body. Experiments have shown that the cue both for identification of the host and release of the parasite's offspring is provided by the toad's skin secretions (Van der Lande and Tinsley, 1976).

There are also some intriguing examples amongst monogenean helminths of the response by the parent parasite to restricted transmission opportunities, and these provide important insight into the intimacy of the relationship between host and parasite.

THE CONTROL OF OVIPOSITION IN MONOGENEANS

It is well-established in the literature that the exact synchrony which enables *Polystoma integerrimum* to release its eggs at the same time as its host, *Rana temporaria*, is due to the control of parasite reproductive biology by host sex hormones. This hypothesis, first proposed by Gallien (1935), is supported by the close correlation of all events in the seasonal cycles of gonadal preparation, oviposition and recovery for both host and parasite. Actually, however, there has never been satisfactory experimental confirmation and the studies which are sometimes cited (Miretski, 1951; Stunkard, 1959; Gorskhov, 1964) are inconclusive and lack adequate controls.

The opportunistic transmission strategy of *Pseudodiplorchis americanus* parallels that of *Polystoma* but with even more exact synchronization. The period of oviposition for both host and parasite is confined to just a few hours when the toads enter temporary ponds following torrential desert rains. The stimulus for parasite oviposition must be very precise: if the gravid adults discharge their offspring just an hour too early or too late then the oncomiracidia may be passed out onto land and lost. The environmental events accompanying the entry of toads into water induce major changes in host physiology and these could provide distinct signals for parasite recognition.

Spadefoot toads can survive severe dehydration during hibernation, and urea accumulates in the liver and blood. When the host enters water, a parasite in the urinary bladder would experience major osmotic and chemical changes and these could provide reliable cues for transmission involving an aquatic infective stage. However, simple experiments (including transfer of infected toads or isolated parasites into water) show that the parasites do not respond to a trigger factor linked with rehydration. Of course, an osmotic cue guarantees that water is available, but it does not provide information that there would be other hosts to infect and the toads are solitary animals except during spawning. A trigger linked specifically with host mating would guarantee both the appropriate medium for transmission and the mass assembly of usually hundreds of potential hosts.

Detailed experiments have established a precise correlation between the release of the infective stages of *P. americanus* and the sexual activity of the host (figure 6.9) (Tinsley, in preparation). During the process of spawning, which may take 3 or 4 hours, the parasites inside the male begin to discharge their larvae as soon as their host begins to clasp the female, and a steady discharge continues throughout amplexus. At this stage there is no response from the parasites within the female, but these begin to release oncomiracidia more or less exactly as the female starts to release spawn and they stop as spawn output stops. This difference in the timing of the parasite response in the two host sexes has important implications for transmission. In natural spawning assemblies, males greatly outnumber females (often by 5 to 1), there is intense competition, and the majority of males never succeed in mating (Tinsley, 1990). If parasite oviposition were linked to gamete release in males, as appears to be the case in females, then a major part of the parasite population would fail to contribute to transmission. However, experiments both in the laboratory and field show that the parasites in male toads are stimulated to release their accumulated infective stages by intense sexual excitement – either by successful amplexus or by the excitement of chorusing to attract females and struggling with other males prior to securing a mate. It may be that the precise cue is actually the same in both sexes, varying only in the time of its appearance, and experiments involving a range of hormones are currently in progress. Nevertheless, the temporal difference in the parasite's response in males and females clearly serves to maximize the opportunistic strategy of *Pseudodiplorchis*. Within a mating assembly, all females will spawn, so a trigger linked with the discharge of ova is infallible. Only a fraction of males will mate, but a trigger linked with sexual behaviour guarantees that all their parasites can exploit the brief chance for transmission (Tinsley, 1990 and in preparation).

In cases like this, selection of the optimum moment for deposition by the parent worm may automatically guarantee favourable circumstances for host invasion; therefore, the encapsulated infective stages usually hatch more or less immediately after release into the external environment, triggered simply by osmotic change or light. The rapid hatching of newly-deposited egg capsules has been described for *Pseudodiplorchis* and *Neodiplorchis* by Tinsley and Earle (1983) and for *Eupolystoma* by Tinsley (1978b).

Nutrition of infective stages

Typically, the platyhelminth egg is enclosed within a protective shell or capsule together with a complement of vitelline cells which supply nutrients to the developing embryo (elsewhere in the animal kingdom this is provided by yolk which forms an integral part of the egg cytoplasm). During the development of encapsulated embryos in the external environment, the number of vitelline cells gradually declines: thus, for *Protopolystoma xenopodis*, the newly-deposited capsule contains over 100 vitelline cells and few remain by the time the oncomiracidium is ready to hatch, after 20 days at 19 °C (Tinsley and Owen, 1975; Jackson and Tinsley, 1988b). This limits the possibilities for a delay in hatching, but the optimum investment of resources has presumably been selected to meet normal circumstances: the aquatic host (the African clawed toad, *Xenopus laevis*) is available for invasion more or less continuously and the parasite larvae probably hatch soon after they have completed development.

In helminths where the encapsulated infective stage, released into the external environment, is adapted to respond to unpredictable opportunities, the 'egg' must be provided with a larger reserve of nutrients to enable a delay. This is seen in an extreme form in the 'sit-and-wait' strategy of *Acanthocotyle lobianchi* where, although hatching is first possible after 15 days (at 13 °C), larvae are provided with lipid reserves enabling up to 83 days waiting (see above).

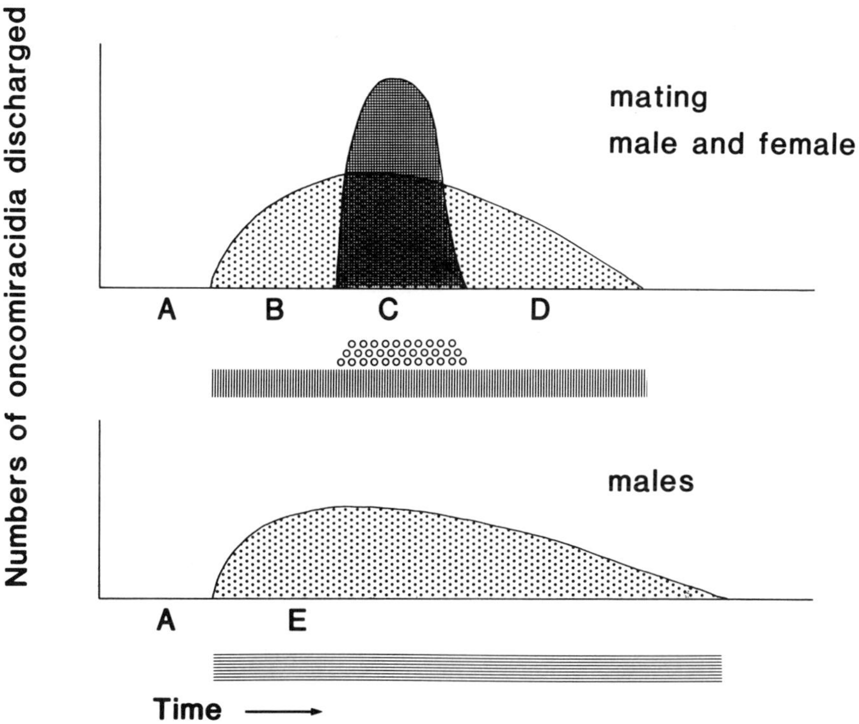

Figure 6.9. *Pseudodiplorchis americanus*: correlation of parasite oviposition with host sexual behaviour.
Upper figure, mating male and female *Scaphiopus couchii* harbouring gravid adult worms.
A. Entry into water and initial rehydration: no oncomiracidia discharged suggesting that an osmotic trigger is not involved. The duration of this phase in laboratory experiments may be 1–4 hours. If the host(s) fail to become sexually aroused this phase continues without release of oncomiracidia regardless of how long the toads remain in water.
B. Initiation of amplexus (||||||||||): virtually simultaneous with the start of sexual activity, gravid parasites in the urinary bladder of the male begin to discharge oncomiracidia (∴∴∴∴), and this continues until amplexus ends (D).
C. Spawning (∘∘∘∘∘∘∘∘∘): in laboratory experiments this occurs a few hours after the start of amplexus.
Characteristically, parasites in the females do not discharge oncomiracidia during the initial period in water (A), nor during the initial period of amplexus by the male (B). Release of infective stages by parasites within female toads (▓▓▓▓) begins only as host oviposition begins (∘∘∘∘∘∘∘∘∘), and ends more or less at the end of spawning (even though amplexus and the period of time spent in water may continue (D)).
Lower figure, group of infected male *S. couchii*, recently emerged from hibernation and put together in water.
A. As above.
E. Start of sexual arousal (☰☰☰☰), males wrestling and attempting amplexus with each other. Discharge of oncomiracidia from their urinary bladder parasites (∴∴∴∴) begins when these males begin to interact and continues throughout the period of sexual arousal.
(Diagrams based on laboratory experimental data (Tinsley, in preparation), the numbers of adult worms is typically greater in males than in females (Fig. 6.9, above), hence oncomiracidial discharge is greater.)

In helminths where embryos complete development *in utero* there is little direct information on the mode of nutrition. In some cases, it may be assumed that the 'eggs' are nutritionally self-contained units which develop within the parent parasite using the vitelline cell provision made before encapsulation. This is probably the case for *Metapolystoma cachani* which can accommodate 200–300 egg capsules *in utero*. These are capable, if deposited, of completing their development in the external environment, but if retained they can develop and hatch within the parent (Murith, 1981b). However, in a number of helminths where embryo development invariably occurs *in utero* (particularly digeneans), there is evidence that nutrients are supplied during development by the parent parasite. Thus, the egg shell of *Halipegus eccentricus* is perforated by narrow channels which may permit nutrients to pass from parental tissues to the developing encapsulated embryo, and studies with radiolabelled carbohydrates and amino acids have demonstrated the uptake and incorporation of parental nutrients by encapsulated embryos in several other digeneans (Tinsley, 1983 and references therein).

The circumstances of embryo production by *Pseudodiplorchis americanus* also suggest adaptations involving direct maintenance of uterine offspring by parental nutrients. There is an extremely reduced vitelline system, and instead of a relatively rigid 'shell', as in other polystomatids, the larva is enclosed in a thin flexible sac. The requirements of the life cycle are very demanding. Although the summer rains, which trigger the spawning of the spadefoot toad host, generally reach Arizona in early July, rain can occur at any time between early June and mid-August. *Pseudodiplorchis* must be prepared for the earliest opportunity, yet the gravid parasites may need to maintain their offspring in an instantly-infective condition for a further 2 months. Recent observations have revealed adaptations which are entirely without precedent elsewhere amongst platyhelminths (Cable, 1989; Cable and Tinsley, 1990). The egg capsule is a multi-lamellate structure composed of concentric layers of membranes. This forms a transparent bag which closely invests the embryo throughout its development and expands continuously as the larva inside it grows. The innermost membranes are modified to form placenta-like processes which 'plumb in' to the tegument of the developing oncomiracidium. These connections are packed with glycogen, and nutrients pass directly from parent worm to offspring. The intestine of the parent parasite is highly branched, with diverticula passing between the folds of the uterus. This ensures close proximity between the sites of digestive uptake and the sites of delivery of nutrients to the developing embryos.

These unique developments provide a remarkable adaptation to the demands of this opportunistic life cycle. The parasites mature quickly, soon after migration to the bladder, and begin to produce embryos whilst they are still relatively small, so accumulation of offspring proceeds at the same time as adult growth. The offspring are produced more or less continuously, so the longer the period before transmission the greater the complement of infective stages. The flexible egg capsule ensures that maximum numbers of oncomiracidia can be packed together within the storage capacity of the uterus. The direct transfer of nutrients from parent to offspring means that resources can be produced and supplied gradually over an extended period of time, instead of having to be stockpiled in the adult reproductive system before egg assembly can begin. Finally, nutrient supply can be related precisely to the period of need, and infective larvae can be maintained in constant readiness for transmission even if the opportunity is delayed.

Reproductive potential in opportunistic life cycles

Factors restricting biotic potential

The strategy of targeting transmission into periods particularly favourable to host invasion may greatly restrict the overall reproductive output of some parasites. Part of this restriction may be determined by the spatial constraints of stockpiling either the components for egg assembly or encapsulated offspring within the body of the parent parasite. Comparison of the total annual egg production by *Polystoma integerrimum* (generally around 3000) with the numbers of infective stages accommodated by *Pseudodiplorchis americanus* (rarely more than 300), suggests that the storage of embryos and the continuous diversion of resources for their development and maintenance permits only about 10% of the reproductive output achieved by storage of components for instant assembly (even allowing for differences in worm size).

Amongst polystomatids which stockpile offspring *in utero*, the reproductive capacity is probably closely determined by body size with records in the range of 200–400 larvae per parasite for *Eupolystoma*, *Metapolystoma*, *Parapolystoma*, *Diplorchis* and several others (Tinsley, 1983). Various modifications may maximize capacity: thus, the large oncomiracidium of *Eupolystoma anterorchis* develops 'doubled up' inside a capsule only about half its length, and the uterus of the relatively small adult worm can accommodate up to 350 infective stages (Tinsley, 1978a,b). In those species (such as *P. americanus*) where it is known that this small number of offspring represents the total reproductive output for the entire year, intrauterine development greatly restricts the numbers of infective stages. However, some polystomatids actually exploit their highly intermittent transmission to transform the efficiency of their life cycles.

Adaptations for boosting reproducing populations

In parasites with direct life cycles, there is the potential for the offspring of an established parasite to be retained and attach alongside its parent without ever leaving the host. This autoreinfection is a characteristic part of the life cycle of viviparous gyrodactylid monogeneans, but among oviparous monogeneans it is relatively uncommon (reviewed by Tinsley, 1983). In polystomatids, however, ovoviviparity has evolved into a dual strategy alongside the adaptations for opportunistic transmission.

Following the brief annual period of egg-laying by *Polystoma integerrimum*, a single egg may remain behind in the uterus, complete development and hatch, and the larva emerges from the parent worm into the urinary bladder (Combes, 1967).

This process can double the parasite burden within each host and has fundamental significance since there is, otherwise, no mechanism by which the frog can be infected after metamorphosis. Similar adaptations occur in tropical *Polystoma* species. *P. grassei* infects a tree frog, *Leptopelis ocellatus*, which spends all its post-metamorphic life in the branches. Mating frogs produce a foam nest in which the tadpoles initially develop before dropping into pools below. Transmission follows the typical *Polystoma* pattern with tadpole infection even though there is no host entry into water to permit oviposition. It must be assumed that the eggs of *P. grassei* drop into the pools from the nest along with the tadpoles. In addition, however, some larvae can apparently hatch internally, increasing the infections in adult hosts which never physically enter transmission sites (Dupouy and Combes, 1977), and this may be responsible for burdens of up to 116 parasites per host recorded by Murith *et al.* (1978).

One of the most remarkable and specialized transmission strategies involving ovoviviparity occurs in *Eupolystoma*. This bladder parasite of tropical anurans is equipped with an extended uterus within which developing larvae accumulate. The life cycle may follow two pathways (figure 6.7). At the end of their development, some encapsulated oncomiracidia are discharged into water, hatch, and are immediately infective to toads which enter temporary pools to spawn. However, some larvae hatch within the parent worm, emerge and attach to the host's urinary bladder without passing into the external environment. Fournier and Combes (1979) demonstrated that these two roles are accomplished by two different types of larvae. The oncomiracidia which hatch *in utero* are not equipped with cilia nor the sense organs involved in host invasion; these alone are able to short-circuit transmission and complete an 'internal cycle'. The oncomiracidia which hatch only after deposition are ciliated and have a complete sensory system, and these are involved solely in host-to-host transmission. Remarkably, the relative numbers of the two types of larvae appear to be controlled in some way by the size of the parasite population in the host individual. If the numbers of established parasites are low, then a majority of non-ciliated larvae is produced and these boost the reproducing population within that host. If the host is heavily infected, ciliated larvae are produced, precluding the internal cycle, and these are released to infect new hosts (figure 6.10). Since the two types of larvae differ structurally, the density-dependent regulation of the parasite population is programmed into the development of *Eupolystoma alluaudi*.

This process has a major impact on the reproductive biology of *Eupolystoma* species: once a host individual is infected, its parasite burden can grow progressively at a time when transmission is impossible. Quite exceptionally amongst monogeneans, intensities of nearly 3000 worms per host can occur (Salami-Cadoux, 1975; Tinsley, 1978a,b), and each adult parasite may produce over 300 immediately-infective larvae. This gives an unprecedented potential for transmission when an opportunity eventually arises. The same principle is seen in some other direct life cycles such as that of the pentastomid *Reighardia sternae* where autoreinfection supplements a life cycle characterized by low *per capita* fecundity (Riley, 1983).

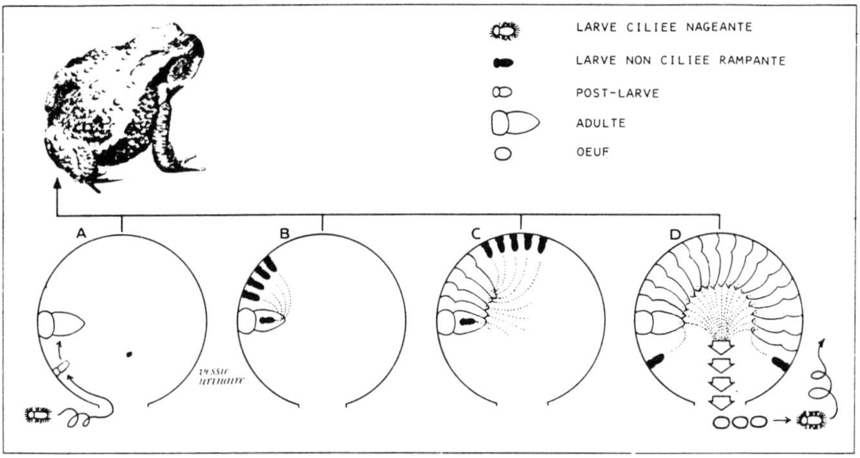

Figure 6.10. The relationship between the cycles of internal auto-infection and external host-to-host transmission in *Eupolystoma alluaudi* (parasite of the urinary bladder of *Bufo regularis*).
A. Host invasion by a swimming oncomiracidium followed by development through juvenile to adult stage.
B. Adult parasite reproduces by ovoviviparity, producing non-ciliated larvae which attach alongside on the bladder wall and develop into further adults.
C. These continue to multiply *in situ* until a threshold is reached, then –
D. The large population of reproducing parasites releases numerous eggs which are expelled with the urine and hatch outside to give ciliated larvae responsible for locating new hosts.
(The detour of invading larvae to the kidneys is not shown. Diagram reproduced with permission from Fournier and Combes (1979), *Comptes Rendus Hebdomadaires des Seances de l'Academie des Sciences*, Paris, **289** Serie D, p.746.)

CONCLUSIONS

Parasites clearly show a range of life cycle strategies designed to meet the problems arising from the irregular distribution of favourable 'patches' within a hostile environment. However, for some parasites host availability is restricted to an extreme degree, so that the opportunities for invasion seem almost impossible to exploit. Presumably, the freedom of some host species from particular groups of parasites may, in part, be the final product of the extreme difficulty of life cycle completion. However, there are also instances where, against all expectations, infection occurs in spite of environmental constraints, and these are of greatest interest within the subject of opportunism.

The evolutionary outcome of opportunism is recorded in the complex multi-host life cycles of many parasites, where the parasite exploits a succession of different host species, each linked to the next by ecological and, particularly, behavioural ties. There are many examples of complex interspecific interactions, such as that recently unravelled by Riley and co-workers (see above) involving a pentastomid parasite (*Raillietiella gigliolii*), leaf-cutting ants, larvae of rhinoceros beetles which inhabit the refuse dumps within the ants' nests, and worm-lizards which are also inquilines of the ant colonies. Summaries of life cycles such as these sometimes present a stereotype sequence of events, deceptive in their diagrammatic simplicity, but behind this must lie complex evolutionary processes. A moment's reflection makes one marvel at the life cycle adaptations to behavioural and trophic links which must, initially, have been chance events. Moreover, these multi-host life cycles require even more marvellous adaptations at a physiological level: tolerance of different osmotic, ionic, and sometimes temperature regimens in abruptly-changing host microenvironments; adaptations to counter immune attack from successive phylogenetically-distinct armouries; developmental flexibility and sensory perception to exploit the conditions characteristic of each host, and so on.

Apart from the inferred evidence of the role of opportunism in the evolution of parasite life cycles, this chapter has focused on the observed role of opportunism in transmission. For this analysis it has highlighted one platyhelminth group, the monogeneans. Because these parasites have direct life cycles, the role of host behaviour in transmission can be identified more clearly, and there is much more detailed information on features such as hatching factors and life cycle synchronization than for most animal parasites. None of the monogeneans exhibits the very high reproductive output typical of endoparasites in mammals, but their life cycles are characterized by a close correlation with host behavioural ecology. They are highly host-specific. Polystomatids are particularly interesting in this analysis because of one characteristic: whilst these parasites are transmitted by a swimming larva, their hosts show varying degrees of independence from aquatic habitats. Indeed, it is in those hosts which enter water most rarely and briefly that we find some of the best examples of opportunism influencing the dynamics of transmission.

A spectrum of factors reduces the possibility of contact between parasite infective stages and their prospective hosts. Some are attributes principally of the environment: the flushing action of tides on rocky shores constrains the transmission of *Parorchis acanthus* despite the more-or-less static gastropod target. Mountain torrents promote instant separation of mobile infective stages from river-dwelling fish and tadpoles. Host activity produces equivalent strong, unidirectional water currents in the case of continuously-swimming sharks and their commensal sucker-fish. Monogenean gill flukes (*Dionchus* species) infect the sucker-fish attached to this mobile substrate by a suite of adaptations which parallel those employed by parasites transmitted in mountain torrents (Tinsley, 1983). Host behaviour may determine that there is only periodic entry into active foci; thus, the emission of schistosome cercariae from infected snails is strongly correlated with periods when specific hosts are likely to enter water. Where parasite eggs are released into the general environment to complete development, host mobility is one of the most important factors influencing contact with infective stages. Where the life cycle involves active invasion in an aquatic environment, the parasite larva is commonly equipped for a short-range search and identification role. A variety of opportunistic strategies has evolved to focus release of these mobile stages into periods when the host is within range, employing indirect cues (circadian rhythms, temperature differences, light–dark cycles, etc.) or more direct (chemical) identification of host occurrence. It is, therefore, a measure of supreme confidence in an opportunistic strategy that the infective stages of some monogeneans (including *Acanthocotyle*) have given up the possibility of final homing-in on the target and literally wait for potential hosts to sit on top of them. Many of the problems involving spatial separation can be overcome if there is an instantaneous response to fleeting opportunities, eliminating the time delays of most infection processes. The ability to wait for these opportunities and then respond rapidly requires specific preparation, and this is often costly in terms of resource-investment and a reduction in total reproductive output. A major reduction in fecundity in order to achieve precision may be a more feasible strategy in direct life cycles, for instance in opportunistic monogeneans (described above) and in some cephalobaenid pentastomids (described by Riley, 1983). Here, a single well-aimed shot may complete the life cycle. However, in multi-host life cycles, where a specific opportunistic strategy achieves only one link in the chain, the overall success rate may be determined by hazards affecting other links. Thus, the oviposition response of *Dracunculus medinensis* is a highly effective mechanism to exploit brief contact with aquatic habitats, but 'food chain' transmission via the copepod host may be far less assured; perhaps to compensate for this, each female *Dracunculus* releases around a million offspring.

A combination of features emerges in these examples which characterize opportunistic transmission: a period of preparation, a response involving speed and precision, and a highly efficient outcome.

By definition, a vital element in opportunism is timing, to synchronize invasion with brief periods of host vulnerability. This is exemplified at an individual level by *Acanthocotyle lobianchi* and *Entobdella diadema*, whose infective stages hatch within 2 or 3 seconds of receiving the essential host signal, and at a population level by *Pseudodiplorchis americanus* in which the entire annual reproductive output is focused into a transmission episode of only 7 hours. In these examples, two strategies have evolved to meet different demands. Where the external environment is relatively stable but the availability of potential hosts is uncertain, parasites may release eggs more or less continuously to develop in the environment, and opportunities for host invasion are exploited by a larval hatching response. Where the environment is inherently unstable and unpredictable, infective stages may develop and be retained within the parent parasite, and exploitation of host availability is achieved by a parental 'oviposition' response. The effects of spatial and temporal constraints on transmission may be closely interlinked in these examples.

Paradoxically, one relationship which demonstrates extreme restriction, both in time and space, is also one of the most efficient. Tinsley and Jackson (1988) showed that, at the end of the short mating season of one population of the spadefoot toad (*Scaphiopus couchii*) in Arizona, every male toad had been infected by *Pseudodiplorchis americanus*: a 100% prevalence and a mean of over 100 worms per host. Probably one of the most remarkable aspects of this strategy is that high levels of infection are achieved with so few offspring per parasite, generally about 100 larvae per year. Not only is this in extreme contrast to the 20 000 eggs per parasite *per day* produced by *Fasciola hepatica*, it represents a very low reproductive potential for most free-living invertebrates. Paradoxically in *P. americanus*, the constraints which so severely restrict the possibility of transmission, in space and time, actually maximize its efficiency when brief opportunities occur. Field data suggest that individual oncomiracidia have a 30% chance of successful host invasion (Tinsley and Jackson, 1988). The reproductive biology of *P. americanus* reveals another paradox: that annual *per capita* egg production by this parasite is less than 10% that of its host (each female *Scaphiopus couchii* deposits around 3000 eggs during spawning, i.e. about 1500 per toad, assuming an equal sex ratio). This contradicts the general relationship between host and parasite reproductive output (see beginning of the chapter), which serves normally to distinguish parasitism from predation. The explanation lies first in the greater efficiency of this life cycle compared with most others, and second in the fact that transmission is targeted on adult hosts which have survived all the hazards and population losses of the tadpole and juvenile stages. (Actually, when the generation time is considered, the parasite's fecundity is much greater than that of the host.) Above all else, the key to this success is the precise integration of two essential processes in host and parasite biology: the desert-dwelling amphibian must return to water to breed and its monogenean parasite is entirely dependent on an aquatic environment for invasion involving a ciliated oncomiracidium. Both partners concentrate their annual reproductive investment into the same very restricted opportunity, created briefly by desert rains. *P. americanus* ensures precise synchronization by employing, as an oviposition cue, the sexual activity of its host. Indeed, all aspects of parasite transmission are closely controlled by host behaviour, and the mass congregation of spawning toads provides a maximum concentration of susceptible targets for parasite invasion.

Host reproduction often provides periods of special vulnerability to parasite invasion. The parasite may exploit a range of associated opportunities: the brief availability of a favourable environment for transmission, the mass assembly of potential hosts (either the mating adults or, subsequently, their offspring), a mechanism for host-to-host transfer, the depression of immune defences (particularly in mammals).

In many of the parasite life cycles which exploit complex interspecific associations, such as that of the pentastomid *Raillietiella*, it is easy to appreciate how the behavioural links between successive hosts increase the efficiency of transmission. It is possible that the life cycles could be completed fortuitously through relatively random events, but regularly-occurring host interactions greatly improve the chances of linkage between the corresponding life cycle stages. However, for parasites such as the monogenean *Pseudodiplorchis*, the essential adaptations are not concerned with improving the efficiency of a process which could occur, albeit inefficiently, by chance. Targeting invasion exclusively into a 7-hour window each year requires absolute precision and efficiency. Many of the adaptations associated with an opportunistic strategy of transmission are quite uncharacteristic of the more general view of parasitism. The case study of *Pseudodiplorchis americanus* presents the complete antithesis of the life cycle of *Fasciola hepatica* which introduces many accounts of parasitology – and this chapter. It is intriguing to consider how, despite seemingly overwhelming constraints, these life cycle strategies represent beautifully-designed solutions to the problems created by host behaviour, environmental instability and fleeting opportunities for transmission.

REFERENCES

Brooke, M. de L. and Davies, N. B., 1988, *Nature*, **335**, 630–632.
Cable, J., 1989, PhD Thesis, University of London.
Cable, J. and Tinsley, R. C., 1990, *in press*.
Calow, P., 1983, *Parasitology*, **86**, 197–207.
Chubb, J. C., 1977, *Advances in Parasitology*, **15**, 133–199.
Combes, C., 1967, *Bulletin de la Societe Zoologique de France*, **92**, 129–133.
Combes, C., 1968, *Memoires du Museum National d'histoire naturelle, Serie A, Zoologie* **51**, 1–195.
Combes, C., 1972, *International Journal for Parasitology*, **2**, 233–238.
Combes, C., Bourgat, R. and Salami-Cadoux, M.-L., 1973, *Zeitshrift fur Parasitenkunde*, **42**, 69–75.
Combes, C., Bourgat, R. and Salami-Cadoux, M.-L., 1976, *Bulletin d'Ecologie*, **7**, 207–214.
Davies, N. B., Bourke, A. F. G. and Brooke, M. de L., 1989, *Trends in Ecology and Evolution*, **4**, 274–278.
Davies, N. B. and Brooke, M. de L., 1988, *Animal Behaviour*, **36**, 262–284.
Davies, N. B. and Brooke, M. de L., 1989a, *Journal of Animal Ecology*, **58**, 207–224.
Davies, N. B. and Brooke, M. de L., 1989b, *Journal of Animal Ecology*, **58**, 225–236.
Dimmitt, M. A. and Ruibal, R., 1980, *Journal of Herpetology*, **14**, 21–29.
Duellman, W. E., 1985, *South African Journal of Science*, **81**, 174–178.
Dupouy, J. and Combes, C., 1977, *Annales des Sciences Naturelles (Paris), Zoologie*, **19**, 397–400.
Fournier, A. and Combes, C., 1979, *Comptes rendus de l'Academie des sciences, (Paris) D* **289**, 745–747.
Gallien, L., 1935, *Travaux du Station de Zoologie, Wimereux*, **12**, 1–181.
Gorshkov, P. V., 1964, *Zoologicheskii zhurnal*, **43**, 272–274.
Jackson, H. C. and Tinsley, R. C., 1988a, *Parasitology*, **97**, 115–128.
Jackson, H. C. and Tinsley, R. C., 1988b, *International Journal for Parasitology*, **18**, 585–589.
Kearn, G. C., 1973, *Parasitology*, **66**, 101–122.
Kearn, G. C., 1974a, *Parasitology*, **68**, 161–172
Kearn, G. C., 1974b, *Parasitology*, **68**, 173–188.
Kearn, G. C., 1975, *Parasitology*, **70**, 87–93.
Kearn, G. C., 1982, *Journal for Parasitology*, **68**, 171–172.
Kearn, G. C., 1986, *Advances in Parasitology*, **25**, 175–273.
Kearn, G. C. and Macdonald, S., 1976, *International Journal for Parasitology*, **6**, 457–466.
Lees, E., 1962, *Parasitology*, **52**, 95–102.
Llewellyn, J., 1984, *Journal of the Marine Biological Association of the United Kingdom*, **64**, 285–302.
Macdonald, S., 1974, *Parasitology*, **68**, 331–338.
Macdonald, S. and Combes, C., 1978, *Chronobiologia*, **5**, 277–285.
Macdonald, S. and Jones, A., 1978, *Journal of Helminthology*, **52**, 23–28.
Miretski, O. Y., 1951, *Doklady Akademii nauk SSSR*, **78**, 613–651.
Muller, R., 1971, *Advances in Parasitology*, **9**, 73–151.
Murith, D., 1981a, *Monitore Zoologico Italiano* (NS) Suppl **15**, 359–365.
Murith, D., 1981b, *Revue suisse de Zoologie*, **88**, 475–533.
Murith, D., Gassmann, M. M. and Vaucher, C., 1978, *Revue suisse de Zoologie*, **85**, 681–698
Ozaki, Y., 1935, *Journal of Science of the Hiroshima University*, B, **3**, 193–225.
Riley, J., 1983, *Parasitology*, **86**, 59–83.
Riley, J., Winch, J. M., Stimson, A. F. and Pope, R. D., 1986, *Journal of Natural History*, **20**, 459–470.
Salami-Cadoux, M. L., 1975, *Bulletin de la Société Zoologique de France*, **100**, 283–292.
Stunkard, H. W., 1959, *Journal of Parasitology*, **45**, 389–394.
Tinsley, R. C. 1974, *Journal of Natural History*, **8**, 355–367.
Tinsley, R. C. 1978a, *Journal of Helminthology*, **52**, 291–302.
Tinsley, R. C. 1978b, *Parasitology*, **77**, 121–132.
Tinsley, R. C. 1983, *Parasitology*, **86**, 161–196.

Tinsley, R. C. 1989, *Parasitology Today*, **5**, 190–195.
Tinsley, R. C. 1990, *American Zoologist*, in press.
Tinsley, R. C. and Earle, C. M., 1983, *Parasitology*, **86**, 501–517.
Tinsley, R. C. and Jackson, H. C., 1986, *Parasitology*, **93**, 451–469.
Tinsley, R. C. and Jackson, H. C., 1988, *Parasitology*, **97**, 437–452.
Tinsley, R. C. and Owen, R. W., 1975, *Parasitology*, **71**, 445–463.
Tocque, K., 1990, PhD Thesis, University of London.
Van der Lande, V. M. and Tinsley, R. C., 1976, *Journal of Zoology*, **180**, 537–563.
Vande Vusse, F. J., 1976, *Journal of Parasitology*, **62**, 552–555.
Whitfield, P. J. and Evans, N. A., 1983, *Parasitology*, **86**, 121–160.
Wilson, R. A., Smith, G. and Thomas, M. R., 1982, In *The Population Dynamics of Infectious Diseases: Theory and Applications* edited by R. M. Anderson (London, New York: Chapman and Hall), pp. 262–319.
Winch, J. M. and Riley, J., 1985, *Parasitology*, **91**, 471–481.

7. A phylogenetic perspective on the evolution of altered host behaviours: a critical look at the manipulation hypothesis

J. Moore and N. J. Gotelli

INTRODUCTION

The study of behavioural changes in parasitized hosts, bemoaned as non-existent two decades ago (Szidat, 1969, p. 784), has flowered as investigators have explored topics as diverse as hormonal interactions between parasitoids and insects (e.g. Beckage, 1985), the mechanics of vector feeding (e.g. Jenni *et al.*, 1980) and effects of parasitized mothers on offspring (Rau, 1985). Parasites have even been hypothesized to affect human mental abilities (e.g. Hay and Aitken, 1983; Kvalsvig, 1988), and affect susceptibility to modern human hunting (Rau and Caron, 1979). Even as this literature has grown, investigators have encountered some of the same challenges and problems that have beset the general fields of ecology and evolution. Studies are often not directly comparable, even when the same parasite is used; methods and traits chosen for study vary among investigators (e.g. Meakins and Walkey, 1975). In addition, despite the fact that examination of some traits may support the hypothesis that altered host behaviours are currently advantageous to the parasite, alternative hypotheses are rarely tested. The existence of the adaptation is accepted as evidence of the historical process by which the trait evolved under natural selection, favouring the parasite with which the adaptation is now associated (but see Gould and Lewontin, 1979; Smith Trail, 1980). With few comparative studies and a limited comprehension of historical constraints, it is difficult to develop an evolutionary framework for understanding and predicting behavioural interactions between parasites and hosts.

The emergence of such a situation is understandable, for the study of behaviour and behavioural alterations in parasitized animals in particular, is complex and may encompass several levels of investigation, including the behaviour itself, the underlying mechanism(s), the ecological result and the evolutionary history. Although not all parasite–host associations lend themselves readily to multiple levels of study, knowledge of one level does not invariably provide insight into others.

For instance, parasites that are transmitted in the food chain have yielded numerous examples of hosts with altered behaviours. Such a parasite is immature in an intermediate host, which must be eaten by a definitive host before the parasite can reach maturity in the latter host. We may reasonably assume that the behavioural repertoire of most prey animals is largely adaptive and is in part the result of selection favouring predator avoidance. Thus, any change in behaviour associated with parasitism may be seen as maladaptive for the host, and possibly predisposing the host to increased predation (but see Milinski and Lowenstein, 1980). At this point, it is not difficult to hypothesize that the altered behaviour is the result of natural selection favouring increased transmission efficiency on the part of parasites that are passed through the food chain (e.g. Moore 1984b).

Indeed, this is a reasonable hypothesis but such manipulation does not always result in transmission. Brassard *et al.* (1982) demonstrated that although parasitized guppies (*Poecilia*) were more susceptible to increased predation, the predator was not necessarily a suitable host for the parasite. They suggested that this might be an unavoidable cost of altered behaviours that accelerate transmission. On the other hand, Bethel and Holmes (1977) showed that differences among behavioural alterations in several host–parasite associations increased vulnerability to appropriate definitive hosts. Without predation tests, it is not clear that altered behaviours are always advantageous to the parasite. Given this uncertainty, speculation about evolution of an alteration becomes even more tenuous.

Even in cases where altered behaviour results in parasite transmission, and we suspect that there are many examples although few have been carefully examined, it is possible that such behaviour is not entirely a direct result of natural selection favouring transmission efficiency but may be, at least in part, a result of other aspects of the host–parasite association. For instance, parasitized animals may themselves have special requirements as a result of being parasitized. In meeting these requirements, and thus benefiting themselves, the hosts may behave differently from uninfected conspecifics. Some of these requirements are reviewed below and may include resources such as oxygen (Smith and Kramer, 1987; and others), heat (Boorstein and Ewald, 1987), energy (Minchella, 1985; Hart, 1988) and food (Milinski, 1985). Meeting these requirements may entail risks, including increased exposure to predators (e.g. Milinski, 1985; and Chapter 4).

Holmes and Bethel (1972) recognized differences among altered behaviours when they explored the ways in which behavioural changes would affect ecological relationships between predators and their prey. They delineated four types of intermediate host behavioural changes that would enhance transmission: (1) decrease in stamina, (2) increase in conspicuousness (often a colour or size alteration), (3) disorientation, and (4) altered responses to environmental stimuli. They noted that the last category was especially interesting, in part because such alterations were the results of changed responses to stimuli, and not necessarily reltated to gross pathology or morphological changes (see Chapter 2). Thus, they recognized that some altered behaviours may be the result of pathology inherent in that particular host–parasite association (e.g. *Trichinella* larvae encysting in muscles and muscular dysfunction).

One might argue, of course, that such pathology is itself a product of natural selection that favours parasite transmission. Such a contention is certainly within reason. Brooks (1980), however, pointed out that parasite site selection, for instance, may have a historical component that can be independent of biotic selective influences (in this case, competition; but see Holmes and Price, 1980). Szidat (1969) hypothesized that site selection in strigeoid metacercariae resulted from natural selection that favoured increasingly protected developmental sites for parasites culminating in the invasion of the lens of the eye. In such a scenario, behavioural alterations associated with infection sites may not be solely the result of natural selection favouring those alterations as transmission mechanisms, but may be at least partially a secondary, coincidental result of developmental site. Regardless of the historical role of natural selection, the contemporary result is a fish with an impaired sensory system. (For discussion of the potential role of definitive host predation efficiency see Szidat, 1969.)

Other alternatives have been considered. Smith Trail (1980) hypothesized that parasitized hosts could benefit from seemingly self-destructive behaviour if it increased inclusive fitness. Dawkins (1982) argued that intermediate host behaviour is a 'shared phenotype', controlled by both parasite and host genes, and that behavioural alterations are an evolutionary compromise between conflicting host and parasite strategies. Minchella (1985) considered a variety of host responses to parasitism (emphasizing gigantism and castration) and noted that a response could: (1) benefit the parasite (possibly increase host survivorship and retain energy that would have gone into reproduction), (2) benefit the host (limit resources available to the parasite), or (3) simply be a side-effect of infection. These alternative explanations may be applied to most host–parasite phenomena.

To date there have been relatively few rigorous tests of such hypotheses (but see Stamp, 1981; McAlister and Roitberg, 1987). In fact, although there have been many demonstrations of altered behaviour, the effect of such behaviour on predation is rarely examined, and behavioural alterations are often simply assumed to result in increased likelihood of predation (e.g. Dobson, 1988). Moreover, even if there were more information on the current beneficiary of altered behaviour, this would not necessarily define the origin of such an alteration.

In Chapter 2, Holmes and Zohar consider the pathology that can be associated with behavioural alterations (see also Vinson and Iwantsch, 1980; Thompson, 1983a; Helluy, 1988; Kavaliers and Podesta, 1988). In this chapter, we shall review some of the literature on altered behaviour and suggest one approach to understanding the evolution of behavioural alterations. We have attempted to include representatives from several major endoparasitic groups in this review (emphasizing helminths), and a number of transmission modes including, but not limited to, passage through the food chain. We have considered a variety of ways in which these parasites might affect their hosts, ranging from the impact of disease on host social behaviour to altered taxes. This review is therefore broad, and necessarily incomplete. (We urge reference to the original literature for details.) We will demonstrate that alterations of host behaviour are widespread, and that in some cases, similar behavioural changes in different host–parasite associations may reflect a variety of underlying mechanisms and could result in a wide range of outcomes for parasite and host. Use of Latin binomials and names of developmental stages is necessary; we have attempted to assist non-specialists with two appendices linking nomenclature with taxonomic affinity, life cycle information and representative references (Appendix 1 and 2).

Because different studies are not parallel and do not always include the same or analogous behavioural traits or alteration, we will discuss one type of alteration at a time. Such an approach is not entirely satisfactory, for in some parasite–host associations several behaviours may be altered, and the relative contribution of each to the parasite or to the host may be difficult to ascertain. We adopt this approach, however, for the sake of comparison.

We wish to emphasize that we suspect the altered behaviours we report here will frequently affect predator–prey relationships. In some, perhaps many, cases this could be shown to benefit the parasite in a way that may be the result of natural selection favouring enhanced transmission. However, at least some of these relationships do not afford parasite transmission and some other alterations may not enhance predation. We believe such ironies call into question the assumption that altered host behaviour is simply and purely a result of natural selection favouring parasite manipulation. Multiple influences are possible, and we suggest some alternative hypotheses and ways to investigate them.

ALTERED HOST RESPONSES: PARASITIC MANIPULATION?

Altered activity

Changes in *host activity* are among the more widespread behavioural alterations. In some cases, activity is reduced by parasitism. B.L. Hart (1988) noted that lethargy and depression are common symptoms of disease. He argued that such behaviour was adaptive for a febrile host, given the metabolic cost of fever, which was in turn beneficial in battling disease organisms. Such behaviour may, however, also benefit the parasite. Mosquitoes engorged more successfully on mice infected with *Plasmodium* spp. or St. Louis encephalitis virus than on controls as a result of a reduction in antimosquito behaviour in the former groups (Day and Edman, 1983, 1984; see also Waage and Nondo, 1982; Turell *et al.*, 1984; Rossignol *et al.*, 1985). (Hart (submitted) has reviewed host behaviours associated with parasite avoidance.)

Upon finding that mice (*Mus musculus*) harbouring the nematode *Trichinella spiralis* exhibited a dose-related reduction in ambulatory and exploratory open field behaviour as the infection progressed, Rau (1983a) hypothesized that infected laboratory mice would be less familiar with the environment and therefore at greater risk of predation. Speed and distance travelled were also reduced in *T. spiralis*-infected laboratory mice (Rau and Putter, 1984). Anthelminthic treatment of parasitized mice resulted in resumption of activity (Zohar and Rau, 1986). Rau and co-workers are currently testing a hypothesis of increased predation by the American kestrel (*Falco sparverius*) on wild strains of *Peromyscus* and *Microtus* infected with a wild strain of *Trichinella* (Rau, personal communication). Likewise, Hutchison *et al.* (1980a) suggested that laboratory mice infected with *Toxoplasma*, a coccidian, might be more subject to predation because they were less active in a Y-maze and less interested in novel stimuli. In cages, they exhibited more general movement and less rearing and digging (Hutchison *et al.*, 1980b) Learning and activity may be linked in infected rodents (Stretch *et al.*, 1960a,b, but see Stretch *et al.*, 1960c). American kestrels (*Falco sparverius*) exhibited reduced flying (increased walking) and preening when infected with *Trichinella pseudospiralis* (Saumier *et al.*, 1988). (See also Dolinsky *et al.* (1981) for diminished activity in *Toxocara*–mouse associations.) Mice infected with the pinworm *Syphacia obvelata* were also less exploratory, but the pinworm *Aspiculuris tetraptera* had no effect (McNair and Timmons, 1977). Pinworms are not transmitted by predation, however, and if their host were eaten, the pinworms would also die.

Activity may also be increased in parasitized mammals. Quinn *et al.* (1987) hypothesized that increased exploratory activity in varying lemmings (*Dicrostonyx richardsoni*) parasitized by the protozoan *Sarcocystis rauschorum*, would increase the risk of predation by snowy owls (*Nyctea scandiaca*) because the animals would be more exposed and noticeable. The animals also groomed less, indicating less fear. In the field, *Microtus arvalis* infected with *Sarcocystis cernae* were over-represented in the diets of kestrels (*Falco tinnunculus*) compared to a snap/trap census (Hoogenboom and Dijkstra, 1987). Both *Sarcocystis* and *Trichinella* invade muscles, although Quinn *et al.* (1987) reported neurological lesions as well. Seed and Khalili (1971) noted that voles (*Microtus montanus*) infected with the protozoan *Trypanosoma gambiense* did not retain circadian pattern in activity but were not at all lethargic. (*Trypanosoma* is transmitted by a vector, and does not depend on predation.)

Other vertebrates can also show the effect of parasites in activity patterns. Fukase *et al.* (1986, 1987) hypothesized that snakes (*Elaphe quadrivirgata*) infected with pathogenic plerocercoids of the cestode *Spirometra erinacei*, which may impede locomotion, may be more easily preyed upon by dogs or cats, the definitive hosts. Skinks (*Sphenomorphus quoyii*) acting as transport hosts for the acanthocephalan *Sphaerechinorhynchus rotundocapitatus*, however, showed no effect on stamina and could dive for longer periods of time than could uninfected animals (Daniels, 1985). This, too, has been supposed to be advantageous for parasite transmission to the next host, which can forage underwater. Daniels (1985) also pointed out that such hydrophilic skinks may have greater exposure to the parasite from aquatic arthropods. Thus, it is possible that skinks with greater diving stamina become infected more frequently, rather than undergoing behavioural changes after infection. Although malaria (*Plasmodium mexicanum*), a protozoan transmitted by a vector, did not affect burst running performance (Schall and Sarni, 1987) in lizards (*Sceloporus occidentalis*), it did reduce running stamina (Schall *et al.*, 1982). *Plasmodium* would not benefit from increased predation on its host, but may benefit from increased vector access (*e.g.* Day and Edman, 1983, 1984). An inspection of broken tails revealed no evidence of increased predator encounters for infected lizards (Schall *et al.*, 1982).

Fish may act as intermediate hosts for a variety of parasites that develop to maturity in piscivorous birds and mammals. Pejerrey (*Basilichthys* spp.) infected with the metacercariae of a trematode, *Diplostomum mordax*, became motionless and tumbled on the surface of the water and could have been easy prey for the cormorant definitive host (Szidat, 1969). Infection with tapeworm plerocercoids (*Schistocephalus solidus*) was associated with the stickleback, *Gasterosteus aculeatus*, moving slowly and near the water surface. Distance travelled did not differ from that of uninfected fish (Meakins and Walkey, 1975; see also Lester, 1971). Sluggish movements also characterized shiners (*Notropis cornutus*) parasitized with plerocercoids of the cestode *Ligula intestinalis* (Dence, 1958). In yet other parasite associations (*e.g.* salmon–trematode metacercariae, Butler and Millemann, 1971), effects on activity were transient and occurred in large part before the parasites were infective for the next host. Increased predation on infected animals was hypothesized in each of these systems. There was no difference, however, in swimming behaviour between juvenile bluegill sunfish (*Lepomis macrochirus*) infected with metacercariae of the trematode *Uvulifer ambloplitis* and those without trematodes, and there was no evidence of differences in predation intensity (Lemly and Esch, 1984). Russell (1980) found that the adult nematode *Truttaedacnitis truttae* did not affect the swimming speed or stamina of rainbow trout (*Salmo gairdneri*). He speculated that this neutrality might be related to the fact that *T. truttae* lives in the intestine and does not damage muscle tissue or grow rapidly enough to impede nutrition. Unlike plerocercoids or mature metacercariae, adult *T. truttae* are not transmitted by predation.

When examined, acanthocephalans often seem to increase the activity levels of their arthropod intermediate hosts (Moore, 1984a; but see Kennedy *et al.*, 1978). The aquatic isopod *Lirceus lineatus* spent more time wandering when infected with *Acanthocephalus jacksoni* and was hypothesized to be more vulnerable to predation (Muzzall and Rabalais, 1975). *Asellus intermedius* infected with *Acanthocephalus dirus* was hyperactive (Camp and Huizinga, 1979). The terrestrial isopod *Armadillidium vulgare* infected with *Plagiorhynchus cylindraceus* tended to move more than uninfected isopods (Moore, 1983a). Although amphipods (*Gammarus pulex*) infected with *Pomphorhynchus laevis* rested on vegetation more than did uninfected animals, they swam differently, spiralling towards the surface (Kennedy *et al.*, 1978). (Metacercariae of the trematode *Microphallus papillorobustus* may also be related to hyperactivity in *Gammarus* spp. (Helluy, 1983a, 1984).) Cockroach (*Periplaneta americana*) hosts of *Moniliformis moniliformis* exhibited increased activity both on running wheels (Moore, 1983b) and in open field (Wilson and Edwards, 1986) situations and were hypothesized to be at greater risk of predation than uninfected cockroaches. Edwards (1987), however, found that this was not related to parasite infectivity for the next host. Infected *A. intermedius, A. vulgare*, and *Gammarus* spp. are more likely than uninfected conspecifics to be preyed upon (laboratory tests: Kennedy *et al.*, 1978; Camp and Huizinga, 1979; Helluy, 1983a, 1984; laboratory and field tests: Moore, 1983a).

Although acanthocephalan cystacanths are transmitted by predation, increased arthropod activity can disseminate parasites in other ways. *Bombus* queens parasitized by the nematode *Sphaerularia bombi* flew almost incessantly, in a disoriented manner, stopping to dig, then moving on. In the process, they deposited nematode larvae in the soil, where they might encounter other *Bombus* (Lundberg and Svensson, 1975; Poinar and Vanderlaan, 1972). On the other hand, there was no difference in flight speed/endurance of the cicada *Magicicada septendecim* infected with the fungus *Massospora cicadina* and that of uninfected conspecifics (White *et al.*, 1983), nor did the flight duration of the bark beetle *Ips typographus* vary with nematode infection (Forsse, 1987), despite differences between the parasite taxa in host-to-host transmission.

As in vertebrate–parasite associations, reduction in invertebrate activity has also been seen as a transmission mechanism. *Gammarus zaddachi* infected with plerocercoids of the cestode *Diplocotyle* were sluggish and simply floated to the water surface where they could remain (Stark, 1965). Stark (1965) cited Dogiel and Volkova (1946) as suggesting that *G. zaddachi* might be subject to increased predation. Larval *Aedes aegypti* containing more than three *Plagiorchis noblei* metacercariae were less active than conspecific mosquitoes (Webber *et al.*, 1987a). In the laboratory, they were more susceptible to vole (*Microtus*, definitive host) predation, but not more vulnerable to guppies (Webber *et al.*, 1987b). The withdrawal response of dogwhelks (*Nucella lapillus*) was not impeded by *Parorchis acanthus* (trematode) parasitism, however. Although infected dogwhelks experience higher predation (see below), other factors contribute to this (Feare, 1971).

A variety of ants infected by dilepidid cestodes and trematode metacercariae move very slowly (cestode *Anomotaenia brevis* in *Leptothorax nylanderi*; Plateaux, 1972; Gabrion *et al.*, 1976; trematode *Brachylecithum mosquensis* in *Camponotus* spp.; Carney, 1969; *Dicrocoelium* spp. in *Camponotus compressicapus* and others; Anokhin, 1966; Frank *et al.*, 1984; citations in Carney, 1969). Parasitized *L. nylanderi* did not avoid disturbance, nor did they forage for the colony, but solicited food instead. Ants with *Brachylecithum* displayed sluggish, circling behaviour, and those with *Dicrocoelium* spp. were immobilized on vegetation, with mandibles locked in the case of some, but not all, species.

All of the ants discussed above must be eaten in order for the parasites to reach the next host. In such a life cycle, reduced ant activity is seen as increasing this probability, unless, as in *Leptothorax*, they rarely leave the nest (Plateaux, 1972). More puzzling is the fact that the snail (*Limicolaria*) first intermediate host of *Dicrocoelium hospes* also moved (and ate) very little. The snail disseminates the slime balls that contain larval trematode states infective for ants (Frank *et al.*, 1984).

Reduced activity on the part of the snail may not serve to disperse the trematode, and if the snail is eaten as a result of higher predation risk, the trematode stages it contains will not be transmitted to the next host. Thus, changes in activity in this association are not so easily linked to enhanced parasite transmission.

Vectors of parasites may also exhibit altered activity levels. The mosquito *Anopheles stephensi* infected with *Plasmodium cynomolgi* exhibited reduced flight distance and speed 12 days postinfection that positively correlated with infection severity (Schiefer *et al.*, 1977). Spontaneous flight activity of *A. stephensi* was reduced by infection with *Plasmodium yoelii* (Rowland and Boersma, 1988). Although different instruments were used, and different parameters measured, we may infer that *Plasmodium*-infected *A. stephensi* do not disperse as readily as do uninfected conspecifics.

Filarial worms also have a negative effect on their vectors. Flight activity was reduced in *A. aegypti* infected with *Brugia pahangi*, proportional to the intensity of infection (Townson, 1970; Berry *et al.*, 1986; Rowland and Lindsay, 1986; see also Husain and Kershaw, 1971). Infection with *Dirofilaria immitis* produced similar results, although unlike *B. pahangi*, this nematode develops in Malpighian tubules and not flight muscles (Berry *et al.*, 1987). Such changes may limit the dispersal of the parasites.

In summary, it is therefore not clear that altered host behaviour always benefits a parasite, despite this frequent suggestion. Other than in work with acanthocephalans, there have been few tests of predation hypotheses in systems where activity is altered (but see Schall *et al.*, 1982; Helluy, 1983a). Activity is measured in different ways by different workers and comparison is difficult; in part, this may be due to the taxonomic variety of the hosts, but even similar hosts are often subjected to different criteria. Both increased and decreased activity in similar hosts have been hypothesized to result in increased predation that benefits the parasite. Reduced activity has also been hypothesized to benefit the host (B. L. Hart, 1988). In some of the preceding instances of altered activity, the parasites (e.g. pinworms, immature metacercariae, adult intestinal nematodes) would be harmed by predation on the host, yet the behavioural changes with which they are associated are similar to those that are alleged to increase predation and favour parasite transmission in other associations. In addition, it is not clear how reduced mobility of vectors enhances parasite dispersal, and the tendency of *Leptothorax* to remain in the nest may not increase transmission.

Although a wide range of activity alterations may be envisaged as increasing the probability of predation, they may do so at different times during a predation sequence. For instance a sluggish intermediate host may have difficulty escaping a predator, fitting the Holmes–Bethel (1972) category of 'decreased stamina'. On the other hand, a hyperactive animal may be more *conspicuous* to a predator (Holmes and Bethel, 1972). If changes in activity are closely linked to increased predation, we might expect depressed activity to be important in prey that depend on escape as a primary predator avoidance mechanism (or for predator definitive hosts that chase), and we might expect hyperactivity in prey that depend upon crypsis (see also Holmes and Bethel, 1972).

Conspicuous behaviour/appearance

Alterations in colour, size, and preferences for light or high places are a few of many other ways that an animal can become more conspicuous. Examples of changes in visual conspicuousness are most abundant, perhaps reflecting human sensory bias. There are, however, examples of changes in sound. *Chironomus* in flight produced a different sound when infected with the mermithid nematode *Limnomermis* (Wulker, 1985; see also Lundberg and Svensson, 1975, for *Bombus* example). Scent may also be affected by parasites. Nesting red grouse hens (*Lagopus lagopus*) were more easily located by dogs when infected with the caecal nematode *Trichostrongylus tenuis*. This may result from inability to reduce caecal emissions, and could make them more vulnerable to predators using olfactory cues (Hudson and Newborn, 1985). Neither *Limnomermis* nor *Trichostrongylus* are transmitted by hosts that are preyed upon and predation would be disadvantageous to the parasite. As with examples of altered activity, if the alteration in conspicuousness occurs in the host of a parasite that must be transmitted by being eaten, then it is tempting to speculate that this is a result of natural selection acting on the parasite inducing the change. However, as with altered activity, such changes may occur in other parasite–host associations as well, where predation does not play a role in transmission, and may be the consequence of other aspects of the parasite–host interaction.

Moving to high places or swimming high in the water column is conspicuous behaviour and might result in increased predation (but see Bethel and Holmes, 1977; Daniels, 1985). Most of these behaviours have been observed in aquatic animals, particularly crustaceans and fish, and have involved parasites transmitted through the food chain. For instance, gammarids (*Gammarus zaddachi*) parasitized by *Diplocotyle* plerocercoids floated upwards and remained on the water surface (Stark, 1965), and *G. pulex* infected with *P. laevis* cystacanths tended to be in open water and on surface vegetation rather than burrowing in gravel (Brown and Thompson, 1986). Gammarids (*Gammarus lacustris*) infected with *Polymorphus minutus* (Acanthocephala), were likely to be in the top area of a tank (Hindsbo, 1972), as were ostracods (*Cypridopsis vidua, Physocypria pustulosa*) infected with *Octospiniferoides chandleri* (Acanthocephala: DeMont and Corkum, 1982) and mosquitoes (*A. aegypti*) infected with *P. noblei* metacercariae (Webber *et al.*, 1987a). In the ocean, chaetognaths (*Sagitta elegans*) infected with trematode metacercariae were more likely to be near the surface of the water. Not only may they be subject to different predation risks, but they themselves may also forage differently from uninfected conspecifics (Pearre, 1979).

Burrowing bivalves infected with trematode metacercariae may be closer to the surface and at greater risk of avian (definitive host) predation. *Venerupis aurea* containing *Gymnophallus fossarum* often reversed its position in the sand, uncovering its ventral side (Bartoli, 1976–78). *Macoma balthica* infected with *Parvatrema affinis* usually occurred closer to the surface and even left tracks in the sand as it moved (Swennen, 1969; Swennen and Ching, 1974). (Hulscher (1982) found that an oystercatcher (*Haematopus ostralegus*) could discriminate against and reject heavily infected *Macoma* (see also Keymer *et al.*, 1983).) Lauckner (1984) reviewed the impact of trematode parasitism on a number of molluscs. The effects ranged from a reduction of byssal threads in *Mytilus edulis* to reduced heat tolerance and possibly inefficient feeding. The estuarine gastropod *Ilyanassa obsoleta* changed its vertical location when infected with the sporocysts of *Gynaecotyla adunca* and was found stranded high on beaches, especially during nocturnal low tides (Curtis, 1987). This host does not have to be eaten for parasite transmission (see also Rothschild, 1940), but the behaviour may none the less function in transmission of the trematode, for it may allow the liberation of the next infective stage in an area where there is increased probability of encountering the next host, nocturnally active beach crustaceans. This behaviour is modified in gastropods with multiple-species trematode infections, but since the snails are castrated by the parasite, it is difficult to hypothesize a selective benefit for the snail that could accrue from the behaviour, short of increasing inclusive fitness in some way (Curtis, 1987). In contrast, *M. papillorobustus* metacercariae must be infective before eliciting behavioural changes in gammarid intermediate hosts (Helluy, 1983b; see also Bethel and Holmes, 1974, Lucius, *et al.*, 1980).

Among aquatic vertebrates, plerocercoids of *Schistocephalus* are thought to increase the conspicuousness of its fish intermediate host, the three-spined stickleback (*Gasterosteus aculeatus*). Lester (1971) observed that infected fish were more likely to frequent shallow water than were uninfected fish. He hypothesized that this could result from increased oxygen demand, and found that infected fish required more oxygen, especially if active. He suggested that this might be due, in part, to the decreased streamlining of infected fish. (The parasite accounts for 25% of the host body mass by September (McPhail and Peacock 1983).) Meakins and Walkey (1975) generally confirmed Lester's result with some discrepancies that might be attributed to methods and measurement techniques (see also Holmes and Bethel, 1972). They suggested that pressure on the swim bladder by the parasite could also alter behaviour. McPhail and Peacock (1983) claimed that the tendency of infected sticklebacks to be closer to the surface had not been well documented, and they suggested that the distended parasitized fish may simply be more conspicuous to the human observer, who may then 'see' more of them near the surface. Lester (1971), however, reported a much higher proportion of infected fish in catches near the shore than in pelagic samples. (Differential susceptibility might also account for this (Lester, 1971).) Smith and Kramer (1987) found that nine-spined sticklebacks (*Pungitius pungitius*) in hypoxic conditions swam closer to the surface than did uninfected conspecifics, and when threatened by a model of an avian predator, they spent less time away from the surface. Smith and Kramer noted that infected three-spined sticklebacks also swam closer to the surface in well-oxygenated water (see above) and suggested that other factors such as temperature and food could play a role in this effect. Dace (*Leuciscus leuciscus*) with heavy infections of *Diplostomum spathaceum* also spent more time near the surface of the water (Crowden and Broom, 1980). In an apparent reversal of this general tendency, Poulin and FitzGerald (1989a) found that juvenile three-spined sticklebacks swam near the surface and avoided ectoparasitic branchiurans (*Argulus canadensis*) that were present in the otherwise preferred microhabitat (vegetation, bottom). (See also Downes *et al.*, 1986, for an example of terrestrial animal habitat choice and possible ectoparasite influence.) Although parasites are often cited as causative agents in strandings of marine mammals, Geraci and St. Aubrin (1987) in a review of that literature concluded that few studies provided firm evidence. *Nasitrema*, a trematode in the cerebrum is thought to have caused some cetacean strandings in southern California (Dailey and Walker, 1978).

Reports of height-seeking in terrestrial animals are less abundant. The ant intermediate hosts of *Dicrocoelium* often crawl up on vegetation before becoming motionless (Romig *et al.*, 1980; Anokhin, 1966, and references therein) as do ants infected with the fungus *Entomophthora* (Loos-Frank and Zimmerman, 1976, cited in Romig *et al.*, 1980). In the case of *Dicrocoelium*, accidental predation on ants transmits the parasite; this is not true of the fungus, but spores may be more widely dispersed from heights. Of five species of gastropods infected with protostrongylid nematodes, only one became negatively geotactic with increasing parasite load (Cabaret, 1984). This behaviour could increase transmission of the nematode to the next herbivorous host. Unfortunately, behaviour of uninfected animals was not reported, and the only species to show this response was also the most heavily infected.

In most of these systems, aquatic and terrestrial, height seeking behaviour has been hypothesized to lead to increased transmission to the next host. Tests of increased transmission, though rare, have confirmed this hypothesis (e.g. Hindsbo, 1972; Brown and Thompson, 1986; Webber *et al.*, 1987b).

Shapiro (1976) noted that some caterpillars that are hosts to parasitoids climb up vegetation. If this resulted in increased predation and thus destroyed the parasitoid, Shapiro suggested that the behaviour could benefit the caterpillar's kin. Smith Trail (1980) expanded this idea to suggest a kin-selected origin for altered host behaviours. She suggested many responses currently seen as adaptive for the parasite might have arisen as host suicide, possibly adaptive for host kin and increasing the inclusive fitness of the host. Stamp (1981) found that height-seeking did not benefit *Euphydryas phaeton* (Lepidoptera) parasitized by *Apanteles euphydryidis*, a braconid wasp. High locations benefitted the parasitoid, which was less subject to hyperparasitism in those locations and which could have found mates more easily upon emergence in high places (see also Moore, 1984a).

Height-seeking behaviour can be confounded with *photophilia* or positive phototaxis in many situations and some studies seek to separate these influences. Hindsbo (1972), Kennedy *et al.* (1978) and Brown and Thompson (1986) used ventral lighting to show that positive phototaxis, not negative geotaxis, was being expressed in parasitized animals. Hindsbo (1972) varied the direction of the light source and observed consistently positive phototaxis in infected gammarids. DeMont and Corkum (1982) found similar results with infected ostracods by varying lighted areas in a petri dish. Gammarids infected with *M. papillorobustus* also moved up towards light (Helluy,1983a, 1984). The response to light at the bottom of the chamber was not as strong, indicating a negative geotactic influence as well.

Bethel and Holmes (1973) found a variety of responses to light in several host–parasite systems. Amphipods infected with cysticercoids of *Lateriporus* spp. or acanthocephalan acanthellae behaved like uninfected amphipods; they were photophobic and negatively phototactic. Infective acanthocephalan cystacanths were associated with changes in these responses. In general, photophilia was increased, but responses to light intensity and disturbance varied among the host–parasite combinations in a way that could contribute to predation by a variety of definitive hosts. This was supported by laboratory predation tests (Bethel and Holmes, 1977).

Moore and Lasswell (1986) found that *A. vulgare* infected with the juvenile nematode *Dispharynx nasuta* were observed more often in the lighted side of a behavioural arena than were uninfected conspecifics. The same host infected with cystacanths of *P. cylindraceus* showed no change in response to light (figure 7.1) (Moore, 1983a). When suddenly exposed to light, American cockroaches infected with cystacanths of *M. moniliformis* were more likely to move toward a light source

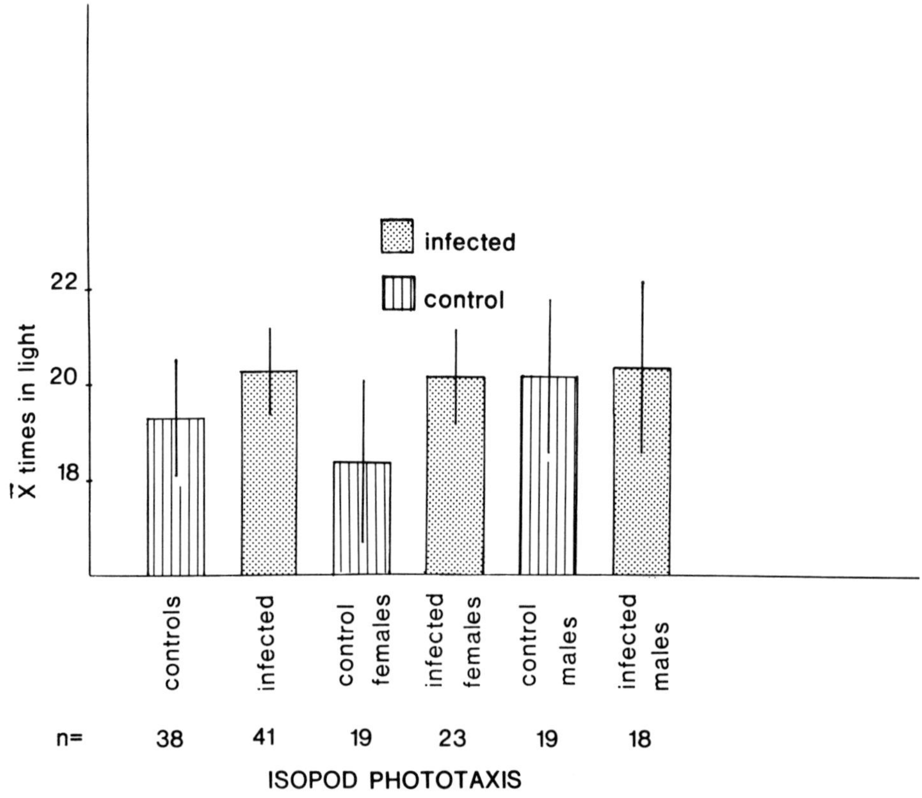

Figure 7.1. Number of times *Armadillidium vulgare* infected with *Plagiorhynchus cylindraceus* and uninfected isopods were observed in the lighted half of a behavioural arena. Total observations per isopod = 40. Data from Moore (1983a).

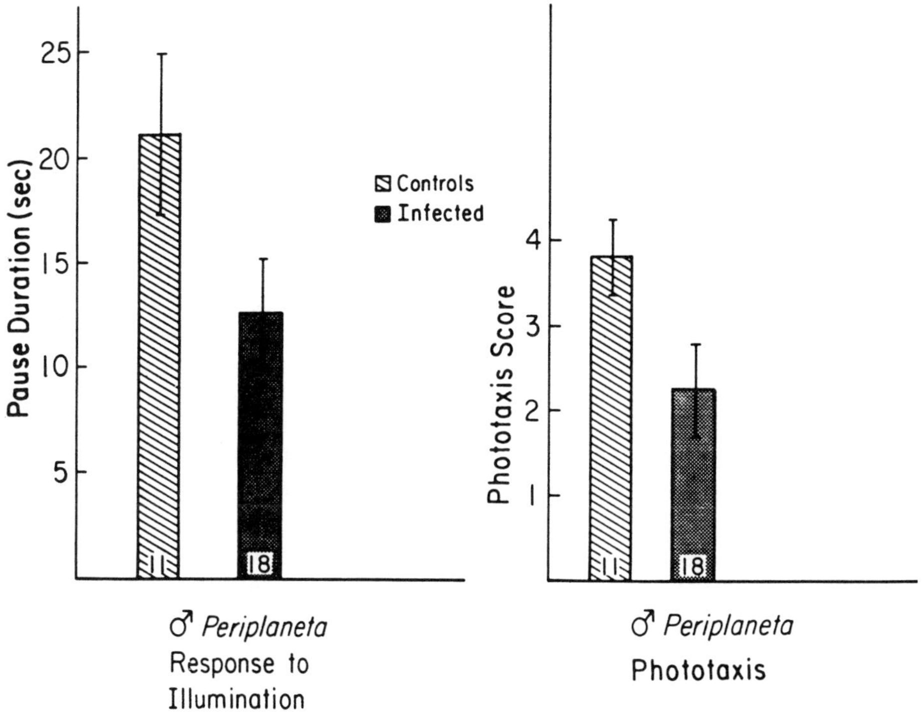

Figure 7.2. Responses of *Periplaneta americana* infected with *Moniliformis moniliformis* and uninfected cockroaches to sudden illumination. Phototaxis score has a range of 0–6: 0 = directly toward the light; 6 = directly away from the light. Data from Moore (1983b).

than were uninfected animals (figure 7.2). They also paused longer before doing so (Moore, 1983b).

The sporocysts of the trematode *Leucochloridium* in the tentacles of the snail *Succinea putris* pulsated in response to light and ceased movement in the dark (Wesenberg-Lund, 1931; see also Lewis, 1977). Although this is thought to attract the attention of bird predators, Wesenberg-Lund watched birds in the vicinity of *Leucochloridium* and saw no predation. Lewis (1977) said that there was no experimental verification of increased predation in this system. Some of the other host–parasite associations have been shown to be more subject to predation, however, supporting the hypothesis that alterations in those associations benefit the parasite (e.g. some gammarid–acanthocephalan associations, *Armadillidium–Plagiorhynchus*).

Some scientists have investigated increased exposure of parasitized hosts in ways that are not easily categorized. For instance, infected *Lirceus lineatus* spent less time under leaves than did uninfected isopods and tended to crawl over vegetation (Muzzall and Rabalais, 1975). Unlike gastropods infected with *Leucochloridium* or *Neoleucochloridium*, which do not move to more visible locations (Lewis, 1977;

but see Ulmer, 1971), dogwhelks infected with the trematode *P.acanthus* delayed forming winter aggregations in the fall and therefore incurred a higher risk of oystercatcher predation on the open shore (Feare, 1971). A similar inhibition may exist in the gastropod *Nassarius obsoletus* when infected with trematodes (Stambaugh and McDermott, 1969). These snails were shedding cercariae, however, and therefore behavioural alterations may not enhance parasite transmission. *Littorina littorea* infected with *Cryptocotyle lingua* sporocysts appeared to be more sedentary than uninfected conspecifics in a Nova Scotia study (Lambert and Farley, 1968; see also Sindermann and Farrin, 1962). Infected snails were generally lower in the intertidal zone, but responded to freezing air more slowly, if at all. Thus during freezing conditions, the infected animals remained relatively high in the intertidal zone. These snails were shedding cercariae on occasion, however, and therefore were probably not infective for predators. Likewise, *Nucella* is the first intermediate host for *Parorchis*, and in most cases, predation on the first intermediate host of a trematode does not continue the life cycle. Although lamellibranchs have been reported as second intermediate hosts for *Parorchis* (Lebour and Elmhirst, 1922; Angel, 1954, cited in Schell, 1985), the cercariae may encyst on the shells of molluscs, including *Nucella* (Stunkard and Cable, 1932; Rees, 1937 as cited by Dawes, 1946). They would be infective in this state. It is questionable, however, whether small metacercariae encysted on the outer surface of a mollusc shell could influence winter aggregation behaviour. Sporocysts growing internally are more likely to have influence, and they are not usually infective for birds.

Ants containing the metacercariae of *B. mosquensis* moved in circles in open spaces and stayed out of the nest when temperatures declined in the autumn. They did not respond to changes in light (Carney, 1969). *Armadillidium vulgare* infected with *P. cylindraceus* responded randomly to overhanging shelter and were more likely to be in areas of relatively low humidity; uninfected isopods were likely to be under the shelter (figure 7.3) and chose more humid environments (figure 7.4) (Moore, 1983a). Light responses were not altered in this association and could therefore not account for the behavioural change, There was no difference in the shelter responses of isopods infected with *D. nasuta* (Moore and Lasswell, 1986). Increased risk of predation was predicted for the parasite–host associations that demonstrated these alterations.

Conspicuous behaviour such as photophilia may be confounded with *heat-seeking behaviour*. Behavioural fever has been observed in vertebrates (B. L. Hart, 1988) and in invertebrates (Boorstein and Ewald, 1987; McClain *et al.*, 1988). Moving towards heat may be advantageous for the host (D. Horton, unpublished, but see Vernberg, 1969).

Parasitized hosts may become conspicuous because of the way they look as well as by the way they behave, and changes in colour can affect the heat exchange of the host, especially in terrestrial animals. Sometimes the parasite itself is part of these *changed appearances*. *Rickettsiella* (Rickettsiaceae) appears as white masses visible through the exoskeleton of terrestrial isopods (El-Al and Holdich, 1987). Metacercariae form visible white cysts in chaetognaths (Pearre, 1979), as do

Schistocephalus plerocercoids in sticklebacks (Milinski, 1985). Many cystacanths can be seen through the exoskeleton of the host, appearing as coloured dots (Bethel and Holmes, 1977; Brown and Thompson, 1986). Moreover, *Asellus aquaticus* infected with cystacanths of *Acanthocephalus lucii* have darker respiratory operculae (Brattey, 1983). Some acanthocephalan infections result in pigmentation dystrophy (Oetinger and Nickol, 1982) of aquatic hosts, which may then be more conspicuous in their customary habitats (e.g. Hindsbo, 1972; Seidenberg, 1973; Muzzall and Rabalais, 1975; Camp and Huizinga, 1979, 1980; Oetinger and Nickol, 1981). Isopods infected with *P. cylindraceus* do not alter their colour, but do alter choice of substrate colour. The dark isopods, when infected, are more likely to be found on light-coloured substrate (figure 7.5) where they are easily seen (Moore, 1983a). In some cases, parasites may actually mimic food items (Jarecka, 1961).

If more than one behaviour is altered, then most predation tests cannot reveal the relative contribution of each alteration to the ecological result (predation). Some alterations in appearances are more amenable to such investigation. By applying paint to the carapaces of uninfected amphipods in a way that mimicked

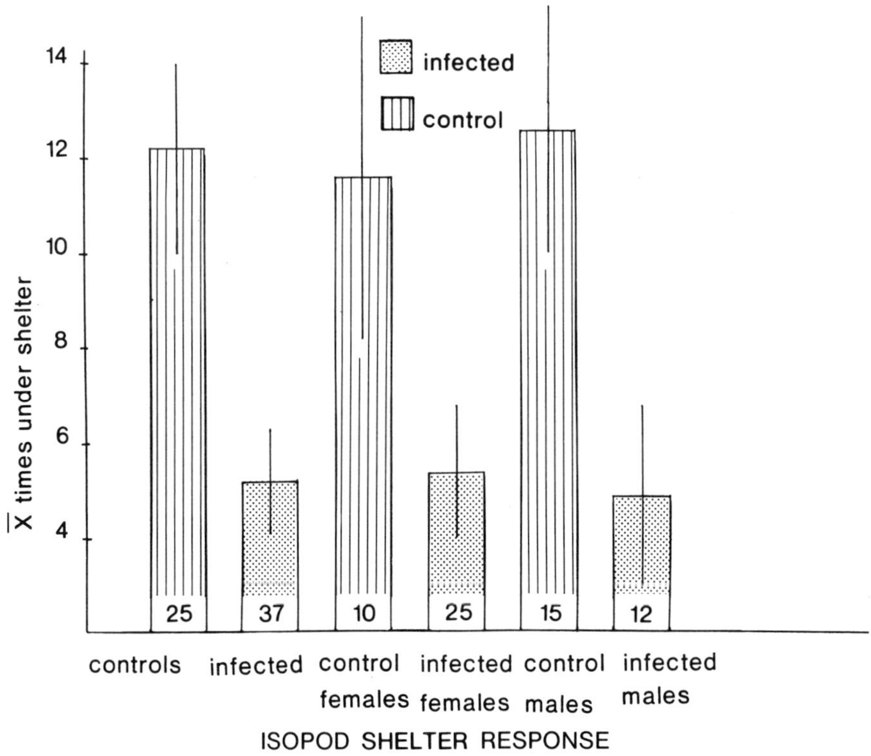

Figure 7.3. Number of times *A. vulgare* infected with *P. cylindraceus* and uninfected isopods were observed under a shelter. Total observations per isopod = 30. Data from Moore (1983a).

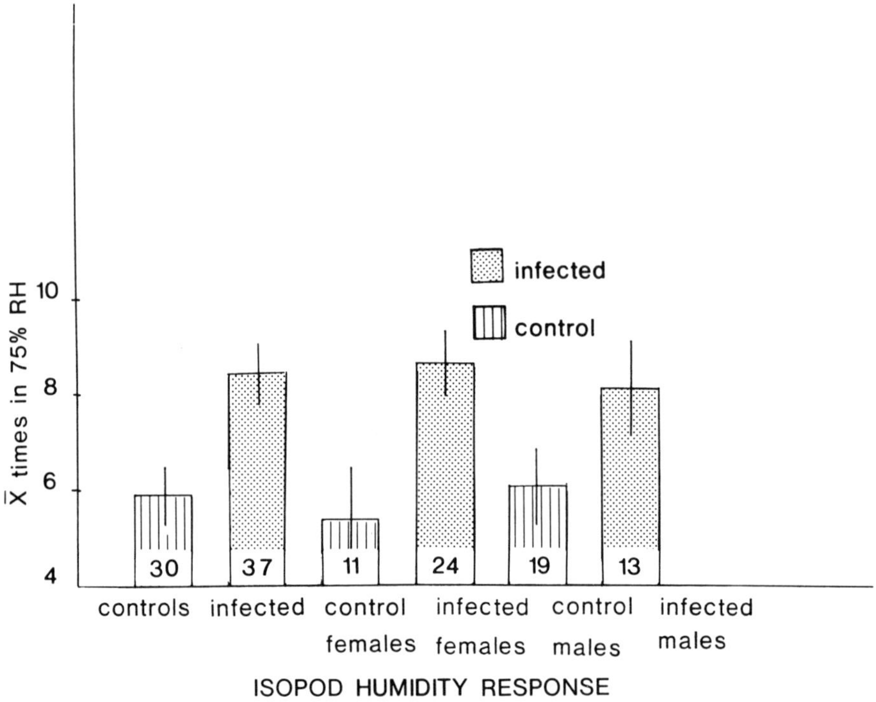

Figure 7.4. Number of times *A. vulgare* infected with *P. cylindraceus* and uninfected isopods were observed in 'low' (75% relative humidity at 24–25°C) humidity. 'High' = 96% relative humidity at 24–25°C. Total observations per isopod = 30. Data from Moore (1983a).

the appearance of the cystacanths, Bethel and Holmes (1977) and Brown and Thompson (1986) were able to show that altered appearance alone did not influence predation frequency. Camp and Huizinga (1979) showed that isopod contrast with substrate colour influenced predation. The infected, light-coloured isopods were eaten in disproportionately high numbers when against a dark substrate similar to their natural environment. This susceptibility decreased in an aquarium with light substrate. (These isopods were not given a substrate choice.)

The distension of the gaster in *Camponotus* infected with *B. mosquensis* resulted in a striped appearance, caused by the exposure of the intersegmental membranes (Carney, 1969). Perhaps the most dramatic alteration of all, however, was that observed by Plateaux (1972) in dilepidid-infected *Leptothorax nylanderi*. Size and shape of body parts were changed, as well as colour (yellow) and patterns. The differences were so great that infected ants were thought to be a new species, and the parasite association was discovered during study prior to the species description! Plateaux (1972) noted that not all ants are so affected by their parasites, but he wondered how many inquiline species had been described that were actually conspecifics of their 'hosts'.

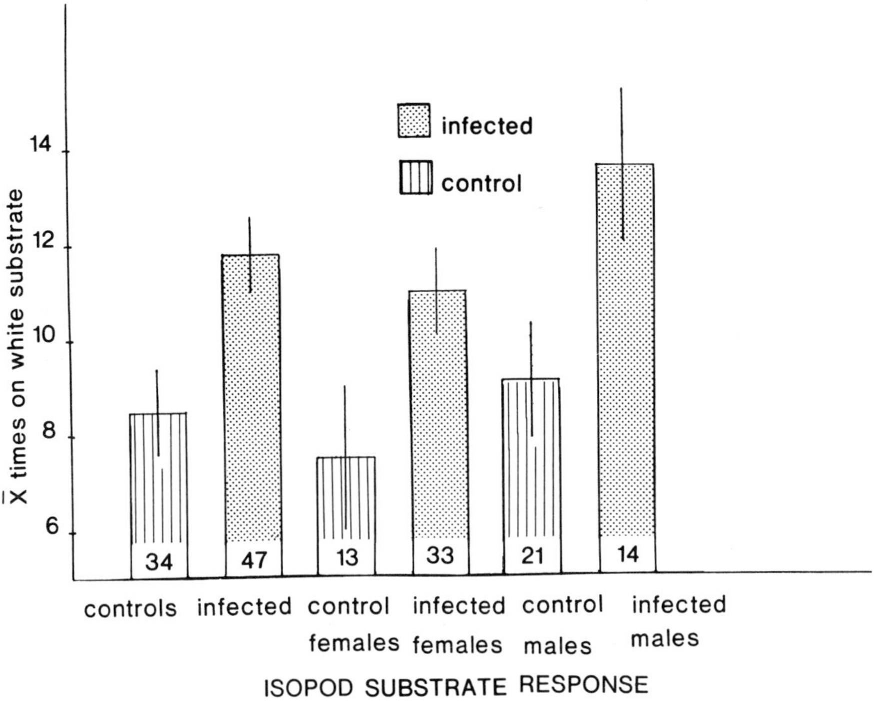

Figure 7.5. Number of times *A. vulgare* infected with *P. cylindraceus* and uninfected isopods were observed on the light substrate half of a behavioural arena. Total observations per isopod = 30. Data from Moore (1983a).

Changes in size may also increase conspicuousness and affect heat balance. Of course, many parasitic infections result in weight loss along with other symptoms of general illness. On the other hand, some parasitic infections result in the host's becoming larger. Perhaps best-known among these is the effect of *Spirometra mansonoides* plerocercoids on rats. The plerocercoids release a hormone mimic that causes the rodents to become noticeably large and fat (Mueller, 1974). As mentioned before, *S. solidus* distends the abdomen of its fish intermediate host, as does *L. intestinalis* (Arme and Owen, 1967; Sweeting, 1977). Due to an increase in parenchyma, the gaster of *Camponotus* infected with *B. mosquensis* is also abnormally large (Carney, 1969).

As with altered activity, changes in conspicuousness, whether height seeking or colour change, can be hypothesized to increase predation rates and be advantageous for a parasite. Tests of this hypothesis are rare, though in many cases it is none the less a reasonable assumption. On the other hand, some of the alterations are not associated with transmissible stages of the parasite, and in the case of heat-seeking, may even be detrimental to the parasite.

Changes in social behaviour

Increased size is often associated with *parasitic castration*. This phenomenon is especially noticeable in some snails (see also Stark, 1965). Sousa (1983) hypothesized that the occurrence of gigantism in castrated gastropods may be linked in some circumstances to host life history: young snails do not manifest gigantism because castration releases few resources. On the other hand, short-lived or semelparous species devote large amounts of energy to reproduction; failure to reproduce may result in gigantism in these species more often than in iteroparous, long-lived snails (see also Minchella, 1985).

Pearre (1976) noted that chaetognaths parasitized by hemiurid metacercariae were partially castrated and exhibited gigantism. He commented that the explanations of sporocyst-induced castration based on energetics would be less reasonable with metacercariae, which do not require as much energy for development. Moore (1983a) found that although female isopods were castrated by *P. cylindraceus* cystacanths, they did not grow as rapidly as uninfected females, possibly because of the energy required by the relatively large cystacanth.

Although large animals may be conspicuous and therefore at greater risk of predation, castration itself is often seen as resulting in a lower risk of predation. For instance, there is no investment in competition for mates or provisioning young (Baudoin, 1975). Gastropods are frequently castrated by trematode parasites (e.g. Wesenberg-Lund, 1931; Lauckner, 1984; Curtis, 1985), as are many arthropod hosts of acanthocephalans (Hynes, 1955; Schmidt and Olsen, 1964; Rumpus and Kennedy, 1974; Oetinger and Nickol, 1981; Brattey, 1983; Moore, 1983a) and our previous discussion of altered behaviours has included some associations that render the host infertile (e.g. Palm, 1948; Plateaux, 1972; Pearre, 1976). Rudd (*Scardinius erythrophthalmus*) infected with plerocercoids of *L. intestinalis* did not display spawning behaviour (Orr, 1966) but a similar cestode, *S. solidus*, had no effect on sticklebacks (McPhail and Peacock, 1983). Even male porcellanid crabs (*Petrolisthes cabrilloi*) are induced by a parasitic barnacle to perform 'parental' behaviours for the benefit of the parasite (Ritchie and Hoeg, 1981; see also Bishop and Cannon, 1979)! Although often thought to promote host longevity, castration occurs in many hosts that exhibit altered behaviours thought to shorten their lifespans. As with gigantism, a related phenomenon, the beneficiary is not always clear (Minchella, 1985).

Mating behaviour itself can be adversely affected by parasites. Cade and Wyatt (1984) suggested that the presence of acoustically-orienting parasitic flies (*Euphasiopteryx ochracea*) may limit calling in one species of the cricket *Gryllus campestris*. Wild-caught female *Gryllus* spp. were more likely to be paired with older males that exhibited less gregarine parasitism (Zuk, 1988) and, as a correlate, greater spermatophore production. Female grasshoppers (*Orchelimum gladiator*) infected with nematomorphs were less likely than controls to mate or respond to mating stimuli (Morris *et al.*, 1975). Male *Drosophila testacea* infected with the nematode *Howardula aoronymphium* were less frequently mated than uninfected males (Jaenike, 1988). Female mice infected with *T. spiralis* were less likely to respond favourably to mating attempts (Edwards and Barnard, 1987). Mating behaviour may also be influenced over evolutionary time if mate choice is influenced by genetic disease resistance (Hamilton and Zuk, 1982; see also Freeland, 1976; Borgia, 1986; Read, 1987; Ward, 1988; see Chapter 7). Female rock doves (*Columba livia*) avoided mates with heavy louse infestations. Because these ectoparasites can be transferred to other doves by direct contact, however, rejection of infected males may indicate contamination avoidance instead of or as well as selection for apparent genetic quality (Clayton, in press, see Chapter 5). Male *G. pulex*, however, did not discriminate between control females and those infected with *P. laevis* (Poulton and Thompson, 1987).

Social dominance, important in both competition and mating, has also been linked to parasitic infection. As with competition (see below), in field studies, it is often unclear if parasitism influences social status or *vice versa*. For instance, it was noted that most diseased red grouse were not territorial; however, the cause of this association was not determined (Jenkins *et al.*, 1963). The opposite trend occurred in baboons (*Papio cynocephalus*), with high ranking males yielding more parasite propagules (Hausfater and Watson, 1976). The heaviest reindeer (*Rangifer tarandus*) calves (dominant mothers) were most likely to be infected with protostrongylids, possibly because of different forage (Halvorsen, 1986), and the heavier starling (*Sturnus vulgaris*) nestlings in a brood tended to be the ones infected with *P. cylindraceus*, perhaps because they begged more successfully and therefore had increased probability of exposure to infected prey (Moore and Bell, 1983). In the laboratory, the development of dominance in male mice was impeded by the nematode, *Heligmosomoides polygyrus* (Freeland, 1981). Rau (1982, 1983b, 1984b) showed that *T. spiralis* in mice caused subordinance even on home territory and reversed previously existing dominance. He suggested that subordinate mice may have an increased risk of predation if expelled from a home area. Except for *T. spiralis*, none of these parasite stages is transmitted by predation.

Other types of social interactions are also affected by parasites. Infected (*Plasmodium*) male lizards devoted less time to social interaction than did uninfected lizards (Schall and Sarni, 1987). Edwards (1988) found that mice infected with *T. spiralis* initiated fewer interactions. American cockroach response to aggregation pheromone decreased with the number of *M. moniliformis* cystacanths in the roach (Wilson and Edwards, 1986). On the other hand, abnormal swimming in guppies (*Poecilia reticulata*) induced by pathology associated with the monogenean *Gyrodactylus bullatarudis*, attracted other guppies and enhanced parasite transmission (Scott, 1985). Schools of minnows (*Pimephales promelas*) infected with metacercariae of *Ornithodiplostomum ptychocheilis* were less compact and cohesive and swam closer to the surface of the water. These metacercariae encyst in the brain and may disrupt the anti-predator function of schooling behaviour (Radabaugh, 1980). Shiners (*N. cornutus*) parasitized by *L. intestinalis* also left schools or formed smaller groups (Dence, 1958).

Several authors have suggested that parasites have affected the *evolution of host social behaviour*. Hamilton (1987) suggested that disease influences host social behaviour at several levels, from short-term behaviours (e.g. grooming) to long-term effects (e.g. sexuality; see also Seger and Hamilton, 1988). Sherman *et al.* (1988) hypothesized that multiple mating in social Hymenoptera might contribute genetic variation and favour disease resistance. The elements of primate sociality (e.g. territoriality, xenophobia, etc.) might inhibit the spread of disease (Freeland, 1976). Yellow baboons (*P. cynocephalus*) alternate sleeping trees in a way that is consistent with a hypothesis of parasite avoidance (Hausfater and Mead, 1982). Although increased disease transmission is often seen as a cost of social behaviour, juvenile sticklebacks (*Gasterosteus* spp.) may form larger groups to reduce individual probability of parasitization by the branchiuran, *A. canadensis* (Poulin and FitzGerald, 1989b) and individual horses in larger herds experience fewer tabanid fly bites (Duncan and Vigre, 1979, see also Chapter 10).

The relationship of social behaviour to predation and transmission is unclear in many instances. At least some castrating parasites are not transmitted through the food chain; others are, and the alleged role of castration in prolonging host life is of dubious advantage to these parasites. Social subordination may predispose a host to predation, yet both subordination and dominance have been associated with parasite stages that are not transmitted to predators. While disruption of schooling may well increase predation, other changes in social behaviour are puzzling. Social reticence in *T. spiralis*-infected mice and *M. moniliformis*-infected cockroaches may be construed as affecting predation, but similar behaviour in the lizard host of *Plasmodium* may not enhance transmission, which in this case is mediated by a vector.

Changes in foraging behaviour

The *foraging behaviour* of hosts can be altered by parasites. Over evolutionary time, parasite avoidance may have been one factor contributing to a widespread tendency among some animals to eschew faecally contaminated food (e.g. Freeland, 1980). Depressed feeding is common among parasitized vertebrates (Symons and Jones, 1978; Crompton 1984; but see Crompton *et al.*, 1985) and may be a host defense (B. L. Hart, 1988; but see Symons, 1969). Rat food intake was negatively correlated with dose of *Nippostrongylus brasiliensis* (Ovington, 1985). Adult intestinal *T. truttae* did not affect trout (*S.gairdneri*) food intake (Russell, 1980). Dace (*L. leuciscus*) infected with *D. spathaceum* were inefficient feeders (Crowden and Broom, 1980), as were sticklebacks infected with *Glugea*, a microsporidian, and/or *S. solidus* (Milinski 1984; but see Chapter 4). Moreover, sticklebacks infected with *S. solidus* took more risks in the presence of a predator (not a definitive host) in order to feed (Milinski, 1985). They recovered more quickly from a fright stimulus, and unlike controls, they would often then resume feeding (Giles, 1983).

Invertebrate feeding behaviour is also subject to parasite interference. Parasitic flies (*Apocephalus* spp.) interfered with foraging behaviour in the ant, *Pheidole titanis* (Feener, 1988). In addition, parasites may alter attraction to food. After the breeding season, gastropods (*I. obsoleta*) infected with trematodes shifted their response to carrion in several ways, depending on the trematode species (Curtis,

1985). Snails (*Helisoma trivolvis*) infected with the trematode *Echinostoma revolutum* were not as attracted to lettuce as were uninfected snails (Boland and Fried, 1984; see also Etges, 1963). *Biomphalaria glabrata* (Gastropoda) infected with *Schistosoma mansoni* (Digenea) fed more often than controls and did so at different times of day (Williams and Gilbertson, 1983).

The parasitoid *Hyposoter exiguae* depressed feeding in the noctuid *Trichoplusia ni*, but food assimilation was increased (Thompson 1982a,b, 1983b, 1985; see also Hunter and Stoner, 1975). With different host instars, food and methods, Beegle and Oatman (1974) found no difference in food consumption between parasitized and control *T. ni.* The protozoan parasite *Nosema* has highly variable effects, ranging from reduced (Johnson and Pavlikova, 1986) to increased consumption (Moffett and Lawson, 1975), with some differences attributable to host sex (Oma and Hewitt, 1984).

Differences are just as pronounced for vectors that transmit parasites when they feed (for a review, see Molyneux and Jefferies, 1986). In almost all cases, infected vectors probe more. Trypanosomes interfered with both mechanoreception and blood flow in the tsetse fly *Glossina* (Molyneux and Jenni, 1981; but see Moloo, 1983; Moloo and Dar, 1985). *Aedes triseriatus* infected with LaCrosse virus may fail to engorge even though they probe more (Grimstad *et al.*, 1980). When *A. aegypti* was infected with *Plasmodium gallinaceum*, mosquito salivary apyrase activity declined and probing increased (Rossignol *et al.*, 1984). *Phlebotomus doboscqi* (sand fly) infected with the protozoan *Leishmania major* began to probe more (failing to engorge) when the parasite had spread anteriorly (Beach *et al.*, 1985; see also Killick-Kendrick *et al.*, 1977; Killick-Kendrick and Molyneux, 1981). Probing also increased in the triatomine bug, *Rhodnius prolixus*, infected with *Trypanosoma rangeli* (Añez and East, 1984). Although multiple probing usually enhances parasite transmission, the vector itself may experience reduced fecundity if increased probing is associated with failure to engorge (Rossignol *et al.*, 1986), and reduced vector mobility (see above) may reduce transmission.

Competitive interactions may also be subject to parasitic influence. The outcome of competition between two species of flour beetle was reversed in the presence of a protozoan parasite (Park, 1948). In a weevil interaction, the presence of a parasitoid promoted coexistence (Utida, 1953). Interspecific ant competition was influenced by the presence of a parasitic phorid fly (Feener, 1981), and wild-caught parasitized sticklebacks took less preferable prey and were poor intraspecific competitors with uninfected conspecifics (Milinski, 1984).

PHYLOGENY AND ALTERED HOST BEHAVIOUR

Much of the evidence reviewed here is consistent with the idea that alterations of host behaviour are an adaptation that increases parasite transmission. These alterations can have important consequences not only for the survival of individual hosts, but also for the dynamics of host and parasite populations (Dobson, 1988) and are among several 'strategies' observed in parasite transmission (Combes, 1980).

Most evolutionary considerations of altered behaviour have emphasized the roles of natural selection and adaptation (of either parasite or host) as the key to understanding. In many cases, host and parasite can be defended equally well as the beneficiary of the alteration. Are these responses beneficial to the host? The parasite? Perhaps both, depending on circumstances? And regardless of current beneficiary, how did they evolve? Yet another hypothesis, often ignored, is that 'non-adaptive' mechanisms may have led to the same behavioural modifications (Gould and Lewontin, 1979; Antonovics, 1987). This does not necessarily imply that alterations are currently of little adaptive significance or maladaptive, although these are possibilities. Rather, in some cases, alterations may have arisen during the evolution of host or parasite taxa, and therefore may reflect the history of the lineage, rather than particular adaptations for parasite transmission in a particular host (see also Pagel and Harvey, 1988).

Thus, an important alternative to the transmission hypothesis is that behavioural modifications of intermediate hosts are phylogenetically constrained. That is, certain behavioural correlates of infection by a parasite characterize related hosts or parasite taxa in a clade. Of course, the behavioural responses *may* have been adaptive in the ancestral association in which they arose but regardless of their origin, they now characterize descendants of that association and *may* or *may not* have current adaptive significance. Brooks (1980, 1985, 1988), Dobson (1985), McLellan *et al.* (1988) and Carpenter (1989) discuss the importance of a phylogenetic perspective in the analysis of parasite or behavioural traits, and phylogenetic constraint may be considered an appropriate 'null hypothesis' in evolutionary studies (e.g. Kochmer and Handel, 1986, see also Coddington, 1988).

A phylogenetic perspective is often missing from ecological analyses (Brooks, 1980, 1985; Lambshead and Paterson, 1986), but has recently been incorporated into studies of behavioural and developmental traits (Dobson, 1985, Moore and Brooks, 1987). For example, gregariousness of butterfly larvae (Sillen-Tullberg, 1988), insect phytophagy (Mitter *et al.*, 1988) and wasp social behaviour (Carpenter, 1989) have been analyzed in the context of phylogeny. Because of possible coevolutionary history, phylogenetic constraints are particularly important in any analysis of parasite–host associations (Brooks, 1987, 1988; Glen and Brooks, 1986; Brooks *et al.*, 1981; J. A. Hart, 1988; Hafner and Nadler, 1988).

Ideally, the existing data on parasite alterations of behaviour could simply be compared to a known host phylogeny to see if there are phylogenetic patterns to the alterations. Unfortunately, there are at least two obstacles to this approach. First, the existing host–parasite behavioural data have been collected with a variety of different methods for disparate host and parasite taxa. Consequently, these data are not suitable for a comparative analysis. To date, there has not been a proper comparative study of behavioural alterations in several related host species, yet there is ample reason to believe a comparative study would prove illuminating.

For instance, Bethel and Holmes (1973) reported that the behaviours they investigated with *G. lacustris* infected by the cestode *Lateriporus* were not altered. These amphipods exhibited a wide range of altered behaviours when infected with *Polymorphus paradoxus* and *P. marilis*. Some, but not all, of the alterations were similar for the two parasites. In another system, Rau (1983a, 1984a) found that unlike *Trichinella spiralis, T. pseudospiralis* in mice had no effect on exploratory behaviour and a reduced effect on ambulatory behaviour. Moore and Lasswell (1986) found no similarity in the behavioural results of nematode infection *versus* acanthocephalan infection of *Armadillidium vulgare*. They concluded that in at least one of those associations, therefore, the response was not a generalized host response to infection.

Using the same parasite in different species of *Gammarus*, Helluy (1983a) found that infection sites and correlated behavioural changes could vary with host species. Jones (1985) suggested that all species of the braconid parasitoid, *Chelonus*, induce precocious host metamorphosis and that this could be viewed as a property of the subfamily. The ant-dicrocoeliid combinations reviewed above hold promise as a behaviourally interesting system of related host taxa and related parasite taxa.

A second obstacle to testing the phylogenetic constraints hypothesis is the lack of appropriate statistical methods. Although there have been some important recent advances in the analysis of comparative phylogenetic data (e.g. Clutton-Brock and Harvey, 1984; Cheverud *et al.*, 1985; Brooks, 1988; Pagel and Harvey; 1988), none of these techniques is precisely suited to the analysis of altered host behaviours (but for a promising new approach, see Grafen, in press).

In the following section, we outline some statistical tests for phylogenetic constraints. Before presenting these tests, we note two caveats. First, none of the host–parasite data we have reviewed are adequate for a quantitative analysis of the phylogenetic constraints hypothesis. Therefore, we have been forced to illustrate our methods with a hypothetical data set. We hope this exercise will encourage parasite ecologists to collect these sorts of comparative data.

Second, the statistical tests are tests of pattern, not mechanism. That is, each test reveals whether a particular pattern of altered host behaviours is likely to have arisen independently of host phylogeny. If there is evidence of non-independence, a phylogenetic constraint is certainly one potential mechanism. But one could also argue that an association between phylogeny and behaviour is adaptive and maintained by natural selection on all of the host–parasite associations. No statistical test can directly answer this question, but considerable insight can be gained by examining the ecological circumstances of the altered host behaviour.

For example, if a suite of related intermediate host species shows altered behaviour, but some of the hosts do not coexist with the predator final host, the argument for a phylogenetic constraint would seem compelling. On the other hand, if only some of those intermediate host species show altered behaviour, and these also happen to be the only ones to coexist with the predator final host, the selection argument would be supported. Finally, if all the intermediate host species show altered behaviour and all coexist with the final host, both arguments would be strengthened. There is the possibility that more than one mechanism may be influential, in which case the interpretations will not be so clear-cut (Hilborn and Stearns, 1982).

With this work in mind, we propose three successive tests for analyzing behavioural alterations in host taxa with known or hypothesized phylogeny. Each test is increasingly more sophisticated and precise, but also entails more underlying assumptions about the data.

Qualitative similarity analysis

Figure 7.6 shows the hypothetical cladogram for ten host taxa that will be analyzed. Such a cladogram might be derived from a quantitative phylogenetic analysis or from more traditional methods. Our analysis assumes that comparable behavioural tests, using both parasitized animals and appropriate controls, have been conducted for all ten host taxa. Clutton-Brock and Harvey (1984) discuss several considerations in the design of comparative experiments of this sort.

For a given variable, such as activity or photophilia, suppose we have measured a significant response to parasitism in taxa H, I, and J, but not in taxa A–G (figure 7.6). Could this pattern represent a phylogenetic constraint, or might it arise by chance? We provide a preliminary answer to this question with a simple combinatorial analysis (Feller, 1968), as follows.

First, note that there are $\binom{10}{3} = 120$ unique three-membered combinations that can be derived from a ten-host cladogram, regardless of its branching pattern. In general, if there are x host taxa, of which y show the altered response, then there are $\binom{x}{y}$ unique combinations, where $\binom{x}{y} = x!/y!(x-y)!$. By inspection, we find that only one of these combinations represents a monophyletic grouping (HIJ). So, the probability of observing a response to parasitism by three taxa, all in the same clade, is $1/120 = 0{\cdot}0083$. Since this is unlikely to occur by chance, we conclude that there is probably a phylogenetic constraint in operation. On the other hand, suppose parasitic alterations were only found in taxa A and B. In this case, there are three other taxon pairs that form monophyletic groups (CD, EF, and IJ), so the probability is $4/45 = 0{\cdot}089$, which is at best marginally non-significant. Table 7.1 gives all the other combinations with their respective probabilities for this cladogram.

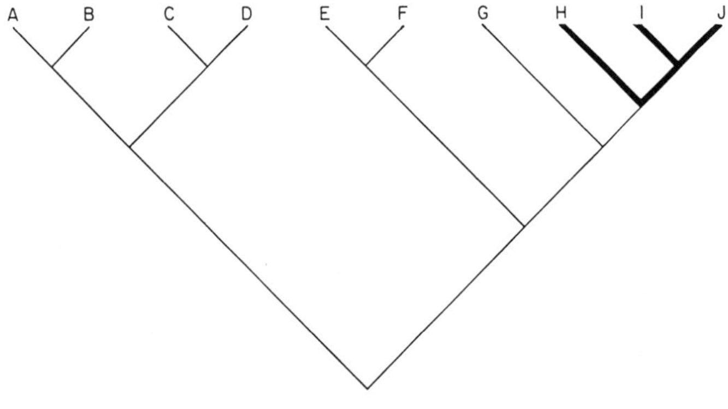

Figure 7.6. Hypothetical cladogram with ten host taxa. Thick bars indicate taxa that show behavioural alterations due to parasite infection.

Table 7.1. Results of a combinatorial analysis on the probability of observing monophyletic groups for the cladogram in figure 7.6.

Number of species in group	Number of monophyletic groups	Number of combinations	Probability
2	4 (AB, CD, EF, IJ)	45	0·089
3	1 (HIJ)	120	0·008
4	2 (ABCD, GHIJ)	210	0·009
5	0	252	—
6	1 (EFGHIJ)	210	0·005

This method quantifies the procedure followed by Dobson (1984), who searched for phylogenetic constraints in the mating systems of North American sciurid rodents. For a cladogram of twelve marmotine taxa, seven exhibited male defense (territorial) polygyny (figure 10.1 in Dobson, 1984). However, these seven taxa did not form a monophyletic group, so Dobson (1984) rejected the hypothesis of phylogenetic constraints (historical origin) and went on to consider alternative (adaptive) hypotheses for mating behaviour.

In a second example, two taxa of ground squirrels showed a characteristic 'run jump' fleeing behaviour (Dobson, 1985). When compared to the best-resolved ground squirrel phylogenies (figures 1b and c in Dobson 1985), the 'run jump' behaviour appears to have evolved once, in the common ancestor to *Spermophilus* and *Cynomys*. For this particular cladogram (figure 1c in Dobson (1985)), the probability of two randomly chosen taxa forming a monophyletic group is only 0·048, suggesting a phylogenetic constraint on the possession of this trait.

Although the combinatorial approach generates a probability value for patterns of altered host behaviour, the analysis is compromised because of non-independence of taxa that are 'nested' on the cladogram. This same criticism applies, to a lesser extent, to the similarity analysis described below and to any statistical test that does not explicitly incorporate the hierarchical structure of the cladogram into the null hypothesis (Felsenstein, 1985). An additional problem is that the test is sample-size dependent. In particular, if there are fewer than seven host taxa, it is impossible to reject the null hypothesis of no phylogenetic constraints, regardless of the pattern. With so few taxa, even the most extreme pattern cannot be rejected at $P<0{\cdot}05$.

Quantitative similarity analysis

A more sophisticated analysis takes into account the magnitude of the response to parasites. The phylogenetic constraints hypothesis can be rephrased as follows: the more closely related two species are the more similar their behavioural response to parasite infection. Let each taxon pair represent a point on a scatterplot. The x axis represents the phylogenetic 'distance' between the two taxa and the y axis represents the *difference* in the behavioural response to parasites between the two taxa. If there are phylogenetic constraints on altered behaviours, the points on the graph should describe a positive relationship: the more distantly related two taxa are (large x), the more dissimilar their responses to parasites (large y). On the other hand, if phylogeny is not important, there should be no association between the degree of relationship and the similarity of the response to parasitism.

In this example, a standard regression analysis is not appropriate because the difference data are not independent of one another, i.e. the relationship between taxa A and B depends on the relationship between A and C and between B and C. Fortunately, Mantel (1967) developed a simple test of association for data of exactly this sort. Schnell *et al.* (1985) describe the use of this test in behavioural studies and analyze several examples of data sets from the literature. We will not describe the mathematical details of the Mantel test, but we will show how the test can be used in the analysis of a hypothetical data set.

The Mantel test begins with the construction of two square difference matrices, one for phylogeny, and one for behaviour. For the phylogeny matrix of n taxa, each element in the $n \times n$ symmetric matrix represents the phylogenetic distance between each taxon pair. Diagonal entries are not included in the test, because these represent the distance of each taxon from itself. There are many ways to measure phylogenetic distance (Sneath and Sokal, 1973); for our purposes, we will use the cladistic difference, that is the number of nodes separating two taxa in the cladogram (Farris, 1969). Table 7.2 gives the distance measures for the hypothetical cladogram in figure 7.6.

Table 7.3 shows three hypothetical response variables that have been measured for each species. The variables might be any behavioural trait that is potentially altered by parasitism such as activity or photophilia. The scale of the responses is not important here, but we do note that the values given in table 7.3 represent the average increase in the behaviour due to parasitism. In other words, the data represent the average difference between parasitized animals and uninfected controls, so that a 0 indicates no change caused by parasitism. We assume that the underlying data are analyzed by appropriate statistical methods to determine the significance of individual behaviours for each species; our intent here is to show how the data can be analyzed in a phylogenetic framework.

The three variables we have chosen represent three potentially different responses to parasitism. Variable 1 exhibits the graded response that the Mantel test

Table 7.2. Symmetric distance matrix for taxa in figure 7.6. Distances are the number of nodes separating each species pair in the cladogram (the cladistic difference measure of Farris (1969)).

	A	B	C	D	E	F	G	H	I	J
A	—	1	3	3	5	5	5	6	7	7
B	1	—	3	3	5	5	5	6	7	7
C	3	3	—	1	5	5	5	6	7	7
D	3	3	1	—	5	5	5	6	7	7
E	5	5	5	5	—	1	3	4	5	5
F	5	5	5	5	1	—	3	4	5	5
G	5	5	5	5	3	3	—	2	3	3
H	6	6	6	6	4	4	2	—	2	2
I	7	7	7	7	5	5	3	2	—	1
J	7	7	7	7	5	5	3	2	1	—

Table 7.3. Three hypothetical variables illustrating the response of each taxon to parasites. '0' indicates no alteration of host behaviour by parasites.

Taxon	Variable 1	Variable 2	Variable 3
A	5	0	4
B	5	0	5
C	3	0	5
D	3	0	−2
E	1	0	3
F	1	0	0
G	0	0	7
H	−2	1	5
I	−3	1	−1
J	−3	1	0

is designed to detect: closely related taxa show the most similar response. Variable 2 is a dichotomous variable, in which only taxa H, I, and J show a response to parasitism (1); all other taxa show no response (0). This is the same scenario we used in the qualitative analysis, so it provides an opportunity to compare the results of the Mantel test with those of the combinatorial test. Finally, variable 3 was created with a random number generator and should reveal no association by the Mantel test.

The Mantel test statistic is calculated as the sum of the products of corresponding elements of the phylogeny and behaviour distance matrices. Schnell *et al.* (1985) describe how the Mantel test statistic can be compared with a normal distribution to generate a probability value. They caution that the approximation to a normal distribution may be poor if the sample size is less than 20. In these cases, they recommend a Monte Carlo simulation to assess significance. For each response variable, we randomized the elements in the phylogeny matrix and calculated the Mantel test statistic for this null association. We repeated the procedure 1000 times and compared the resulting distribution to the observed value of the Mantel test statistic to estimate significance. Finally, we calculated the correlation coefficient between the two difference matrices. This provided a familiar index of the degree of association between the two variables.

For variable 1, there was a strong linear response between behaviour and phylogeny (figure 7.7(a)), so the Mantel test statistic was highly significant, with a large correlation coefficient (table 7.4). Variable 2 also gave a significant result, although the relationship was not very linear (figure 7.7(b)). This variable

Table 7.4. Results of the Mantel test for association between phylogeny and behaviour.

Response variable	Mantel test statistic	r	P_1	P_2
1	1792	0·898	<0·001	<0·001
2	232	0·532	<0·001	<0·001
3	1444	−0·071	0·681	0·619

For each of three hypothetical variables (table 7.3), P_1 is the probability value calculated from the approximation to the normal distribution, and P_2 is the probability value determined by Monte Carlo simulation. r is the correlation coefficient between the square difference matrices for behaviour and phylogeny (table 7.1). See figure 7.7.

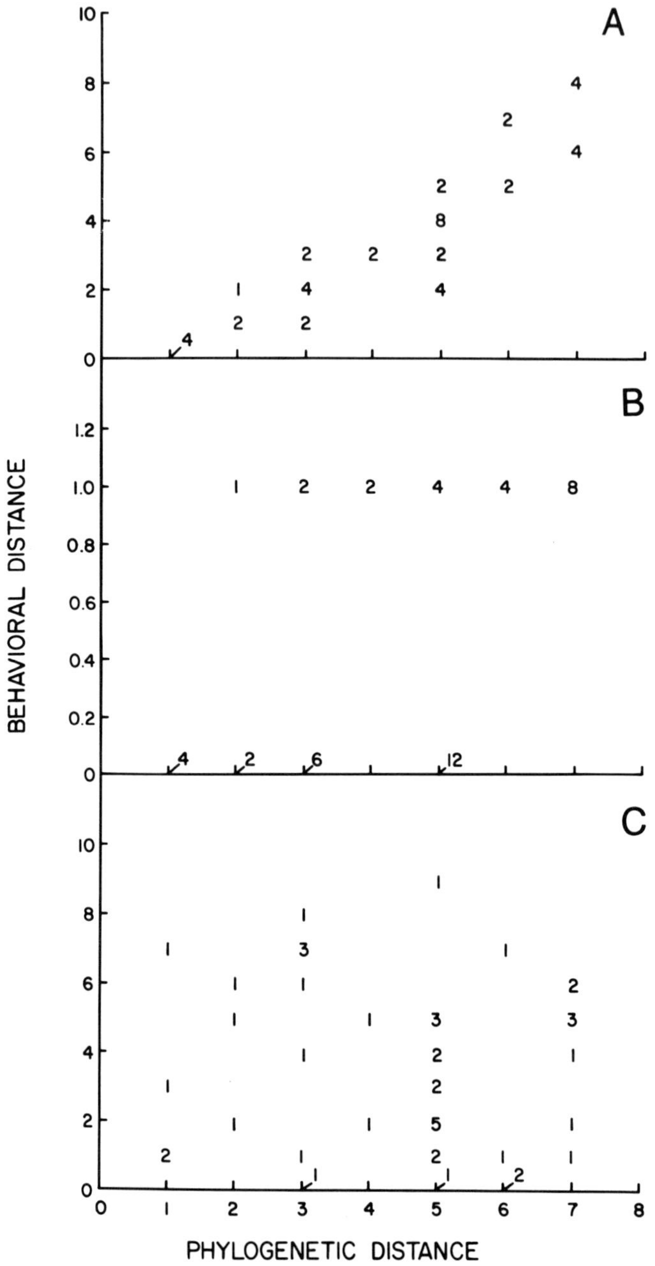

Figure 7.7. Associations between behaviour and phylogeny. Each point represents a different species pair ($n = 45$). Numerals represent multiple points. The x-axis shows the phylogenetic distance (= cladistic difference) between each species pair (table 7.2) and the y-axis shows the difference in the response to parasitism of each species pair. a–c correspond to response variables 1–3 respectively (table 7.3). See table 7.4 for statistical analyses.

corresponds to the pattern in figure 7.6. Thus, the Mantel test appears to give reliable results even in cases where the phylogenetic pattern is qualitative, rather than quantitative. Finally, variable 3 was non-significant and showed no pattern with phylogeny (figure 7.7(c)). Even though only ten species were used in the analysis, the probability values for the large scale approximation were very similar to those generated by the Monte Carlo simulation. These findings all suggest that the Mantel test is a powerful technique for revealing phylogenetic constraints on altered behaviours.

One potential flaw with using cladistic differences in the Mantel test is that this measure is sensitive to variation in low-level clusters in the cladogram (Farris, 1973). However, a distortion coefficient for cladistic differences should reveal the level at which the behavioural and distance matrices disagree (Farris, 1973).

Quantitative cladistic analysis

Ideally, a comparative analysis should not simply show that altered behaviours are correlated or associated with host phylogeny, but should reveal the way that behaviours changed in concert with speciation events in a lineage. In theory, this could be done by incorporating behavioural data as 'characters' in a cladistic analysis of host phylogeny. Carpenter (in press) takes just such an approach to test a scenario for the evolution of wasp social behaviour.

Such an analysis is beyond the scope of this present review, but we can at least outline the steps that would be involved. First, continuous behavioural data would have to be converted to a set of discrete behavioural 'states'. Thorpe (1984) and Archie (1985) describe several techniques for gap-coding continuous data for a cladistic analysis. Next, and far more difficult, is the determination of ancestral and derived character states. Unless an appropriate outgroup can be found, it is most parsimonious to assume that the unaltered (control) behavioural state is plesiomorphic, that is, the 'original, pre-existing character' (Wiley, 1981, p. 9) in evolutionary terms.

Both the behavioural characters and traditional morphological characters would be used in a cladistic analysis to reconstruct phylogeny. The Wagner tree method of Farris (1970), for example, could be used to construct a parsimonious tree that incorporates both morphological and behavioural changes during host evolution.

The final step would be to determine whether behavioural alterations are 'cladistically reliable characters' in the terminology of Farris (1969). Minimum Wagner trees can be constructed with and without behavioural characters included in the analysis. If there is no change in the structure of the most parsimonious tree, this would suggest that behavioural alterations overlay host phylogeny, even if some homoplasies are present in the data. (Homoplasies are two similar characters in two species that are not found in the ancestor. They have evolved independently and are 'parallel' or 'convergent' (Wiley, 1981).) On the other hand, if the tree changes radically with the inclusion of behavioural characters, the correspondence between phylogeny and behaviour would be properly de-emphasized.

Farris (1969) describes a character weighting iteration that rapidly converges on the 'true' (underlying) phylogeny. These character weights also reveal the reliability of behavioural characters in the analysis; if behaviour is concordant with phylogeny, then weights for behavioural characters should be high and similar to weights for morphological characters. On the other hand, if behaviour does not change concordantly with morphology (i.e. they are mostly homoplastic), then behavioural characters will receive small weights and not have a great influence on the final cladogram. Simulation studies (Farris, 1969) demonstrate that this technique is insensitive to the relative numbers of cladistically reliable and unreliable characters in the analysis.

An alternative approach would be to construct two cladograms, one based on morphological (or biochemical) characters and one based on behavioural characters. For example, McLennan *et al.* (1988) created a highly consistent phylogenetic tree for stickleback fishes based entirely on behavioural characters. This cladogram was then used, qualitatively, to resolve ambiguities in cladograms constructed from traditional morphological and biochemical characters. Mickevich (1978) discusses quantitative methods for measuring congruence between cladograms (note that the Mantel test, when used for two matrices of cladistic differences, also provides a measure of congruence between behavioural and morphological cladograms).

We stress that these cladistic analyses are provisional, and cannot provide a definitive test. It is difficult to analyze branching, hierarchical structures such as cladograms statistically (Felsenstein, 1985). The cladistic analyses we have proposed are exploratory, rather than definitive tests of the relationship between phylogeny and behavioural alterations.

Further considerations

Although the methods we have discussed will provide considerable insight into the evolution of altered behaviours, there are many potential problems and pitfalls that can arise even in a careful comparative study. In this section, we briefly outline potential complications that may arise in investigating the phylogenetic constraints hypothesis.

First, our analysis has considered evolutionary change only from the perspective of the host. But there are many examples in which host and parasite evolve together in a lineage (e.g. Brooks *et al.*, 1981; Glen and Brooks, 1986). If co-speciation has occurred, then phylogenetic patterns of altered behaviour may not be apparent from an analysis of host responses to a single species of parasite. Indeed, when co-speciation is strong, the important phylogenetic constraints may not be on which hosts show altered behaviours, but on the hosts in which the parasite can successfully develop.

Second, the analysis is only as good as the phylogeny upon which it is based (Dobson, 1985). Phylogenetic analyses themselves have underlying assumptions that are beyond the scope of this review (e.g. see Felsenstein, 1982, 1983). In particular, a cladistic analysis of altered behaviours will be flawed if it is based on a paraphyletic (nonnatural; see Wiley, 1981) grouping. Yet, it may not be practical to test all the species in a taxon. 'Representative' species will have to be used, and the results will depend on how accurately their behavioural responses characterize the entire taxon.

Finally, even if a phylogenetic constraint is discovered, there will be questions about the precise mechanism. If parasitic alterations characterize only certain taxa, is this due to a developmental constraint (Maynard Smith *et al.*, 1985), a structural or physiological constraint, perhaps imposed by limitations of body size (Clutton-Brock and Harvey, 1984; Stearns, 1983), or some other ecological correlate of host phylogeny? In theory, it is possible to control statistically for the effects of phylogeny (Felsenstein, 1985), but in practice there will always be difficulties of interpretation in comparative analyses. None the less, if a comparative analysis indicates that phylogenetic constraint is likely to have played a role in the evolution of alterations, then the investigation of mechanisms may be enhanced, as *a priori* hypotheses may emphasize physiological, developmental or ecological traits that are shared by members of the clade under scrutiny.

The methods we propose will use non-behavioural data and phylogenetic analyses to examine the importance of phylogenetic influences in the evolution of behavioural alterations. Several alterations may occur in a given parasite-host association, and we have no reason to suspect that they all evolved in response to the same influences. For any given alteration, the relative importance of phylogeny, host/parasite responses or current ecological influences can be evaluated. These are certainly not mutually exclusive, but to the extent that phylogenetic constraint on the part of parasite or host appears influential, we can look for evolutionarily shared mechanisms. Failure to demonstrate a phylogenetic pattern will lead to increased consideration of non-phylogenetic influences (e.g. current ecology). Despite the scarcity of precise tests, we nonetheless expect that many cases of altered behaviour or appearance will be associated with increased predation. We will examine those altered behaviours from a perspective that is phylogenetic and historical as well as adaptationist and ecological, and for the first time, place them in the predictive context that the study of evolution allows.

Lest we appear to be too certain, we leave you with the words of that most venerable scientist, Spock, from the Starship Enterprise:

> A truly successful parasite is commensal, living in amity with its host, or even giving it positive advantages, as, for instance, the protozoans who live in the digestive system of your termites and digest for them the wood they eat. A parasite that regularly and inevitably kills its hosts cannot survive long, in the evolutionary sense, unless it multiplies with tremendous rapidity . . . It is not pro-survival. (Carabatsos, 1968).

With all due respect to Spock, much remains to be learned about how parasites 'succeed', and the constraints on that evolution.

ACKNOWLEDGEMENTS

We thank D. Hough for performing the Mantel tests and Monte Carlo simulations and C. McCallister for drafting the figures, A. Grafen and B. Hart for sharing unpublished manuscripts, and P. Harvey and A. Keymer for discussion. Portions of the manuscript benefitted from comments by J. Carpenter, G. Graves, and D. Simberloff. This research was supported by NSF BSR-8452076 to J. M. and NSF BSR-8817495 to J. M. and N. J. G.

REFERENCES

Añez, N. and East, J. S., 1984, *Acta Tropica*, **41**, 93–95.
Anokhin, I. A., 1966, *Doklady Biological Sciences*, **166**, 93–94.
Antonovics, J., 1987, *The American Naturalist*, **129**, 321–331.
Archie, J. W., 1985, *Systematic Zoology*, **34**, 326–345.
Arme, C. and Owen, R. N., 1967, *Parasitology*, **57**, 301–314.
Bartoli, P., 1976–78, *Haliotis*, **7**, 23–28.
Baudoin, M., 1975, *Evolution*, **29**, 335–352.
Beach, R., Kiiku, G. and Leeuwenburg, J., 1985, *American Journal of Tropical Medicine and Hygiene*, **34**, 278–282.
Beckage, N. E., 1985, *Annual Review of Entomology*, **30**, 371–413.
Beegle, C. C. and Oatman, E. R., 1974, *Journal of Invertebrate Pathology*, **24**, 188–195.
Berry, W. J., Rowley, W. A. and Christensen, B. M., 1986, *Journal of Medical Entomology*, **23**, 441–445.
Berry, W. J., Rowley, W. A. and Christensen, B. M., 1987, *Journal of Medical Entomology*, **24**, 699–701.
Bethel, W. M. and Holmes, J. C., 1973, *Journal of Parasitology*, **59**, 945–956.
Bethel, W. M. and Holmes, J. C., 1974, *Journal of Parasitology*, **60**, 272–274.
Bethel, W. M. and Holmes, J. C., 1977, *Canadian Journal of Zoology*, **55**, 110–115.
Bicik, V. and Lysek, H., 1970, *Parasitologica Polonica*, **18**, 99–105.
Bishop, R. K. and Cannon, L. R. G., 1979, *Journal of Fish Diseases*, **2**, 131–144.
Boland, L. M. and Fried, B., 1984, *The Journal of Parasitology*, **70**, 436–439.
Boorstein, S. M. and Ewald, P. W., 1987, *Physiological Zoology*, **60**, 586–595.
Borgia, G., 1986, *Behavioral Ecology and Sociobiology*, **9**, 355–358.
Brassard, P., Rau, M. E. and Curtis, M. A., 1982, *Parasitology*, **85**, 495–501.
Brattey, J., 1983, *The Journal of Parasitology*, **69**, 1172–1173.
Brooks, D. R., 1980, *Systematic Zoology*, **29**, 192–203.
Brooks, D. R., 1985, *Annals of the Missouri Botanical Gardens*, **72**, 660–680.
Brooks, D. R., 1987, *International Journal for Parasitology*, **17**, 291–297.
Brooks, D. R., 1988, *Annual Review of Ecology and Systematics*, **19**, 235–259.
Brooks, D. R., Thorson, T. B. and Mayes, M. A., 1981, In *Advances in Cladistics: Proceedings of the First Meeting of the Willi Hennig Society*, edited by V. A. Funk and D. R. Brooks (New York: New York Botanical Garden), pp. 145–175.
Brown, A. F, and Thompson, D. B. A., 1986, *Journal of Biological Education*, **20**, 121–127.
Butler, J. A. and Millemann, R. E., 1971, *The Journal of Parasitology*, **57**, 860–865.
Cabaret, J., 1984, *Annales des Parasitologie Humaine et Comparée*, **59**, 529–530.
Cade, W. H., 1984, *Canadian Journal of Zoology*, **62**, 226–228.
Cade, W. H. and Wyatt, D. R., 1984, *Behaviour*, **88**, 61–75.
Camp, J. W. and Huizinga, H. W., 1979, *The Journal of Parasitology*, **65**, 667–669.
Camp, J. W. and Huizinga, H. W., 1980, *The Journal of Parasitology*, **66**, 299–304.
Carabatsos, S. W., 1968, In *Star Trek 2* adapted by J. Blish (New York: Bantam Books), pp. 71–88.
Carney, W. P., 1969, *American Midland Naturalist*, **82**, 605–611.
Carpenter, J. M., 1989, *Cladistics*, **5**, 131–144.
Cheverud, J. M., Dow, M. M. and Leutenegger, W., 1985, *Evolution*, **39**, 1335–1351.
Clayton, D. H., 1989, *American Zoologist*, in press.
Clutton-Brock, T. H. and Harvey,, P. H., 1984, In *Behavioral Ecology: an Evolutionary Approach*, edited by J. R. Krebs and N. B. Davies (Oxford: Blackwell), pp. 7–29.

Coddington, J. A., 1988, *Cladistics*, **4**, 3–22.
Combes, C., 1980, *Vie Milieu*, **30**, 55–63.
Crompton. D. W. T., 1984, *Federation Proceedings*, **43**, 239–245.
Crompton, D. W. T., Arnold, S. E., Walters, D. E. and Whitfield, P. J., 1985, *Parasitology*, **90**, 449–456.
Crowden, A. E. and Broom, D. M., 1980, *Animal Behaviour*, **28**, 287–294.
Curtis, L. A., 1985, *Biological Bulletin*, **169**, 377–390.
Curtis, L. A., 1987, *Science*, **235**, 1509–1511.
Dailey, M. D. and Walker, W. A., 1978, *Journal of Parasitology*, **64**, 593–596.
Daniels, C. B., 1985, *Journal of Herpetology*, **19**, 160–162.
Dawes, B., 1946, *The Trematodes* (Cambridge: Cambridge University Press).
Dawkins, R., 1982, *The Extended Phenotype: The Gene As the Unit of Selection* (London: Oxford University Press).
Day, J. F. and Edman, J. D., 1983, *The Journal of Parasitology*, **69**, 163–170.
Day, J. F. and Edman, J. D., 1984, *Comparative Biochemistry and Physiology*, **77A**, 447–452.
DeMont, D. J. and Corkum, K. C., 1982, *The Journal of Parasitology*, **68**, 125–130.
Dence, W. A., 1958, *The Journal of Parasitology*, **44**, 334–338.
Dobson, A. P., 1988, *The Quarterly Review of Biology*, **63**, 139–165.
Dobson, F. S., 1984, In *Biology of Ground-Dwelling Squirrels*, edited by J. O. Murie and G. R. Michener (Lincoln: University of Nebraska Press), pp. 229–249.
Dobson, F. S., 1985, *Evolution*, **39**, 1384–1387.
Dolinsky, Z. S., Burright, R. G., Donovick, P. J., Glickman, L. T., Babish, J., Summers, B. and Cypress, R. H., 1981, *Science*, **213**, 1142–1144.
Downes, C. M., Theberge, J. B. and Smith, S. M., 1986, *Canadian Journal of Zoology*, **64**, 622–629.
Duncan, P. and Vigne, N., 1979, *Animal Behaviour*, **27**, 623–625.
Edwards, J. C., 1987, Parasite-induced changes in host behaviour, PhD Thesis, University of Nottingham.
Edwards, J. C., 1988, *Animal Behaviour*, **36**, 529–540.
Edwards, J. C. and Barnard, C. J., 1987, *Animal Behaviour*, **35**, 533–540.
El-Aal, M. A. and Holdich, D. M., 1987, *Journal of Invertebrate Pathology*, **49**, 252–258.
Etges, F. J., 1963, *American Journal of Tropical Medicine and Hygiene*, **12**, 696–700.
Farris, J. S., 1969, *Systematic Zoology*, **18** 374–385.
Farris, J. S., 1970, *Systematic Zoology*, **19**, 83–92.
Farris, J. S., 1973, *Systematic Zoology*, **22**, 50–54.
Feare, C. J., 1971, *Bird Study*, **18**, 121–129.
Feener, D. H., Jr., 1981, *Science*, **214**, 815–817.
Feener, D. H., Jr., 1988, *Behavioral Ecology and Sociobiology*, **22**, 421–427.
Feller, W., 1968, *An Introduction to Probability Theory and Its Applications* (New York: Wiley).
Felsenstein, J., 1982. *The Quarterly Review of Biology*, **57**, 379–404.
Felsenstein, J., 1983, *Annual Review of Ecology and Systematics*, **14**, 313–333.
Felsenstein, J., 1985, *The American Naturalist* **125**, 1–15.
Forsse, E., 1987, *Journal of Applied Entomology*, **104**, 326–328.
Frank, W., Lucius, R. and Romig, T., 1984, In *Recent German Research on Problems of Parasitology, Animal Health and Animal Breeding*, edited by H. Markl and A. Bittner (Metzingen: Georg Atauser), pp. 38–58.
Freeland, W. J., 1976, *Biotropica*, **8**, 12–24.
Freeland, W. J., 1980, *Ecology*, **61**, 1297–1303.
Freeland, W. J., 1981, *Science*, **213**, 461–462.
Fukase, T., Itagaki, H., Wakui, S., Kano, Y., Goris, R. C. and Kishida, R., 1986, *Japanese Journal of Herpetology*, **11**, 86–95.
Fukase, T., Itagaki, H., Wakui, S., Kano, Y., Goris, R. C. and Kishida, R., 1987, *Japanese Journal of Herpetology*, **12**, 39–44.
Gabrion, C., Plateaux, L., and Quentin, C., 1976, *Annales de Parasitologie Humaine et Comparée*, **51**, 407–420.
Geraci, J. R. and St. Aubrin, D. J., 1987, *International Journal for Parasitology*, **17**, 407–414.
Giles, N., 1983, *Animal Behaviour*, **31**, 1192–1194.
Glen, D. R. and Brooks, D. R., 1986, *Biological Journal of the Linnean Society*, **27**, 331–354.
Gould, S. J. and Lewontin, R. C., 1979, *Proceedings of the Royal Society of London B*, **205**, 581–598.
Grafen, A., 1989, *Philosophical Transactions of the Royal Society*, in press.
Grimstad, P. R., Ross Q. E., Craig, G. B., Jr., 1980, *Journal of Medical Entomology*, **17**, 1–7.

Hafner, M. S. and Nadler, S. A., 1988, *Nature*, **332**, 258–259.
Halvorsen, O., 1986, *Oikos*, **47**, 71–74.
Hamilton, W. D., 1987, In *Animal Societies: Theories and Facts*, edited by Y. Ito, J. L. Brown and J. Kikkawa (Tokyo: Japan Scientific Society Press), pp. 81–102.
Hamilton, W. D. and Zuk, M., 1982, *Science*, **218**, 384–387.
Hart, B. L., 1988, *Neuroscience and Biobehavioral Reviews*, **12**, 123–137.
Hart, J. A., 1988, *Cladistics*, **4**, 339–366.
Hausfater, G. and Meade, B.J., 1982, *Primates*, **23**, 287–297.
Hausfater, G. and Watson, D. F., 1976, *Nature*, **262**, 688–689.
Hay, J. and Aitken, P. P., 1983, *Ecology of Disease*, **2**, 203–209.
Helluy, S., 1982, *Annales de Parasitologie Humaine et Comparée*, **57**, 263–270.
Helluy, S., 1983a, *Review Ecologie* (Terre Vie), **38**, 211–223.
Helluy, S., 1983b, *Annales de Parasitologie Humaine et Comparée*, **58**, 1–17.
Helluy, S., 1984, *Annales de Parasitologie Humaine et Comparée*, **59**, 41–56.
Helluy, S. A., 1988, On the mechanisms by which a larval parasite (*Polymorphus paradoxus*, Acanthocephala) alters the behavior of its intermediate host (*Gammarus lacustris*, Crustacea), PhD Thesis, University of Alberta, Edmonton.
Hilborn, R. and Stearns, S. C., 1982, *Acta Biotheoretica*, **31**, 145–164.
Hindsbo, O., 1972, *Nature*, **238**, 333.
Holmes, J. C. and Bethel, W. M., 1972, In *Behavioural Aspects of Parasite Transmission* edited by E. U. Canning and C. A. Wright (New York: Academic Press), pp. 123–149.
Holmes, J. C. and Price, P. W., 1980, *Systematic Zoology*, **29**, 203–213.
Hoogenboom, I. and Dijkstra, C., 1987, *Oecologia*, **74**, 86–92.
Hudson, P. and Newborn, D., 1985, *The Game Conservancy Annual Review*, **17**, 121–128.
Hulscher, J. B., 1982, *Ardea*, **70**, 89–152.
Hunter, K. W., Jr. and Stoner, A., 1975, *Environmental Entomology*, **4**, 381–382.
Husain, A. and Kershaw, W. E., 1971, *Transactions of the Royal Society of Tropical Medicine and Hygiene*, **65**, 617–619.
Hutchison, W. M., Aitken, P. P. and Wells, B. W. P., 1980a, *Annals of Tropical Medicine and Parasitology*, **74**, 145–150.
Hutchison, W. M., Bradley, M., Cheyne, W. M., Wells, B. W. P. and Hay, J., 1980b, *Annals of Tropical Medicine and Parasitology*, **74**, 337–345.
Hynes, H. B. N., 1955, *Journal of Animal Ecology*, **24**, 352–387.
Jaenike, J., 1988, *The American Naturalist*, **131**, 774–780.
Jarecka, L., 1961, *Acta Parasitologica Polonica*, **9**, 409–426.
Jenkins, D., Watson, A. and Miller, G. R., 1963, *Journal of Animal Ecology*, **32**, 317–376.
Jenni, L., Molyneux, D. H., Livesey, J. L. and Galun, R., 1980, *Nature*, **283**, 383–385.
Johnson, D. L. and Pavlikova, E., 1986, *Journal of Invertebrate Pathology*, **48**, 232–238.
Jones, D., 1985, *Annals of the Entomological Society of America*, **78**, 141–148.
Kavaliers, M. and Podesta, R., 1988, *Canadian Journal of Zoology*, **66**, 2653–2657.
Kennedy, C. R., Broughton, P. F. and Hine, P. M., 1978, *Journal of Fish Biology*, **13**, 265–275.
Keymer, A., Crompton, D. W. T. and Sahakian, B.J., 1983, Parasitology, **86**, 455–460.
Killick-Kendrick, R., Leaney, A.J., Ready, P. D. and Molyneux, D. H., 1977, *Proceedings of the Royal Society of London B*, **196**, 105–115.
Killick-Kendrick, R. and Molyneux, D. H., 1981, *Transactions of the Royal Society of Tropical Medicine and Hygiene*, **75**, 152–154.
Kochmer, J. P. and Handel, S. N., 1986, *Ecological Monographs*, **56**, 303–325.
Kvalsvig, J. D., 1988, *Parasitology Today*, **4**, 206–208.
Lambert, T. C. and Farley, J., 1968, *Canadian Journal of Zoology*, **46**, 1139–1147.
Lambshead, P.J. D. and Paterson, G. L. J., 1986, *Journal of Natural History*, **20**, 895–909.
Lauckner, G., 1984, *Helgoländer Meeresuntersuchungen*, **37**, 185–199.
Lebour, M.V. and Elmhirst, R., 1922, *Journal of the Marine Biological Association of the United Kingdom*, **12**, 829–832.
Lemly, A. D. and Esch, G. W., 1984, *The Journal of Parasitology*, **70**, 475–492.
Lester, R.J. G., 1971, *Canadian Journal of Zoology*, **49**, 361–366.
Lewis, P. D., Jr., 1977, *Proceedings of the Montana Academy of Sciences*, **37**, 70–81.
Lucius, R., Romig, T. and Frank, W., 1980, *Zeitschrift für Parasitenkunde*, **63**, 271–275.
Lundberg, H. and Svensson, B. G., 1975, *Norwegian Journal of Entomology*, **22**, 129–134.
McAlister, M. K. and Roitberg, B. D., 1987, *Nature*, **328**, 797–799.
McClain, E., Magnuson, P. and Warner, S.J., 1988, *Journal of Insect Physiology*, **34**, 279–284.

McLennan, D.A., Brooks, D.R. and McPhail, J.D., 1988, *Canadian Journal of Zoology*, **66**, 2177–2190.
McNair, D.M. and Timmons, E.H., 1977, *Laboratory Animal Science*, **27**, 38–42.
McPhail, J.D. and Peacock, S.D., 1983, *Canadian Journal of Zoology*, **61**, 901–908.
Mantel, N., 1967, *Cancer Research*, **27**, 209–220.
Maynard Smith, J., Burian, R., Kauffman, S., Alberch, P., Campbell, J., Goodwin, B., Lande, R., Raup, D. and Wolpert, L., 1985, *The Quarterly Review of Biology*, **60**, 265–287.
Meakins, R.H. and Walkey, M., 1975, *Journal of Fish Biology*, **7**, 817–824.
Mickevich, M.F., 1978, *Systematic Zoology*, **27**, 143–158.
Milinski, M., 1984, *Behavioral Ecology and Sociobiology*, **15**, 35–37.
Milinski, M., 1985, *Behaviour,* **93**, 203–216.
Milinski, M. and Lowenstein, C., 1980, *Zeitschrift für Tierpsychologie*, **53**, 325–340.
Minchella, D.J., 1985, *Parasitology*, **90**, 205–216.
Mitter, C., Farrell, B. and Wiegmann, B., 1988, *The American Naturalist*, **132**, 107–128.
Moffett, J.O. and Lawson, F.A., 1975, *Journal of Economic Entomology*, **68**, 627–629.
Moloo, S.K., 1983, *Parasitology* **86**, 51–56.
Moloo, S.K. and Dar, F., 1985, *Transactions of the Royal Society of Tropical Medicine and Hygiene*, **79**, 119.
Molyneux, D.H. and Jefferies, D., 1986, *Parasitology*, **92**, 721–736.
Molyneux, D.H. and Jenni, L., 1981, *Transactions of the Royal Society of Tropical Medicine and Hygiene*, **75**, 160–163.
Moore, J., 1983a. *Ecology*, **64**, 1000–1015.
Moore, J., 1983b, *The Journal of Parasitology*, **69**, 1174–1176.
Moore, J., 1984a, *The American Naturalist*, **123**, 572–577.
Moore, J., 1984b, *Scientific American*, **250**, 108–115.
Moore, J. and Bell, D.H., 1983, *The Journal of Parasitology*, **69**, 387–390.
Moore, J. and Brooks, D.R., 1987, *Evolution*, **44**, 882–891.
Moore, J. and Lasswell, J., 1986, *The Journal of Parasitology*, **72**, 186–189.
Morris, G.K., Kerr, G.E., and Gwynne, D.T., 1975, *Canadian Journal of Zoology*, **53**, 1127–1130.
Mueller, J.F., 1974, *The Journal of Parasitology*, **60**, 3–14.
Muzzall, P.M. and Rabalais, F.C., 1975, *Proceedings of the Helminthological Society of Washington*, **42**, 116–118.
Oetinger, D.F. and Nickol, B.B., 1981, *The Journal of Parasitology*, **67**, 672–684.
Oetinger, D.F. and Nickol, B.B., 1982, *The Journal of Parasitology*, **68**, 270–275.
Oma, E.A. and Hewitt, G.B., 1984, *Journal of Economic Entomology*, **77**, 500–501.
Orr, T.S.C., 1966, *Nature*, **212**, 736.
Ovington, K.S., 1985, *Parasitology*, **91**, 157–167.
Pagel, M.D. and Harvey, P.H., 1988, *The Quarterly Review of Biology*, **63**, 413–440.
Palm, N.-B., 1948, *Opuscula Entomologica Supplementum*, **7**, 3–101.
Park, T., 1948, *Ecological Monographs*, **18**, 265–308.
Pearre, S., Jr., 1976, *Journal of the Marine Biological Association, UK.*, **56**, 503–513.
Pearre, S., Jr., 1979, *Internationale Revue des Gesamten Hydrobiologie*, **64**, 193–206.
Plateaux, L., 1972, *Annales des Sciences Naturelles, Zoology, Paris*, **14**, 203–220.
Poinar, G.O. and Vanderlaan, P.A., 1972, *Nematologica*, **18**, 239–252.
Poulin, R. and FitzGerald, G.J., 1989a, *Canadian Journal of Zoology*, **67**, 14–18.
Poulin, R. and FitzGerald, G.J., 1989b, *Behavioral Ecology and Sociobiology*, **24**, 251–255.
Poulton, M.J. and Thompson, D.J., 1987, *Animal Behaviour*, **35**, 1577–1579.
Quinn, S.C., Brooks, R.J. and Cawthorn, R.J., 1987, *Journal of Parasitology*, **73**, 265–271.
Radabaugh, D.C., 1980, *Journal of Fish Biology*, **16**, 621–628.
Rau, M.E., 1982, in *Aspects of Parasitology* edited by E. Meerovitch (Montreal: The Institute of Parasitology, McGill University), pp. 335–360.
Rau, M.E., 1983a, *Parasitology*, **86**, 311–318.
Rau, M.E., 1983b, *Parasitology*, **86**, 319–322.
Rau, M.E., 1984a, *Parasitology*, **88**, 371–373.
Rau, M.E., 1984b, *Parasitology*, **88**, 415–419.
Rau, M.E., 1985, *Journal of Parasitology*, **71**, 774–778.
Rau, M.E. and Caron, F.R., 1979, *Canadian Journal of Zoology*, **57**, 2466–2478.
Rau, M.E. and Putter, L., 1984, *Parasitology*, **89**, 579–583.
Read, A.F., 1987, *Nature,* **328**, 68–70.
Ritchie, L.E. and Hoeg, J.T., 1981, *Journal of Crustacean Biology*, **1**, 334–347.
Romig, T., Lucius, R. and Frank, W., 1980, *Zeitschrift für Parasitenkunde*, **63**, 277–286.

Rossignol, P. A., Ribeiro, J. M. C., Jungery, M., Turell, M. J., Spielman, A. and Bailey, C. L., 1985, *Proceedings of the National Academy of Sciences*, **82**, 7725–7727.
Rossignol, P. A., Ribeiro, J. M. C. and Spielman, A., 1984, *American Journal of Tropical Medicine and Hygiene*, **33**, 17–20.
Rossignol, P. A., Ribeiro, J. M. C. and Spielman, A., 1986, *American Journal of Tropical Medicine and Hygiene*, **35**, 277–279.
Rothschild, M., 1940, *Journal of the Washington Academy of Sciences*, **30**, 437–448.
Rowland, M. and Boersma, E., 1988, *Parasitology*, **97**, 221–227.
Rowland, M. W. and Lindsay, S. W. L., 1986, *Physiological Entomology*, **11**, 325–334.
Rumpus, A. E. and Kennedy, C. R., 1974, *Parasitology*, **68**, 271–284.
Russell, L. R., 1980, *Canadian Journal of Zoology*, **58**, 1220–1226.
Saumier, M. D., Rau, M. E. and Bird, D. M., 1988, *Canadian Journal of Zoology*, **66**, 1685–1692.
Schall, J. J., Bennett, A. F. and Putnam, R. W., 1982, *Science*, **217**, 1057–1059.
Schall, J. J. and Sarni, G. A., 1987, *Copeia*, **1987**, 84–93.
Schell, S. C., 1985, *Handbook of Trematodes of North America North of Mexico* (Moscow: University of Idaho Press).
Schiefer, B. A., Ward, R. A. and Eldridge, B. F., 1977, *Experimental Parasitology*, **41**, 397–404.
Schmidt, G. D. and Olsen, W. O., 1964, *The Journal of Parasitology*, **50**, 721–730.
Schmidt, G. D. and Roberts, L. S., 1985, *Foundations of Parasitology*, 3rd ed. (St. Louis: Times Mirror/Mosby).
Schnell, G. D., Watt, D. J. and Douglas, M. E., 1985, *Animal Behaviour*, **33**, 239–253.
Scott, M., 1985, in *Ecology and Genetics of Host-Parasite Interactions* edited by D. Rollinson and R. M. Anderson (New York: Academic Press), pp. 21–38.
Seed, J. R. and Khalili, N., 1971, *Journal of Interdisciplinary Cycle Research*, **2**, 91–99.
Seger, J. and Hamilton, W. D., 1988, in *The Evolution of Sex* edited by R. E. Michod and B. R. Levin (Sutherland, Massachusetts: Sinauer Assoc. Inc.), pp. 176–193.
Seidenberg, A. J., 1973, *The Journal of Parasitology*, **59**, 957–962.
Shapiro, A. M., 1976, *The American Naturalist*, **110**, 900–902.
Sherman, P. W., Seeley, T. D. and Reeve, H. K., 1988, *The American Naturalist*, **131**, 602–610.
Sillen-Tullberg, B., 1988, *Evolution*, **42**, 293–305.
Sindermann, C. J. and Farrin, A. E., 1962, *Ecology*, **43**, 69–75.
Smith, R. S. and Kramer, D. L., 1987, *Canadian Journal of Zoology*, **65**, 1862–1865.
Smith Trail, D. R., 1980, *The American Naturalist*, **116**, 77–91.
Sneath, P. H. and Sokal, R. R., 1973, *Numerical Taxonomy* (San Francisco: W. H. Freeman).
Sousa, W. P., 1983, *Journal of Experimental Marine Biology and Ecology*, **73**, 273–296.
Stambaugh, J. E. and McDermott, J. J., 1969, *Proceedings of the Pennsylvania Academy of Science*, **43**, 226–231.
Stamp, N. E., 1981, *The American Naturalist*, **118**, 715–725.
Stark, G. T. C., 1965, *Parasitology*, **55**, 415–420.
Stearns, S. C., 1983, *Oikos*, **41**, 173–187.
Stretch, R. G. A., Leytham, G. W. H. and Kershaw, W. E., 1960a, *Annals of Tropical Medicine and Parasitology*, **54**, 487–492.
Stretch, R. G. A., Stretch, S. J. E., Leytham, G. W. H. and Kershaw, W. E., 1960b, *Annals of Tropical Medicine and Parasitology*, **54**, 376–380.
Stretch, S. J. E., Stretch, R. G. A., Leytham, G. W. H. and Kershaw, W. E., 1960c, *Annals of Tropical Medicine and Parasitology*, **54**, 483–486.
Stunkard, H. W. and Cable, R. M., 1932, *Biological Bulletin*, **62**, 328–338.
Sweeting, R. A., 1977, *Journal of Fish Biology*, **10**, 43–50.
Swennen, C., 1969, *Netherlands Journal of Sea Research*, **4**, 376–379.
Swennen, C. and Ching, H. L., 1974, *Netherlands Journal of Sea Research*, **8**, 108–115.
Symons, L. E. A., 1969, in *International Review of Tropical Medicine* edited by D. R. Lincicome and A. W. Woodruff (New York: Academic Press), **3**, 49–100.
Symons, L. E. A. and Jones, W. O., 1978, *Experimental Parasitology*, **44**, 7–13.
Szidat, L., 1969, *Journal of the Fisheries Research Board of Canada*, **26**, 753–786.
Thompson, S. N., 1982a, *Parasitology*, **84**, 491–510.
Thompson, S. N., 1982b, *The Journal of Parasitology*, **68**, 936–941.
Thompson, S. N., 1983a, *Comparative Biochemistry and Physiology*, **74B**, 183–211.
Thompson, S. N., 1983b, *Parasitology*, **87**, 15–28.
Thompson, S. N., 1985, *Comparative Biochemistry and Physiology*, **81B**, 21–42.
Thorpe, R. S., 1984, *Evolution*, **38**, 244–255.

Townson, H., 1970, *Annals of Tropical Medicine and Parasitology*, **64**, 411–420.
Turell, M.J., Bailey, C.L. and Rossi, C.A., 1984, *American Journal of Tropical Medicine and Hygiene*, **33**, 1232–1238.
Ulmer, M.J., 1971, in *Ecology and Physiology of Parasites* edited by A.M. Fallis (London: Adam Hilger Ltd), pp. 123–159.
Utida, S., 1953, *Ecology*, **34**, 301–307.
VanDobben, W.H., 1952, *Ardea*, **40**, 1–63.
Vernberg, W.B., 1969, *American Zoologist*, **9**, 357–365.
Vinson, S.B. and Iwantsch, G.F., 1980, *The Quarterly Review of Biology*, **55**, 143–165.
Waage, J.K. and Nondo, J., 1982, *Transactions of the Royal Society of Tropical Medicine and Hygiene*, **76**, 119–122.
Ward, P.I., 1988, *Animal Behaviour*, **36**, 1210–1215.
Webber, R.A., Rau, M.E. and Lewis, D.J., 1987a, *Canadian Journal of Zoology*, **65**, 1340–1342.
Webber, R.A., Rau, M.E. and Lewis, D.J., 1987b, *Canadian Journal of Zoology*, **65**, 2346–2348.
Wesenberg-Lund, C., 1931, *Memoires de l'Academie Royale des Sciences et des Lettres de Danemark, Copenhague, Section des Sciences, 9m. series*, **4**, 90–142.
White, A., Ganter, P., McFarland, R., Stanton, N. and Lloyd, M., 1983, *Oecologia*, **57**, 281–286.
Wiley, E.O., 1981, *Phylogenetics: The Theory and Practice of Phylogenetic Systematics* (New York: John Wiley and Sons).
Williams, C.L. and Gilbertson, D.E., 1983, *The Journal of Parasitology*, **69**, 671–676.
Wilson, K. and Edwards, J., 1986, *Animal Behaviour*, **34**, 942–944.
Wülker, W., 1985, *Zeitschrift für Parasitenkunde*, **71**, 409–418.
Zohar, A.S. and Rau, M.E., 1986, *The Journal of Parasitology*, **72**, 464–466.
Zuk, M., 1988, *Evolution*, **42**, 969–976.

Appendix 1. Taxonomic affinities of some parasites, with notes on transmission modes and references. This is *not* complete life cycle information; some stages not under consideration may be transmitted differently. (See also general references, e.g. Schmidt and Roberts, 1985.) 'Predation' refers to consumption of intermediate host containing immature parasite, and not to all methods of ingesting parasites. DH = definitive host.

Parasite genus	Major taxon	Transmission	References
Acanthocephalus	Acanthocephala	Predator DH	Seidenberg, 1973; Muxxall and Rabalais, 1975; Camp and Huizinga, 1979; Oetinger and Nickol, 1981, 1982; Brattey, 1983
Apanteles	Hymenoptera	Oviposition; no predation	Stamp, 1981
Argulus	Crustacea	Direct transmission; no predation	Poulin and Fitzgerald, 1989a, b
Aspiculuris	Nematoda	Direct transmission; no predation	McNair and Timmon, 1977
Brachylecithum	Trematoda:Digenea	Predator DH	Carney, 1969
Brugia	Nematoda:Filariata	Dipteran vector	Berry *et al.*, 1986; Rowland and Lindsay, 1986
Cryptocotyle lingua	Trematoda:Digenea	Cercariae; no predation	Lambert and Farley, 1968
Dicrocoelium	Trematoda:Digenea	Predator DH	Anokhin, 1966; Lucius *et al.*, 1980; Romig *et al.*, 1980; Frank *et al.*, 1984
Diplocotyle	Cestoidea	Predator DH	Stark, 1965
Diplostomum	Trematoda:Digenea	Predator DH	Szidat, 1969; Crowden and Broom, 1980; Brassard *et al.*, 1982
Dirofilaria	Nematoda:Filariata	Dipteran vector	Berry *et al.*, 1987
Dispharynx	Nematoda: Spirurata	Predator DH	Moore and Lasswell, 1986
Echinostoma	Trematoda:Digenea	Predator DH	Etges, 1963; Boland and Fried, 1984
Glugea	Protozoa: Microspora	Direct transmission; no predation	Milinski, 1984, 1985
Gymnophallus	Trematoda:Digenea	Predator DH	Bartoli, 1976–1978

Gynaecotyla	Trematoda:Digenea	Predator DH	Curtis, 1987
Gyrodactylus	Monogenea	Direct transmission; no predation	Scott, 1985
Heligmosomoides	Nematoda:Strongylata	Direct transmission; no predation	Freeland, 1981
Howardula aoronymphium	Nematoda:Allantonematidae	Direct transmission; no predation	Jaenike, 1988
Hyposoter	Hymenoptera	Oviposition; no predation	Thompson, 1982a, b, 1983b, 1985
Lateriporus	Cestoidea	Predator DH	Bethel and Holmes, 1973
Leishmania	Protozoa:Kinetoplastida	Dipteran vector	Killick-Kenrick *et al.*, 1977; Beach *et al.*, 1985
Leucochloridium	Trematoda:Digenea	Predator DH	Wesenberg-Lund, 1931; Lewis, 1977
Ligula	Cestoidea	Predator DH	VanDobben, 1952; Dence, 1958; Orr, 1966; Sweeting, 1977
Limnomermis	Nematoda:Mermithidae	Direct transmission; no predation	Wülker, 1985
Massospora	Fungi	Direct transmission; no predation	White *et al.*, 1983
Microphallus	Trematoda:Digenea	Predator DH	Helluy, 1983a, b
Moniliformis	Achanthocephala	Predator DH	Moore, 1983b; Wilson and Edwards, 1986; Edwards, 1987
Nippostrongylus	Nematoda:Strongylata	Direct transmission; no predation	Keymer *et al.*, 1983; Ovington, 1985
Nosema	Protozoa:Microspora	Direct transmission; no predation	Moffett and Lawson, 1975; Oma and Hewitt, 1984; Johnson and Pavlikova, 1986; Boorstein and Ewald, 1987
Octospiniferoides	Acanthocephala	Predator DH	DeMont and Corkum, 1982
Ornithodiplostomum	Trematoda;Digenea	Predator DH	Radabaugh, 1980
Parorchis	Trematoda;Digenea	Preadtor DH	Feare, 1971
Parvatrema	Trematoda:Digenea	Preadtor DH	Swennen, 1969; Swennen and Ching, 1974; Hulscher, 1982
Plagiorchis	Trematoda:Digenea	Predator DH	Webber *et al.*, 1987a, b
Plagiorhynchus	Acanthocephala	Predator DH	Moore, 1983a

Plasmodium	Prtozoa:Apicomplexa	Dipteran vector	Schiefer *et al.*, 1977; Schall and Sami, 1987; Schall *et al.*, 1982; Day and Edman, 1983, 1984; Rossignol *et al.*, 1984, 1985; Rowland and Boersma, 1988
Polymorphus	Acanthocephala	Predator DH	Hindsbo, 1972; Bethel and Holmes, 1973, 1974, 1977
Pomphorhynchus	Acanthocephala	Predator DH	Kennedy *et al.*, 1978; Brown and Thompson, 1986; Poulton and Thompson, 1987
Sarcocystis	Protozoa:Apicomplexa	Predation	Hoogenboom and Dijkstra, 1987; Quinn *et al.*, 1987
Schistocephalus	Cestoidea	Predator DH	Arme and Owen, 1967; Lester, 1971; Meakins and Walkey, 1975; McPhail and Peacock, 1983; Giles, 1983; Milinski, 1984, 1985; Smith and Kramer, 1987
Schistosoma	Trematoda:Digenea	Direct transmission; no predation	Stretch *et al.*, 1960; Williams and Gilbertson 1983
Sphaerechinorhynchus	Acanthocephala	Predator DH	Daniels, 1985
Sphaerularia	Nematoda:Tylenchoidea	Direct transmission; no predation	Palm, 1948; Poinar and Vanderlaan, 1972; Lundberg and Svenson, 1975
Spirometra	Cestoidea	Predator DH	Mueller, 1974; Fukase *et al.*, 1987
Toxocara	Nematoda:Ascaridata	Predator DH possible	Dolinsky *et al.*, 1981
Toxoplasma	Protozoa:Apicomplexa	Predator DH	Hutchinson *et al.*, 1980a, b; Hay and Aitken, 1983
Trichinella	Nematoda:Trichurata	Predator DH	Rau, 1983a, 1984a; Rau and Putter, 1984; Zohar and Rau, 1986; Edwards and Barnard, 1987; Saumier *et al.*, 1988
Trichostrongylus	Nematoda:Strongylata	Direct transmission; no predation	Hudson and Newborn, 1985
Trypanosoma	Protozoa:Kinetoplastida	Insect vector	Seed and Khalili, 1971; Jenni *et al.*, 1980; Molyneux and Jenni, 1981; Añez and East, 1984
Uvulifer	Trematoda:Digenea	Predator DH	Lemly and Esch, 1984

Appendix 2. A Brief summary of the terminology of parasite developmental stages used in this chapter (see Schmidt and Roberts, 1985)

Cystacanth: an immature acanthocephalan. Transmitted via predation from the intermediate to definitive host.

Cysticercoid: an immature cyclophyllidean cestode. Transmitted via predation from intermediate to definitive host.

Metacercaria: an immature digenean (trematode), usually encysted in second intermediate host. Transmitted via predation to definitive host.

Plerocercoid: penultimate developmental stage of pseudophyllidean and proteocephalan cestodes. Transmitted via predation to definitive host.

Sporocyst: proliferative, sac-like stage of digenean (trematode) that may yield additional stages. In rare cases, sporocyst and contents are consumed by the definitive host.

8 Human behaviour and the epidemiology of helminth infections: cultural practices and microepidemiology

G. S. Nelson

INTRODUCTION

Most public health workers are macroepidemiologists concerned with the broad patterns of the distribution, prevalence and intensity of infection by age and sex together with estimates of morbidity, mortality and economic impact; information which is sometimes supplemented with mathematical models describing the dynamics of transmission for use in the design and evaluation of control programmes. All too often the information is based on objective measurements with little if any concern for the human component in relation to transmission. As a medical biologist studying diseases such as schistosomiasis, trichinosis and hydatid disease, where both wild and domestic animals were involved in transmission, and working in remote communities like Turkana in Kenya, I was fortunate to become involved in studies on both the macro- and microepidemiology of these diseases. We were particularly interested in unravelling the intimate relationships between man, the parasites and the environment and we soon realized that there were enormous gaps in our knowledge.

Microepidemiology is concerned with the variability of the organisms themselves and with the relationship between the parasite and the host, both in the internal and external environments: internally with the genetics of susceptibility and immune responses, externally with the variability of the way in which the parasites are transmitted, with their survival in nature, and with the exact locations and densities of the different stages of the parasites in the environment, and more particularly with human behaviour in relation to the dissemination and acquisition of infection. It is these external aspects of microepidemiology which are the subject of this chapter. Microepidemiology provides the scientific basis for the control of those parasitic diseases where health education and community participation are essential features of the programme. In 1972 at a joint meeting of the British Society for Parasitology and the Linnean Society I was asked to comment on 'the control of schistosomiasis by altering human behaviour patterns.' In my reply I arrogantly maintained that: 'We are likely to have even less success in preventing children in the tropics from bathing in infected water or preventing indiscriminate defaecation than the health educationalists in the Western world have had in preventing people from committing suicide by smoking' (Nelson, 1972). How wrong I was! In recent years we have seen a dramatic decline in smoking-related diseases in the Western world and with the universal access to radio and television, together with the emphasis given to Primary Health Care by WHO, there has been a wide dissemination of knowledge about parasitic diseases, even in remote regions of the world.

The control of parasitic diseases depends on either: (1) killing the parasites by chemotherapy, (2) killing the vectors with insecticides or changing the environment to prevent their development, or (3) changing human behaviour to avoid transmission.

We have excellent drugs for the treatment of most parasitic diseases, and especially the helminths; we know a great deal about the control of vectors but we are much more ignorant of the way in which man disseminates and acquires infection. Our techniques for changing human behaviour are also crude and as a result very little progress has been made in getting the participation of the community in parasitic disease control programmes. This is compounded by what Warren (1988) calls the 'luddite anti-technology type of mentality among specialists in health care delivery in the developing world'. This is unfortunately true and it has discouraged the application of rigorous scientific studies of the anthropology of disease transmission.

As a medical helminthologist I make no apology for selecting trichinosis, hydatid disease and schistosomiasis as the main topics for discussion because of the history of changing ideas about their epidemiology and because they are all problems where changes in human behaviour have markedly reduced transmission. They also represent the main groups of helminths, the nematodes, cestodes and trematodes, and they illustrate the dangers of accepting the dogma enshrined in the stylized life cycles which are a feature of all parasitological textbooks. Also there are fundamental differences in their epidemiology in different geographical regions determined by variations in their host specificity and pathogenicity and differences in their mode of transmission. For example, in Africa and the Arctic *Trichinella* is maintained in nature in wild carnivores and it is not infective to the usual source of infection for man in temperate regions namely the domestic pig. Similar differences in infectivity occur with *Echinococcus granulosus*. For example, in Ireland and England the most prevalent strain of *Echinococcus* produces hydatid cysts in horses but not in man. Schistosomiasis also illustrates the dangers of generalization with variations in the parasites ranging from the 'non-infective'

Schistosoma japonicum in Taiwan to the highly infective *Schistosoma japonicum* in mainland China. In every case these fundamental differences in the properties of the parasites affect the way in which we need to assess the microepidemiology in relation to transmission and it is only when we have local knowledge that we can devise control programmes based on changing human behaviour.

TRICHINOSIS

> And the swine it is unclean to you. Ye shall not eat of their flesh nor touch their dead carcasses.
>
> Moses, Deuteronomy, Chapter 14.
>
> O ye who believe eat of the good things we have provided for you and be grateful to God. He hath only forbidden you to eat meat and blood and flesh of swine.
>
> Koran – Sura Al-Baqarah (2); Ayat 172–173.
>
> Avoid eating carnivorous animals and birds with talons.
>
> Saying of the Prophet Muhamed,
> related by Al-Bukhari and Muslim.

Introduction

Trichinosis deserves prominence in any account of the microepidemiology of parasitic diseases because it was the first generalized infection of man in which the aetiological agent was recognized and where a knowledge of the life cycle led to practical methods of control. It is an ancient infection prevalent throughout the world in almost all meat-eating, cannibalistic or carrion-feeding carnivores including man. Sporadic infections are likely to have affected man from his early days as a hunter–gatherer but the major epidemics which now occur in urban communities are of more recent origin and result from the domestication of the pig. The development of methods of pig husbandry, including the feeding of pigs with garbage containing pork products has made them into cannibals and so perpetuated the infection of the domestic environment. The parasite *Trichinella spiralis* was discovered by a medical student, James Paget, in the muscles of a cadaver in the dissecting rooms of St. Bartholomew's hospital in London in 1835. It is no wonder that he later attained fame and a knighthood because he had the arrogance to say that: 'All men in the dissecting room, teachers included, saw the little specks in the muscles but I believe that I alone looked at them and observed them'. But it was not until 1860 that trichinosis became recognized as a widespread, sometimes fatal human disease. This resulted from the study by Zenker (1860) who diagnosed trichinosis in a housemaid in Dresden in Germany, who had died with what at first was thought to be severe typhoid fever. He recognized the cysts in the muscles at the autopsy and then fed samples to a dog and recovered the tiny adult worms in the intestine. After an arduous mid-winter journey of over 80 miles by coach to the farm where the girl had been working he established that she and several other members of the household had acquired the infection from a pig which had been slaughtered to make a traditional Christmas dish. The domestic pig and the peridomestic rat were soon found infected in many parts of the world and autopsy studies as late as the 1940s revealed infection rates in many countries in Western Europe and the USA as high as 16% (Wright *et al.*, 1944).

Life cycles of Trichinella

Until quite recently almost all life cycle charts showed transmission from rats to pigs, pigs to pigs and pigs to man and differences in infection rates in man were thought to be entirely determined by differences in the way in which pork products were consumed. This is true for most of Europe and the USA where human infections are usually acquired by ingesting the *Trichinella* cysts in undercooked sausages, pork pies and seasoned pork products eaten raw. The ingested cysts develop within 3 days in the small intestine into very small ephemeral worms which mate and produce larvae which then migrate to the muscles where they stimulate an immunopathological response which is responsible for the disease. In most human cases, and in the normal animal host, the reaction is mild and the larvae become encysted in the muscles where they persist for many years in a viable state waiting to be ingested by a new host.

We now know that the true maintenance cycle of *Trichinella* is in wild carnivores and carrion-feeding mammals and that the parasite is not a uniform species. There are such marked differences in the biological properties of the four major forms that they have been designated as different species: *T. nativa* in the Arctic, *T. spiralis* in temperate regions and *T. nelsoni* in the tropics, and sub-tropics (Britov and Boev, 1972). The fourth species, *T. pseudospiralis* has a non-cystic stage in the muscles and has been described from birds and mammals but not from man (Garkavi, 1972) (see figure 8.1 and table 8.1). The taxonomic status of these different forms has been discussed at length by Dick (1983). The maintenance cycles showing the main hosts and the possible ways in which man becomes infected are illustrated in figure 8.2.

Authoritative and extensive reviews of the epidemiology of trichinosis in relation to host range, geographical distribution and prevalence have been produced by Campbell (1983) and Kim (1983). Campbell also speculates on the microepidemiology with particular emphasis on how various species of animals including pigs become infected and he suggests that in the sylvatic cycle scavengers and carnivores contract their infection mainly by eating infected carrion, and that domestic pigs are mainly infected because man has made the domestic pig into a carnivore.

Domestic cycle of Trichinella

In low-income societies the main source of infection for the pig is still from garbage containing pork scraps, but this is no longer the case in Europe and the USA where

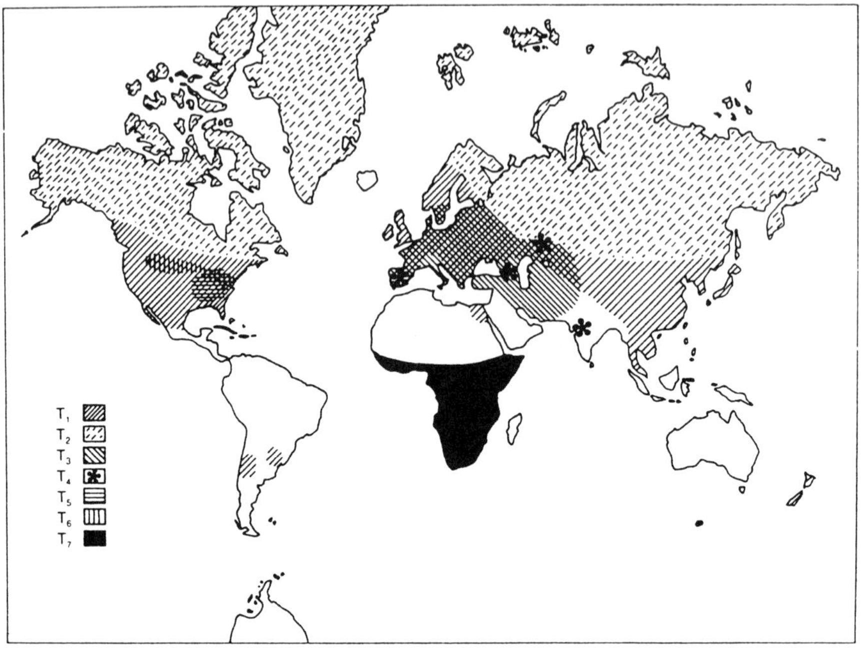

Figure 8.1. Currently estimated global distribution of the seven recognized *Trichinella* gene pools. For further details see table 8.1. Modified from Pozio *et al.*, 1989.

strict laws introduced for the control of hog cholera and swine vesicular exanthema have practically eliminated this form of trichinosis. Here pigs are likely to become infected due to overcrowding where tail-eating and cannibalism are frequent. The omnivorous domestic pig will readily consume a fellow pig which has died and they have no aversion to eating rats or the carcasses of other infected animals. This propensity for eating carrion was responsible for one of the largest outbreaks of trichinosis ever recorded when carcasses of foxes from a fox farm were fed to pigs and by chance the infected pig meat was subsequently consumed over a wide area of Poland, on a particular feast day when it is customary to eat pork as an uncooked 'delicacy'. This Polish outbreak demonstrates the need for meticulous investigations into the source of the infected meat and the way in which it is consumed. An even more bizarre outbreak of severe trichinosis occurred in the Lebanon in a group of strict Muslims who followed the teaching of the Koran. Unfortunately meat purchased as lamb in the market, and served as an uncooked delicacy in the form of 'Kubbeh nayyah', turned out to be adulterated with pork (Matossian *et al.*, 1974). Similar outbreaks have afflicted the faithful in Egypt and an outbreak among German tourists who thought they were eating camel meat was probably due to similar adulteration with pig meat from Egypt.

The decline in pig-borne trichinosis in Europe and the USA is to a large extent due to a reduction in the prevalence of the parasite in domestic pigs as a consequence of better husbandry including strict laws relating to garbage and better

Table 8.1. Taxonomy, distribution and hosts of the seven recognized *Trichinella* gene pools.

Trichinella classifications 1[a]	2[b]	3[c]	Distribution area	Reference strain code[d]	Hosts	Infection in man
T. spiralis s.s. (T_1)	*T. s. spiralis*	*T. spiralis*	Cosmopolitan	MSUS/PO/60/ISS4	Domestic pig but also sylvatic animals	Yes
T. nativa (T_2)	*T. s. nativa* type 1	*T. nativa*	Arctic and sub-arctic regions	MURS/NO/84/ISS10	Sylvatic mammals	Yes
Trichinella type 3 (T_3)	—	*T. nelsoni*	South-west area of Palaearctic region	MVUL/IT/82/ISS2	Sylvatic animals but also domestic animals in contact with sylvatic cycle	Yes
T. pseudospiralis (T_4)	*T. pseudospiralis*	*T. pseudospiralis*	Unknown	MPRO/SU/72/ISS13	Birds and mammals	Unknown
Trichinella type 5 (T_5)	*T. s nativa* type 2	—	Nearctic region	MURS/US/83/ISS34	Sylvatic animals	Unknown
Trichinella type 6 (T_6)	*T. s. nativa* type 3	—	Nearctic region	MURS/US/82/ISS35	Sylvatic animals	Unknown
T. nelsoni (T_7)	*T. s. nelsoni*	*T. nelsoni*	Ethiopic region	MCRO/KE/78/ISS29	Sylvatic animals	Yes

[a] According to *Pozio et al* (1989).
[b] According to Zarlenga and Murrell (in press).
[c] According to Shaikenov and Boev (1983).
[d] Reference strains used by Pozio *et al.* (1989).

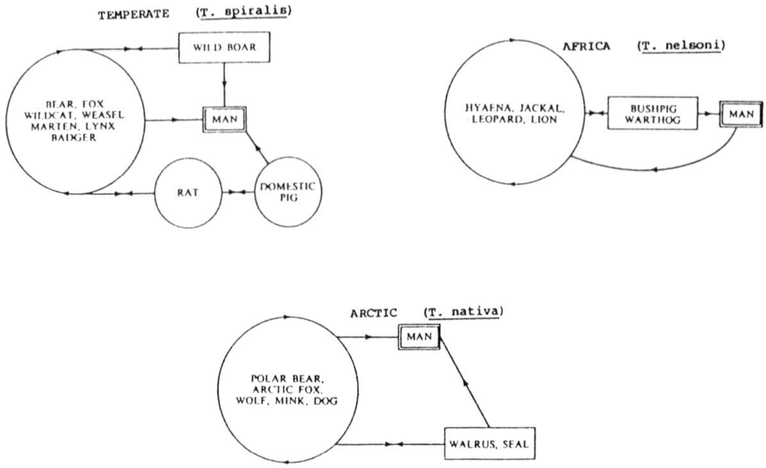

Figure 8.2. Maintenance cycles and transmission to man of *Trichinella spiralis*, *T. nelsoni* and *T. nativa*.

methods of meat hygiene. Other contributing factors include storage methods which ensure that pork products like salamis and smoked hams, which are to be eaten uncooked, are free from infection. The use of home freezers is particularly important, *T. spiralis* derived from the domestic pig can be killed by freezing at a temperature of $-15\,^{\circ}C$ for 20 days or by cooking to an internal temperature of $60\,^{\circ}C$. It can also be killed by gamma irradiation but this excellent method of sterilizing and prolonging the shelf-life of meat has not been widely accepted. The dictum that pork be cooked until it is white ensures that transmission is unlikely from the 'Sunday joint' but the tendency to taste sausage meat to see if it is properly seasoned for stuffing chickens or turkeys and the eating of raw sausages have led to quite severe outbreaks of trichinosis in England and Ireland (Semple *et al.*, 1954; Corridan and Gray, 1969). The danger of eating barbecued pork was dramatically demonstrated in an outbreak of trichinosis amongst the Gurka regiment in Hong Kong (Simon *et al.*, 1986). Although the domestic pig is the most important animal host of *T. spiralis* there have been several outbreaks traced to eating horse flesh. An outbreak in Paris affected 126 people and was traced to eating horse meat from the USA (Ancelle *et al.*, 1985) and a similar outbreak was described in Italy by Bellani *et al.* (1978). The horse, being a herbivore would be unlikely to be infected in nature. It is suspected that the horses involved in those epidemics had either eaten an infected rodent in their food or they had been fed high protein supplements prepared from pig meat, in much the same way as the outbreaks of salmonellosis occur in poultry which are given protein supplements prepared from recycled meat.

Although most human infections in the world are still acquired from domestic pigs there is now increasing evidence of human infection acquired from wild animals. This is particularly true in the Arctic and in Africa which have been used in the following sections to illustrate the way in which man acquires infection from wild animals.

Trichinosis in the Arctic

The first reported outbreak in man in the Arctic was in Greenland where 300 people were affected and 30 died. The source of this outbreak was traced to eating the flesh of the walrus (Thorborg *et al.*, 1948).

As a result of the studies by Merkushev and his colleagues in the northern USSR, Rausch and his colleagues in Alaska and Madsen in Greenland we now know that there is a high prevalence of *Trichinella* in Arctic carnivores, particularly the polar bear, brown bear and arctic fox. There is a secondary cycle in marine mammals which have fed on dead carnivores or indulged in some other form of carrion-feeding or cannibalistic activity (for a review of *Trichinella* in the Arctic see Rausch, 1970). It is likely that trichinosis has long been a serious disease among Eskimos. It has also been a hazard for Arctic explorers, causing the death of members of the fatal Andrée expedition when they tried to cross the North Pole by balloon in 1879 (Sundman, 1970). It is also likely to have been responsible for the severe rheumatism which affected the Norwegian explorer Nansen when he spent the Arctic winter of 1895 subsisting on the meat of polar bears and walrus on Franz Joseph Island. Also it has been recorded among explorers in the Soviet Arctic (Ozeretskovskaia and Upenski, 1957) and the German Garrison in Spitzbergen suffered from a severe outbreak due to eating 'hams' prepared from polar bear meat during the 1939–45 war.

Both brown and black bears have been the source of other outbreaks and the sledge dogs which are often heavily infected may be eaten by man when other sources of food are exhausted.

The first indication that the Arctic parasite differed quite markedly from *T. spiralis* was obtained by Nelson *et al.* (1966) who found that although a strain of *Trichinella* from a brown bear from Alaska was morphologically identical to *T. spiralis* from Europe and the USA, it differed in being of very low infectivity to rats and domestic pigs. Similar differences have been seen in other isolates from the Arctic and as a result of their genetic studies Britov and Boev (1972) named the Arctic parasite *Trichinella nativa*. Other differences have since been described including differences in the isoenzymes and DNA and, of great importance in relation to the epidemiology and control, difference in resistance to low temperatures. For example, *T. nativa* has been shown to remain viable in polar bear meat for as long as 12 months at $-15\,^{\circ}$C. This resistance to freezing together with the longevity of the cysts in the hosts, which in most animals is for the lifetime of the infected animal, has ensured the wide distribution of *T. nativa* even where the host population is of very low density. Man has no role to play in the natural distribution of *T. nativa* but as a result of the farming of foxes and mink for fur and the exhibition of wild arctic animals in zoos the parasite may be more widely disseminated. It is fortunate that the Arctic parasites are not infective to domestic pigs. The next section shows that the same is true for *Trichinella* from carnivores in Africa.

Trichinosis in Africa

Trichinosis was first discovered in Africa south of the Sahara in 1961 when exceptionally heavy infections were recorded in eleven boys on the lower slopes of Mount Kenya by Forrester *et al.* (1961). At the same time both the medical and veterinary authorities assumed that the parasite must have been introduced in pork products for feeding the British troops operating against the Mau Mau, and that the boys must have stolen imported sausages or ham. There were no previous records in man in Kenya and none in domestic pigs even though large numbers had been

Figure 8.3. Carnivore and carrion-feeding hosts of *Trichinella nelsoni* in East Africa. 1 = Lion (*Panthera leo*); 2 = Serval (*Felis serval*); 3 = Leopard (*Panthera pardus*); 4 = Red-river hog (*Potamochoerus poreus*); 5 = Spotted hyaena (*Crocuta crocuta*); 6 = Jackal (*Canis adustus*); 7 = Striped hyena (*Hyaena hyaena*). From Nelson *et al.* (1983).

examined by meat hygienists who were helping with the establishment of a major pig farming industry in Kenya. The clue that the parasite was indigenous came from questioning the infected boys who gave an account of killing and eating a red river hog, a wild bush pig which is only distantly related to the domestic pig (see figure 8.3).

Eventually *Trichinella* was found in a bush pig and a leopard from Mt. Kenya but it was only after high infection rates were seen in hyenas, jackals and lions and other carnivores in more remote areas of the Rift Valley region that the authorities were persuaded that the infection was indigenous (see figure 8.2). Subsequent studies in other parts of the continent showed that carnivores together with secondary scavengers, including bush pigs and warthogs, were infected all over tropical Africa from Senegal to as far south as the Kruger National Park where it was suggested that *Trichinella* might be responsible for limiting the population density of lions and other large carnivores. Of particular significance was that no infections were found in rodents or the small carnivores like mongooses which feed on rodents. Further studies of isolates of *Trichinella* from carnivores in Kenya showed that although the parasite was of high infectivity to man and monkeys it was of exceptionally low infectivity to rats and domestic pigs. Furthermore it could be used to protect rats and pigs from infections with strains of *Trichinella* from Europe and America. It was therefore possible that the widespread distribution of *T. nelsoni* in tropical Africa accounted for the absence of domestic strains of *T. spiralis* in this part of the world.

The young Kenyans who were infected on Mt. Kenya were so hungry that they ate the flesh of an animal which is usually taboo amongst the Kikuyu. Eating the flesh of wild carnivores is also taboo in many parts of Africa although some West Africans will eat dogs. It is likely that these taboos preceded the formalization of the proscriptions against eating the flesh of swine in the Bible and the Koran. Trichinosis is therefore rare in man in Africa because of the taboos against eating carnivores but also because many of the people in West and East Africa are followers of Islam. An outbreak in Senegal was amongst Europeans (Gretillat and Vassilades, 1967) who had barbecued a warthog (*Phacochoerus aethiopicus*). The same species was the source of an outbreak in Tanzania (Bura and Willet, 1977). There have been no outbreaks traced to eating carnivores. The disease therefore can be controlled by reinforcing existing taboos or by health education ensuring that the meat of wild pigs and other potential hosts such as dogs is avoided or well cooked.

[The history of the discovery of *Trichinella* in Africa and an account of the role of 'carrion-feeding, cannibalistic carnivores and human disease in Africa' can be found in more extensive accounts by Nelson (1970, 1982, 1988)].

HYDATID DISEASE

> If a dog drinks from the utensils of anyone of you it is essential to wash it seven times.

Angels do not enter a house where there is a dog.

Sayings of the Prophet Muhamed, related by Muslim, Ahmed, Abu Dawood, Al-Bukhari and Al-Bayhaqi.

Introduction

There are many features in common in the history of hydatid disease and trichinosis. The discovery of life cycles of both *Trichinella* and many species of cestodes originated with the studies of the same group of German parasitologists, more particularly Kuchenmeister, Leuckart, Zenker and Von Siebold, who combined their studies in medical helminthology with studies on similar parasites in domestic animals. Before they carried out their experiments the cystic stages of cestodes in man and animal were regarded as distinct species with no relationship to adult cestodes. The most dramatic experiment of this period was that reported by Kuchenmeister in 1857 when he fed cysts from pigs to 'a convict scheduled to be dispatched from this life to death by the guillotine'. The cysts were mixed in soup and black puddings and were said to have tasted so good that the convict on the evening before his death 'thought it necessary to express his warm thanks to the doctor concerned'! At autopsy the cysts had changed into immature tapeworms with the typical scolices of *Taenia solium*. A similar result was obtained by Naunyn in 1863 who obtained the adult of *Echinococcus* from dogs which he infected with material from a large hepatic hydatid cyst from man. In the same year Leuckart carried out the reverse experiment and successfully produced hydatid cysts in pigs which he infected with *Echinococcus* eggs from dogs. This completed the life cycle but, as with trichinosis, public health workers and veterinarians became so obsessed with the domestic cycle between dogs and sheep, and dogs and man that it was many years before the true maintenance cycles were discovered in wild animals. We now recognize four different species: *E. granulosus*, including several subspecific variants with life cycles involving domestic dogs, livestock, wild carnivores and wild herbivores; *E. multilocularis*, transmitted between foxes and rodents; and two species in South America: *E. oligarthus* in wild cats and rodents and *E. vogeli* in dogs and rodents. The present chapter is illustrated by reference to the main parasite of man, namely *E. granulosus*, with particular emphasis on the studies carried out in Turkana in Kenya where there is the highest prevalence of hydatid disease in man in the world. [More extensive accounts of the history of the discovery of the parasite and the studies in Kenya can be found in Nelson (1988). The global problem of hydatid disease is reviewed in the FAO/UNEP/WHO Guidelines for Surveillance, Prevention and Control of Echinococcosis/Hydatidosis (WHO, 1981) and the recent advances in experimental aspects can be seen in *The Biology of Echinococcus and Hydatid Disease* edited by Thompson, 1986.]

Echinococcus and hydatis disease in the Turkana district of Kenya

The life cycle of *Echinococcus* is more complex than that of *Trichinella* in that there is no direct transfer from host to host. There is a further component in the life cycle in that the taeniid-type eggs containing the hexacanth embryo are passed in the faeces of a definitive host, usually the dog, and then have to survive in the external environment before they are ingested by man or a normal herbivorous intermediate host such as the sheep or camel. There are remarkable differences in the incidence of hydatid cysts in man in different parts of the world, even in different parts of the same country, with no correlation between the human infection rates and the prevalence in dogs and domestic animals; microepidemiological studies are necessary to explain these differences. Observations on the relationship between man and dogs have helped to explain this paradox in Kenya where, following the lead given by Schwabe (1964) who had studied ethnic and cultural practices in relation to the transmission of hydatid disease in Lebanese Christians and Muslims, emphasis was given to both the biological and anthropological aspects of transmission.

The Turkana district covers more than 60 000 square kilometres in the north-west corner of Kenya bordering Ethiopia, the Sudan and Uganda. The people who occupy this semi-arid region are known as 'The Turkana'. They are wandering pastoralists with great herds of cattle, sheep, goats, camels and donkeys always accompanied by numerous dogs. They live in temporary thorn scrub manyattas and their traditional diet consists of a mixture of blood and milk from their animals supplemented with berries and leaves; meat is reserved for ceremonial occasions. The great drought of 1958–61 resulted in the death of many of the people and their livestock and this attracted numerous relief agencies and missionaries to the area. Nearly 30 years later there are still many Turkana congregated around the famine camps where they are provided with regular rations of maize meal, dried milk and vegetable oil but most of the people have reverted to their traditional way of life.

Hydatid disease was first described as a serious problem in the Turkana by Wray (1958) who noted that most of his patients with enormous self-diagnosed hepatic cysts were all Turkana from 300 miles away and not local Masai. He suggested that the unusually high prevalence in Turkana might be caused by a different parasite or a different life cycle from that seen elsewhere in Kenya, where there were very few human infections although sheep and goats were known to be heavily infected. At that time there were no records of adult worms from dogs or wild carnivores in Kenya so Wray's suggestion was quite plausible. However, the cysts from man and animals were typical unilocular cysts of the *E. granulosus* type and subsequent observations by Rausch and Nelson (1963) on adult material from dogs, jackals, hyenas and hunting dogs revealed that the Kenya material was similar to *E. granulosus* from many parts of the world.

The first explanation for the extraordinary high prevalence of hydatid disease in Turkana came from observations showing that there were unusually large numbers of heavily-infected dogs per family and that there was an exceptional degree of intimacy between the people and their dogs. Other unusual features in Turkana were the high prevalence in children and the high female–male ratio with more than twice as many women infected as men.

In general, human infections may result from the direct ingestion of eggs of *Echinococcus* from the faeces of dogs, or indirectly by handling dogs which may have eggs attached to their fur or even around their mouths, due to their habit of licking their own anus. The eggs may also be ingested from contaminated food, water, soil or eating utensils. Under conditions of moisture and shade, as for example on the sheep pastures in Wales or New Zealand, the eggs of *Echinococcus* can survive for many weeks but they are rapidly killed by direct sunshine such as occurs in the arid areas of Turkana. It was therefore suggested that there must be a rapid transfer of eggs from dogs to the people and that the high prevalence in Turkana children was due to the custom of allocating a dog to each child to be used as a 'nurse' to guard them and to clean them when they defaecated or vomited (Nelson and Rausch, 1963). In a country where the women may have to walk many miles to the nearest water hole this custom saves water and for the same reason the women may even use dogs to clean themselves when they are menstruating. Another possible cause of infection was the use of dog faeces for tanning leather, as described by Schwabe and Abou-Daoud (1961) in the Lebanon, or as a medicament. In the first British 'textbook' of helminthology Cobbold (1864) refers to the observation by Leuckart who thought that the unusually high prevalence of hydatid disease in Iceland could be explained because: 'The treatment of the quacks is exactly suitable to keep up the epidemic, for amongst their remedies dog urine and fresh dog excrement play a conspicuous part'. In fact dog's faeces mixed with charcoal are used in Turkana for treating wounds and the women use dog's faeces mixed with fat as a lubricant to prevent damage to their skin under their necklaces; especially their modern bead necklaces which are much heavier than the traditional ostrich egg shell necklaces (French *et al.*, 1982). However there is no evidence of ingestion of the tapeworm eggs in a medicament such as occurred in Iceland or as still occurs in South Africa where witch-doctors use ground-up

Figure 8.4. Turkana woman with dogs.

tapeworms of *Taenia* as a medicament and as a result cause cysticercosis of the brain and epilepsy (Heinz and McNab, 1965).

Another source of infection in Turkana is probably contaminated water. The eggs of *Echinococcus* can survive long periods in water but it has been difficult to prove their presence in waterholes because *Echinococcus* eggs are morphologically identical to the eggs of *Taenia saginata* and *T. solium* of man and of a variety of other species such as *T. hydatigena* which are practically universal in dogs. This problem has now been solved by Craig *et al.* (1988) who have developed an immunodifferentiation technique based on a monoclonal antibody specific for *Echinococcus* eggs. An important spin-off of this work has been the development of a reliable method of diagnosing *Echinococcus* infections in dogs using anal swabs. This avoids the messy business of looking for adult worms in the faeces of dogs after purgation or examining the bowel contents at autopsy. The immunodifferentiation technique has shown that *Echinococcus* eggs are widespread within the homesteads. They have been found contaminating eating bowls, the dried skins used as sleeping mats and the soil in shaded areas of the huts, where the eggs have the best chance of survival. Studies by Watson-Jones and Macpherson (1988) on man/dog contact in Turkana have also shown that the time spent by dogs in the manyattas is strongly correlated with the prevalence of hydatid disease (figure 8.5 and table 8.2).

Although the epidemiological studies suggest that the unusually high prevalence in Turkana is due to the intimacy between man and dog there is still the possibility that the parasite is also different and that it has an unusually high infectivity for man. Elsewhere in the world where *Echinococcus* is prevalent there are strong selection pressures for the parasite to become adapted for transmission in a variety of restricted carnivore–herbivore cycles (Thompson and Lymbery, 1988). In one of these cycles between dog and horses in England there is epidemiological and experimental evidence of very low infectivity to man and monkeys (Smyth, 1977). The reverse may be the case in Turkana because, here man, unlike anywhere else in the world, is a normal component of the life cycle. There are no burial customs, except for respected elders, so dogs and jackals feed on human corpses containing human hydatid cysts (Macpherson, 1983). They also acquire infection from sheep, goats and camels but there is no wildlife cycle such as exists in Masailand between the wild carnivores and wild herbivores. The situation in Turkana may be a fairly recent event due to their possessing enormous numbers of rifles which were originally supplied by the British Government to help fight the Ethiopians and Italians during the 1939–45 war. The rifles are now used for cattle raiding and for killing the few remaining wild game animals which are essential prey of the large carnivores; with their disappearance a true wild carnivore – wild herbivore cycle is no longer possible. However, jackals which feed on small mammals, birds and carrion, including human corpses, still survive in considerable numbers and are excellent hosts for the human Turkana parasite. When cysts from man were fed to dogs and jackals (figure 8.6) the jackals produced even heavier infections than the dogs (Macpherson *et al.*, 1983). It is likely that jackals help to maintain the life cycle among the domestic animals, which will have access to infected jackal faeces on the

Table 8.2. Percentage of time spent by dogs in different parts of the study settlements in relation to the overall incidence of hydatidosis among the inhabitants.

		Location of dog(s)						
	Hydatidosis incidence per 100 000 per year	1 Akai	2 Atabo/ekal	3 Outside akai – own manyatta	4 = 1 + 2 + 3 Total time in own manyatta	5 Neighbouring manyatta	6 Other parts of village	7 Bush/fields
Lokichokio 1986	198	8·6	0·6	27·0	36·2	34·8	19·2	10·1
Lokichokio	198	20·7	15·3	38·5	74·5	10·6	10·9	10·1
Lokitaung	128	0·8	20·9	42·5	64·2	1·6	0·0	34·2
Lokichar	28	0·5	16·2	45·4	62·1	1·7	1·7	34·5
Kapeddo	17	0·0	0·0	52·5	52·5	2·6	0·2	44·7
Longetch	Low	0·0	0·0	35·3	35·3	17·2	5·4	42·1
Kositea	Low	2·2	4·1	12·3	18·6	19·9	11·3	50·2
Komads 1986	?	0·8	4·4	76·5	81·7	—	—	18·3
Spearman's rank correlation coefficient		+0·515	+0·691	+0·406	+0·986[b]	−0·319	−0·899[a]	

[a] Spearman's rank coefficient (rs) is significant at $P<0{\cdot}05$ for $(rs)>0{\cdot}829$ ($n = 6$, one-tailed test, i.e. in the predicted direction).
[b] Spearman's rank correlation coefficient (rs) is significant at $P<0{\cdot}01$ for $(rs)>0{\cdot}943$ ($n = 6$, one-tailed test, i.e. in the predicted direction). Spearman's rank applies to the 1985 values only.
Each settlement comprises a Manyatta enclosed by a fence within which are located huts and shelters for various activities. These are: the Akai, a hut used for sleeping at night/preparation of food; the Ekal, used in the daytime as an eating and resting place; the Atabo, an open structure usually containing a fireplace and used for the preparation of food; and the Anok, a thorn fence enclosure for livestock. Modified from Watson-Jones and Macpherson (1988).

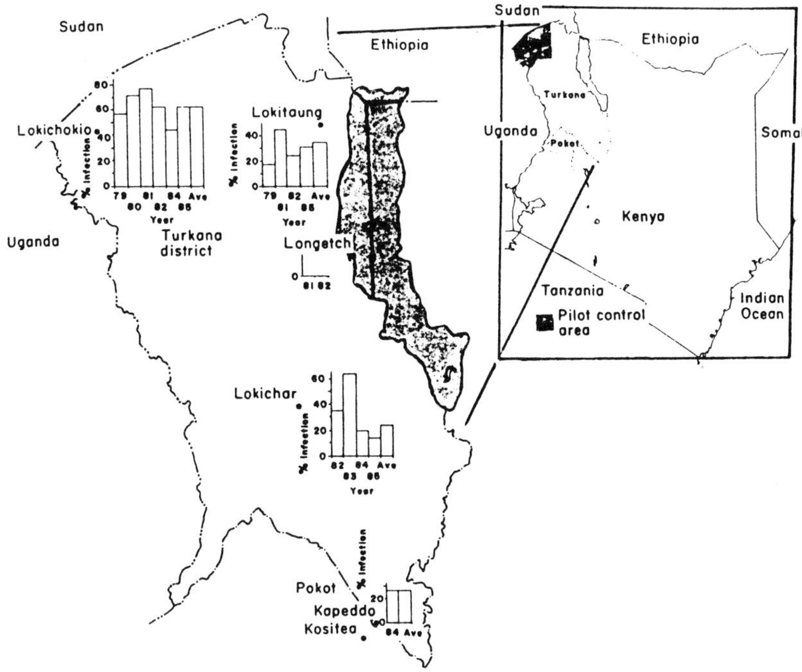

Figure 8.5. Map of the Turkana district showing the prevalence of *Echinococcus granulosus* among dogs in the different locations studied. See also table 8.2.

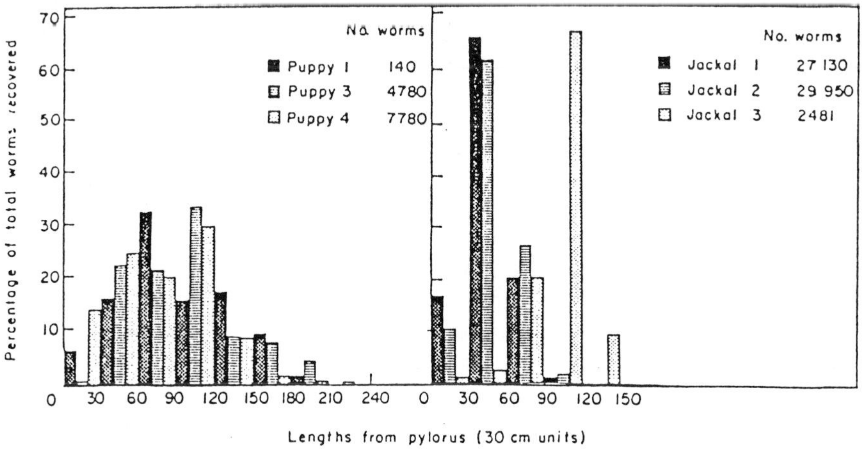

Figure 8.6. Distribution of worms in the small intestines of three puppies (a) and three silver-backed jackals (b) infected with *E. granulosus* of human origin. Three spotted hyaenas were also infected but no worms were recovered at autopsy. All animals were given 160,000 protoscolices packed into a gelatin capsule with a volume of 0.2 ml. The material was obtained from an hydatid cyst removed surgically from a Turkana patient in Lodwar hospital 2 days previously. Modified from Macpherson *et al.* (1983).

pastures but apart from the contamination of the water holes it is unlikely that they are of any significance in transmission to man. Although a great deal of work has been done on the experimental transmission from man to different hosts and from dogs to different hosts and although numerous isolates have been compared for possible differences in their epidemiology, biochemistry and genetic characteristics, the question as to whether the Turkana parasite is different has not been answered. (See McManus and Bryant, 1986 for a general review of this problem.)

Control of hydatid disease in Turkana

A major control programme is underway in Turkana with day to day supervision by the African Medical Research and Education Foundation (AMREF). It is based on the following methods:

1. Mass surveys of the human population using ultrasonography and serology followed by treatment of all cases with either surgery or albendazole.
2. The reduction of dogs by killing all strays and unwanted animals followed by the registration of the remaining dogs together with their repeated treatment with praziquantel.
3. Health education to disseminate knowledge and active participation of the whole community in the programme.

Treatment is very expensive and although during the past 12 years the Flying Doctor Service from AMREF has operated on more than 1700 patients of a total population of only 120 000 people, many more cases are being detected every day. The drug is very effective, and invaluable in complicated cases, but it has to be taken daily for up to 3 months which is impracticable without the special provision of treatment centres. The treatment of dogs is also expensive. A detailed study of the population dynamics of the parasite in the dogs has shown that reinfection occurs within 6 weeks after treatment (Wachira, 1989). Research is in progress to develop a long-acting depot preparation of praziquantel for the control of the infection in dogs and efforts are being made to develop vaccines but at present the only hope for the long-term control of hydatid disease in Turkana is to change human behaviour so as to avoid infection.

A great deal is being done to gain the trust and respect of the people by studying their customs and beliefs and by working through traditional methods of communication. All the health educators are local Turkana; many of them are patients who have been successfully treated for hydatid cysts. Of particular importance are the 'hydatid songs' which they have composed in their own language and which they sing accompanied by dance whenever there is a tribal gathering. There are many songs and they cover all aspects of the life cycle. For example the 'Prevention Song' can be translated as follows:

It is easy to prevent this disease – hydatidosis (repeat)
Those dogs that have owners, take them, take them to be dosed
Those dogs that have no owners, kill them, kill those beasts

And the cyst – don't give it to the dog
Throw it in the fire – don't give it to the dog (repeat)

Another song is entitled 'Playing with the Dog':

Don't play with the dog – it brings bad things (repeat)
Don't play with that animal it brings bad things

Don't let it lick your baby
Don't let it lick your sores
It will bring bad things

The dogs will leave 'eggs' on our hands
The dogs will leave 'eggs' on our skins
The dogs will leave 'eggs' inside our huts
The dogs will leave 'eggs' inside our walls
The dogs will leave 'eggs' on our clothes

We chase the dogs from our dwellings
Our brothers chase away that animal
Our sisters chase away that animal
Our mothers chase away that animal

It brings hydatid to the people
So that we have to be cut open.

The people are also 'educated' through proverbs and poems about the disease and many of the patients who have been treated are proud to wear T-shirts illustrating the life cycle. Before this study began the people recognized hydatid cysts in their livestock but they did not associate these with their own 'big belly disease'. They had no idea of the role of dogs in the life cycle and since dogs are so essential in the way of life of the people it will be many years before their attitudes can be changed. In the meantime there is a greatly increased awareness of hydatid disease in the country. Much is being done by the medical and veterinary officers of the Kenya Government, the medical missionaries, by AMREF and by the research workers in the University of Nairobi and the Kenya Medical Research Institute, but the eventual control of the disease will depend on the enthusiasm of the Turkana themselves to apply the control measures.

SCHISTOSOMIASIS

Introduction

The last group of parasites selected to illustrate the importance of understanding human behaviour in relation to transmission and control are the schistosomes: *Schistosoma mansoni*, *S. haematobium* and *S. japonicum*. Their life cycles are much more complex than those of *Trichinella* and *Echinococcus* because they require highly specific snail intermediate hosts for their development and multiplication and in the case of *S. mansoni* and *S. japonicum* man is not the only definitive host. In many parts of the world schistosomiasis is becoming more widespread and it is increasing in prevalence and intensity of transmission as a result of water resource developments. This is a paradox of development in the developing world; attempts to produce more food or energy through irrigation or the building of dams results in an increase in the snail hosts of schistosomiasis. This has occurred, for example, in Egypt with the building of the Aswan High Dam, in Ghana with the development of the Volta Lake, in the Sudan in the vast Gezira irrigation scheme and in Brazil in numerous large and small irrigation and hydroelectricity schemes. In many parts of South East Asia the enormous human population explosion has been followed by the green revolution with increased irrigation and in some areas, like the Philippines, by an explosive increase in schistosomiasis. There have been several successful control programmes particularly in China, Japan, Iran and the West Indies but despite enormous efforts the disease is still a major cause of morbidity in countries like Egypt, Brazil and many parts of Africa.

The discovery of the life cycles of schistosomes

The great Danish zoologist Steenstrup in his thesis on *Alternation of Generations* in 1845 demonstrated that cercariae and rediae (which previously were thought to be distinct species of helminths parasitic in snails), were the intermediate stages of adult trematodes, but it was not until 1913 that the life cycle of the first schistosome was unravelled. Bilharz had discovered the adult worms in the mesenteric and vesical blood vessels of man in Egypt in 1850 and it was soon realized that these parasites were the cause of the haematuria which had plagued the Egyptians for more than 4000 years. Bilharz recognized that these were unusual trematodes in that they had separate sexes and they were elongated to live inside the blood vessels. He also noted that they produced eggs which when placed in water liberated a typical free-swimming ciliated trematode miracidium. This suggested that there might be a snail intermediate host but despite numerous experiments by Bilharz and his successors none was found and with the dogmatic authority of the great German parasitologist Looss (who had discovered the hookworm life cycle) it became accepted that the infection was acquired by swallowing the miracidia in drinking water. There were some who doubted this conclusion including Brock (1894) in South Africa who claimed that the parasite penetrated the skin. He had observed that it was most prevalent in boys who bathed in the local streams but uncommon in girls who never bathed but who drank the same water; moreover the boys complained of itching after bathing.

The crucial experiment demonstrating skin penetration was carried out by Fujinami and Nakamura (1909) in Japan where *S. japonicum* had been found in man, cats, dogs and cattle. They showed that muzzled cows with 'wellington boots' were not infected when they paddled in the water but control animals without boots became infected. It was also in Japan where the first snail hosts were discovered by Miyairi and Suzuki (1913) who produced a detailed description of all stages of development of the parasite including the infective cercariae.

The Japanese discoveries were confirmed by Leiper and Atkinson (1915) in China, but only with great difficulty, mainly due to a clash of personalities. The story of this 'collaborative' study carried out by the irrascible, ambitious Scottish academic Leiper and the very 'English' naval officer Atkinson, who had been parasitologist on Scott's last expedition to the Antarctic has been told elsewhere (Nelson, 1977) and exemplifies the egotistical race for priority which was a feature of this period of the history of tropical medicine. Leiper later went on to help the army in Egypt to solve the problem of schistosomiasis and as a result of a series of brilliant observations he not only worked out the life cycles of both *S. mansoni* and *S. haematobium* but he also developed effective methods for their control (Leiper, 1915).

Schistosomiasis has always been a disease which threatens armies operating in endemic areas: Napoleon's troops complained that they were menstruating like women; similar complaints were made by British troops in the Boer war in South Africa and later in Egypt, and it has been claimed that the failure of Republican China to conquer Taiwan was because thousands of their troops became ill with acute *S. japonicum* infections whilst training for the invasion on the mainland.

The concentration of research on schistosomiasis in Egypt led to the misconception that the disease was maintained in nature by interhuman transmission. This is true for Egypt and many other parts of the world where *S. haematobium* and *S. mansoni* are 'man-made' problems but it is not true of *S. japonicum* where there is an enormous reservoir of the parasite in wild and domestic animals and it is not true of *S. mansoni* in Africa where baboons and possibly rodents are reservoir hosts. Schistosomiasis is also a zoonosis in the Caribbean and in South America where several species of rodents and aquatic animals are infected with *S. mansoni*. The presence of an animal reservoir has obvious implications for the control of the disease, especially where there is emphasis on mass chemotherapy. (The significance of an animal reservoir host in the epidemiology of schistosomiasis has been reviewed by Nelson (1975) and Rollinson and Southgate (1987).)

The control of schistosomiasis

There are four main methods for controlling schistosomiasis:

1. Kill the parasite with chemotherapy to reduce morbidity and to reduce transmission.
2. Kill the snails with molluscicides or change the environment so that the snails cannot breed.
3. Prevent the contamination of potential transmission sites with excreta containing the eggs of the parasite.
4. Protect people from exposure to the infective cercariae in infected water by providing alternative clean water.

Vaccines are being developed but they have not been tried in man and even in animal models they are only partially effective.

The following account, which is concerned with the factors affecting transmission and the control of transmission by changing human behaviour, deals mainly with contamination and exposure.

Contamination of transmission sites with infective eggs

> The Prophet Muhamed said:
>
> Avoid the three curing tasks –
> Defaecation in water sources
> Defaecation in pathways
> and Defaecation in shaded areas
>
> (Related by Muslim, Ahmed and Dawood)
>
> He also said:
>
> Avoid urination in stagnant water
>
> and
>
> Avoid urination in bathing places
>
> (Related by Ahmed, An-Nasai, Ibn Majah and 'The Five')

If the people of Islam had obeyed the Prophet's injunctions against contaminating the environment with human faeces and urine to the same extent as they have observed his injunctions against eating pig meat, schistosomiasis and many other helminths would be as scarce now in Islamic countries as trichinosis.

The severity of schistosomiasis is proportional to the number of parasites in the body and because the only stage of multiplication of the parasite is in the snail host, the severity of the disease in man is determined by his degree of exposure to infected water. This in turn is determined by the amount of contamination of the water by schistosome eggs and the subsequent concentration of cercariae in the water. In the worst affected communities children between 5 and 14 years old have the highest egg output (figure 8.7) and they may be responsible for most of the contamination. This is particularly true for *S. haematobium* because children of this age group represent a high proportion of the population in developing countries and there is a rapid decline in egg output in adults. The same may be true for many *S. japonicum* and *S. mansoni* foci. (See Jordan and Webbe (1982) and Wilkins (1987) for reviews of this aspect of the epidemiology of schistosomiasis.) However, there are dangers in generalizing from observations in limited areas. For example, there are striking differences in the epidemiology of *S. japonicum* in China, with its small child population, compared with the Philippines with its relatively much greater proportion of children. There are also some areas where there is a very high prevalence and intensity of infection of *S. mansoni* in all age groups. Here adults may be just as important sources of contamination as children, for example on the Nile (figure 8.8) in Uganda (Nelson, 1958; Ongom and Bradley, 1972) or in the Sudan where canal cleaners are particularly heavily infected (figure 8.9).

Also, statistical data on age-specific prevalence and intensity rates may be misleading as pointers to the age groups responsible for contamination in that they represent risks of exposure influenced by possible immunity but not necessarily the groups responsible for infecting snails. This can be seen in dry savanna areas where heavily-infected children often defaecate and urinate around the homesteads, where the eggs in the urine and faeces are rapidly destroyed by desiccation, whereas less heavily-infected adults seek the privacy of shaded banks of rivers or ponds to defaecate. There have been very few detailed scientific studies of modes of contamination because of the obvious rather unsavoury task of studying urination and defaecation. More attention has been given to studying risks of exposure by watching people washing and bathing. The most detailed study of human defaecation behaviour in relation to a medical problem was carried out by Kochar *et al.* (1976) in their studies on hookworm infections in Bengal in India. Here they were able to quantify both the degree of contamination and the risk of exposure by counting hookworm larvae in the soil. In schistosomiasis similar efforts have been made to develop methods for determining miracidial densities but so far they have not proved of value in the field. It is therefore necessary to determine the degree of contamination less directly by using sentinel snails and by estimating snail infection rates.

Contamination by S. mansoni *eggs*

One of the best studies on contamination has been carried out by Ouma (1988) in Kenya where he has made detailed observations on human defaecation in all age groups in relation to potential transmission sites. The area he selected had a high prevalence and intensity of *S. mansoni* infection; a great deal was known about the infection rates of the snails in the diverse but relatively small habitats and there was

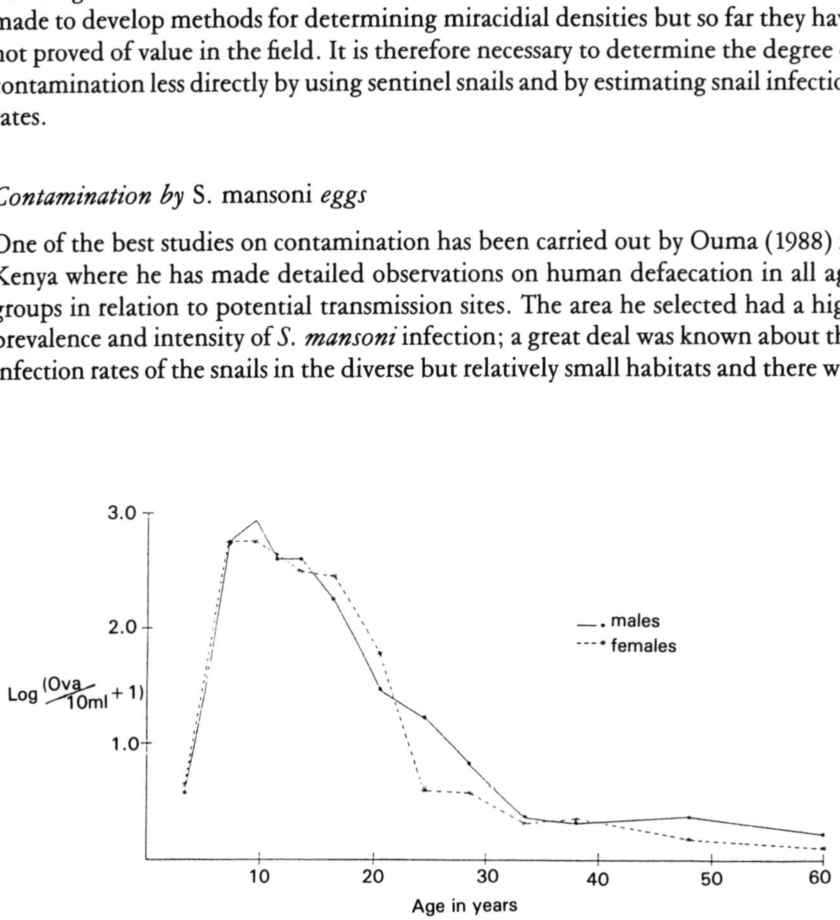

Figure 8.7. Age-related changes in the output of eggs of *Schistosoma haematobium* in male and female inhabitants of villages near Bansang in McCarthy Island Division of The Gambia. Data from Wilkins (1977).

Figure 8.8. Woman heavily infected with *Schistosoma mansoni* on the Nile in Uganda.

Figure 8.9. Canal cleaners heavily infected with *Schistosoma mansoni* in the Gezira irrigation area of the Sudan.

excellent participation in the project by the local people. He concluded that defaecating, even close to the water, was far less important than bathing. By using a saline swab technique a very high proportion of the people were found to have eggs of *S. mansoni* around the anus and his epidemiological studies showed that snail infection rates and cercarial densities were higher in bathing places than in similar water bodies used for washing clothes or for collecting drinking water.

Although important and probably underestimated, this is not the only way in which *S. mansoni* eggs get into the water. During the rainy season the eggs deposited near the water may remain viable and will be washed into the ponds and streams before they desiccate and young children and some adults may defaecate directly into the water. The Moslem custom of washing the anus after defaecation may be important and even washing clothing contaminated with faeces may be responsible for infecting snails. Unprocessed, or even partially processed sewage, may contain large numbers of viable eggs and effluents containing the eggs may infect snails. However, rivers which are heavily polluted by sewage effluent are usually inhospitable to the snail hosts of schistosomes. Transmission along the shores of Lake Victoria is unusual in that here the snail hosts *Biomphalaria choanomphala* are found at considerable depths and they are thought to be infected from the passing lake steamers and lighters which are often crowded with hundreds of people.

Animal reservoirs as sources of contamination

Baboons are particularly important hosts in areas where they are commensals of man, stealing his crops, sharing his water holes and often roosting in trees or rocks where they drop their faeces either directly into the water or alongside the streams and ponds where they are easily washed in by the rain (Nelson *et al.*, 1962; Fenwick, 1969). Baboons are increasing in numbers due to the destruction of their main predator the leopard, and more especially due to conservation and they are now found in very large numbers in the endemic areas of Africa and Arabia. Rodents are not such good hosts and Pitchford and Visser (1962) have shown, by field experiments, that they are unlikely to be important hosts in South Africa. There is however some evidence that they may be good reservoir hosts in Kenya (Kawashima *et al.*, 1978) and their role together with a great variety of other animals is still debatable in South America. The studies by Ouma (1988) in Kenya suggest that coprophagous animals which ingest human faeces could act as transport hosts from man to water and he has observed high spurious 'infection rates' in dogs which eat human faeces.

Contamination by S. haematobium *eggs*

Most of the eggs of *S. haematobium* are passed in the urine and since urination is almost a reflex action when bathing this is likely to be the commonest source of contamination. But not all *S. haematobium* eggs are deposited in the urinary tract. The adult worms are found thoughout the pelvic veins and the eggs are sometimes excreted in the faeces. They are also frequently found around the cervix of the uterus and Hustings (1965, 1970) has suggested that this might be a secondary mode of transmission.

Although chimpanzees, baboons and several species of monkeys are excellent laboratory hosts for *S. haematobium* very few animals have been found naturally infected and there is as yet no evidence of the natural maintenance of *S. haematobium* in non-human primates, as was demonstrated by Fenwick (1969) for *S. mansoni*.

Contamination by S. japonicum *eggs*

Because the eggs of *S. japonicum* are passed in the faeces the transmission from man to snails is similar to that of *S. mansoni*, but also because *S. japonicum* has numerous animal hosts and is confined to Asia there are many biological and cultural differences affecting transmission. The higher standards of hygiene together with mass chemotherapy, and more particularly modern agricultural developments, have practically eliminated *S. japonicum* from the endemic areas of Japan and from the lower Yangste valley in China. Contamination by human faeces was formerly due to the widespread use of human faeces as a fertilizer and also the daily washing of chamber pots in the canals containing *Oncomelania* snails.

Another important source of contamination in South East Asia is the practice of constructing latrines over water. There is also evidence suggesting that the eggs of *S. japonicum* can survive much longer than *S. mansoni* and *S. haematobium* in the environment and so contamination is more likely to occur when *S. japonicum* eggs are washed into the water from the land (Garcia, 1976).

S. japonicum is primarily an animal rather than human parasite and the cercariae emerge from the snails at night as an adaptation to transmission by nocturnal animals. Nocturnal rodents proved to be the 'missing link' in the mathematical model designed by Hairston (1962) to explain the dynamics of schistosomiasis transmission in the Phillipines. Elsewhere other animals are involved in transmission but everywhere man is the most abundant host.

At one time the most important animal reservoir in Japan was the domestic cow. The replacement of the cow by Mr Honda's motorized ploughs was claimed by Yokagawa (1970) to be an important reason for the decline of *S. japonicum* in Japan. In China, where cattle and water buffaloes are sources of contamination, special harnesses with plastic containers are attached to the back end of the animals to collect their faeces.

Acquisition of schistosome infections

Although schistosome infections can be acquired by drinking, as a result of cercarial penetration of the buccal mucosa, the normal method of infection is by cercariae penetrating the skin. There have been several observational studies relating rates of exposure and types of exposure to the prevalence and intensity of infection. These studies have given invaluable information on acquired immunity and they have provided a baseline for assessing control measures.

Important information in relation to the epidemiology and control of schistosomiasis has come from observational studies such as those by Husting (1965, 1970) in Zimbabwe, Polderman (1974) in Ethiopia, Dalton (1976) in St. Lucia, Dalton and Pole (1978) in Ghana and Cheesemond and Fenwick (1981) in the Sudan. Studies which have been particularly important in relation to naturally acquired immunity have been those by Butterworth *et al.* (1987) in Kenya on *S. mansoni* and Wilkins *et al.* (1987) in the Gambia on *S. haematobium*.

Exposure can be correlated with infection rates in snails and in some situations with cercarial densities as measured by cercariometry using the technique first used by Rowan (1958) in Puerto Rico and improved by Theron (1979), Blumenthal and Jewsbury (1983) and Prentice (1984). But there is still the problem with cercariometry of not being able to identify the species of schistosomes and in area where bovine and other animal schistosomes are abundant this can cause confusion. An alternative method for estimating the infectivity of different water bodies is by exposure of sentinel animals. This is a costly and sometimes cruel business but provided the immersed animals subsequently produce identifiable male and female adults the results can be of value.

All of these observations on both contamination and exposure provide essential information for control programmes where serious attempts are made to get the full participation of the community in reducing transmission. They also pin-point the contaminated water bodies which then can be either modified to make them unsuitable for snails or treated by local molluscicide applications.

Control of schistosomiasis based on changing human behaviour

With few exceptions the history of the control of schistosomiasis is a record of undue optimism and enthusiasm followed by failure and pessimism. Often this has resulted from short-term projects when molluscicides have been used for killing the snails or chemotheropy has been used for killing the parasites. The disillusionment has been so profound that in many countries the control of transmission has been abandoned in favour of the concept of reducing morbidity by reducing worm-loads with chemotherapy often unsupported by any attempt to change human behaviour. With drugs such as praziquantel which kills all three parasites this is the easiest and least expensive option. Molluscicides are no longer used on a wide scale but when the transmission sites are known and if funds are available to supplement chemotherapy, focal application is recommended. Even with these methods some community participation is necessary but in the longer term the most successful control schemes have depended on modifying human behaviour so as to avoid transmission by improving sanitation and by the avoidance of snail-infested water. In some areas this has been possible by legislation and by the fencing-off of transmission sites but in the most successful schemes it has been achieved by good health education and the provision of alternative water supplies for bathing, cooking and for washing clothes. The advantages of this approach were clearly demonstrated in the St. Lucia project which also had the added advantage of reducing other water-borne diseases (Jordan, 1985). But St. Lucia is a small island in the Carribbean and the control programme depended on a great deal of external help both financially and with staff. The problem is much more difficult in the poorer countries of Africa. For example in Egypt, where for many years schistosomiasis has been given the highest priority in the public health programme, there have been enormous efforts to control the disease but there are still more people infected with schistosomiasis in the Nile delta now than there were in 1914 when Leiper worked out the life cycles of *S. mansoni* and *S. haematobium*. There has been some successful control against *S. haematobium* in Middle and Upper Egypt, as a result of a World Bank–Ministry of Health project, but even here according to an 'Independent Evaluation Mission' (1985) there has been no detailed analysis of human behaviour in relation to transmission and the Health Education Programmes have been largely ineffective. However, there have been some recent improvements and one of the most popular daily programmes on Egyptian television is a satirical comedy with two leading celebrities discussing ways of contracting and avoiding schistosomiasis and the best ways of curing the disease.

Health education is only one of the facets of 'integrated control programmes' but it is an essential part on which everything else depends. Health education also requires accurate anthropological information. In a stimulating essay entitled 'The behaviour of *Homo sapiens*, the forgotten factor in the transmission of human disease' Gillett (1985) illustrates the failure to study human behaviour in relation to the epidemiology and control of some of the great vector-borne diseases such as malaria, dengue and yellow fever. He goes so far as to say that WHO and relevant University departments should be radically reorganized to give this subject the highest priority because: 'we have, after all, had almost 100 years to try out the old methods and, as far as the Third World is concerned they have for the most part failed'. This is not altogether true; there have been some striking successes in the control of parasitic and vector-borne diseases but mainly as a result of dictatorial or colonial policies which are no longer acceptable. Future control programmes will require the whole-hearted participation of the affected communities and their consent to change traditional behaviour, because they themselves recognize the benefits of such change

In the last analysis all public health control programmes need the political will of National Governments to provide the initiative and necessary funds to carry them through. It is no accident that the most extensive and most successful programme for the control of schistosomiasis was in the Yangtse Valley in China. The programme was initiated and encouraged by Chairman Mao Tse Tung. He said:

> Knowledge begins with practice and
> Theoretical knowledge acquired through practice
> Must then return to practice

This philosophy inspired practitioners, research workers and everyone concerned with the control of schistosomiasis in China. It is a philosophy which can inspire all of us in our efforts to study and control many other parasitic diseases of mankind.

ACKNOWLEDGEMENTS

Over a period of more than 30 years I have had the pleasure of collaborating with many scientists in the study of trichinosis, hydatid disease and schistosomiasis in East Africa. Nothing would have been achieved without the willing participation of the local people. I am particularly grateful to many of my former students and to the field staff of the Division of Vector-Borne diseases and the African Medical Research and Education Foundation in Kenya for their cheerful assistance often under very difficult conditions. I am grateful to the Publishers for permission to use extracts from this chapter for my Presidential Address to the Royal Society of Tropical Medicine and Hygiene in October 1989. (See *Transactions Royal Society of Tropical Medicine and Hygiene*, 1990, **84**, 3–13).

REFERENCES

Ancelle, T., Dupony-Camet, J., Heyer, F., Faurant, C. and Lapierre, J., 1985 *Lancet* **2**, 660.

Bellani, L., Mantovani, A., Pampliglione, S. and Filippini, I., 1978, In *Trichinellosis* edited by C. W. Kim and Z. S. Pawlowski (Hanover: University Press of New England), pp. 535–84.

Blumenthal, U. J. and Jewsbury, J. M., 1983, *Annals of Tropical Medicine and Parasitology* **77**, 441–442.

Britov, V. A., and Boev, S. N., 1972, *Vsesoylnaya Konferentisya (viii) po Prirodnoi ochagovostgi boleznei zhiretnkich i ohrane ikh chrislennosti*, **1**, 83–84.

Brock, G. S., 1894, *Journal of Pathology and Bacteriology*, **2**, 52.

Bura, M. W. T., and Willett, W. C., 1977, *East African Medical Journal*, **54**, 185–193.

Butterworth, A. E., Capron, M., Cordingley, J. S. *et al.*, 1985, *Transactions of the Royal Society of Tropical Medicine and Hygiene*, **79**, 393–408.

Campbell, W. C., 1983, In *Trichinella and Trichinosis* edited by W. C. Campbell (New York; Plenum Press), pp. 425–442.

Cheesmond, A. K. and Fenwick, A., 1981, *Journal of Tropical Medicine and Hygiene*, **84**, 101–107.

Cobbold, T. S. M 1864, *Entoza, An Introduction to Helminthology* (London: Groomsbridge).

Corridan, J. P. and Gray, J. J., 1969, *British Medical Journal*, **2**, 727–730.

Craig, P. S., Macpherson, C. H. L., Watson-Jones, D.L. and Nelson, G. S., 1988, *Transactions of the Royal Society of Tropical Medicine and Hygiene*, **82**, 268–274.

Dalton, P. R., 1976, *Bulletin of the World Health Organization* **54**, 587–595.

Dalton, P. R., and Pole, D., 1978, *Bulletin of the World Health Organization*, **56**, 417–426.

Dick, T. A., 1983, In *Trichinella and Trichinosis* edited by W. C. Campbell (New York: Plenum Press), pp. 31–73.

Fenwick, A., 1969, *Transactions Royal Society of Tropical Medicine and Hygiene*, **63**, 557–567.

Forrester, A. T. T., Nelson, G. S. and Sander, G., 1961, *Transactions of the Royal Society of Tropical Medicine and Hygiene*, **55**, 503–513.

French, C.M., Nelson. G.S. and Wood, M., 1982, *Annals of Tropical Medicine and Parasitology*, **76**, 425–437.

Fujinami, A. and Nakamura, H., 1909, *Kyoto Igaku Zasishi*, **6**, 224–252.

Garcia, E. G., 1976, *South East Asian Journal of Tropical Medicine and Public Health*, **7**, 190–196.

Garkavi, B. L., 1972, *Veterianariya*, **10**, 90–91.

Gillett, J. D., 1985, *Transactions of the Royal Society of Tropical Medicine and Hygiene*, **79**, 12–20.

Gretillat, S. and Vassilades, G., 1967, *Resue Eleve Medicine Veterinaire Pays Tropicale*, **21**, 85–91.

Hairston, N. G., 1962, In *Ciba Foundation Symposium on Bilhaziasis* edited by G. E. W. Wolstenholme and M. O'Connor (London: J. A. Churchill), pp. 36–62.

Heinz, H.J., McNab, G. M., 1965, *South African Journal of Medical Science* **30**, 19–31.

Husting, E. L., 1965, *Central African Medical Journal* **11**, 330–331.

Husting, E. L., 1970, *Central African Medical Journal* **16**, 5–10.

Independent Evaluation Mission (1985) *Report on an Independent Evaluation Mission in the National Bilharzia Control Programme in Egypt, 1985, Transactions of the Royal Society of Tropical Medicine and Hygiene* **81** (Supplement), 1–57.

Jordan, P., 1985, *Schistosomiasis the St. Lucia Project* (Cambridge: Cambridge University Press).
Jordan, P. and Webbe, G., 1982, in *Schistosomiasis* edited by P. Jordan and G. Webbe (London: Heinemann).
Kawashima, K., Datamine, D., Sakamoto, M., Shimada, M., Nojima, H., and Miyahara, M., 1978, *Japanese Journal of Tropical Medicine and Hygiene*, **6**, 195–203.
Kim, C. W., 1983, In *Trichinella and Trichninosis* edited by W.C. Campbell (New York: Plenum Press) pp. 445–500.
Kochar, V. K., Schad, G. A., Chowdhury, A. B., Dean, C. G., and Nawalinski, 1976, In *Medical Anthropology* edited by F. X. Grollig, S. J. Harold and H. B. Haley (The Hague: Mounton Publishers), pp. 288–312.
Leiper, R. T., 1915, *Journal of the Royal Army Medical Corps*, **25**, 48–55; 182–192; 261–267.
Leiper, R. T. and Atkinson, F. L,. 1915, *British Medical Journal* **1**, 201–203.
McManus, D .P ., and Bryant, C., 1986, In *The Biology of Echinococcus and Hydatid Disease* edited by R. C. A. Thompson (London: Allen and Unwin), pp. 114–142.
Macpherson, C. N. L., Karstad, L., Stevenson, P., and Arundel, J. H., 1983, *Annals of Tropical Medicine and Parasitology* **77**, 61–73.
Macpherson, C. N. L., 1983, *American Journal of Tropical Medicine and Hygiene* **32**, 397–404.
Matossian, R. M., Rebeiz, J., and Stephan, E., 1974, *Lebanon Medical Journal* **27**, 267–273.
Miyairi, K., and Suzuki, M., 1913, *Tokyo Iji Shinshi*, **1836**, 1–5.
Nelson, G. S., 1958, *East African Medical Journal* **35**, 311–319; 336–344; 543–546; 581–586.
Nelson G. S., 1970, In *Trichinosis in Man and Animals* edited by S. E. Gould (Springfield: Charles C. Thomas), pp. 473–492.
Nelson G. S., 1972, In *Behavioural Aspects of Parasite Transmission* edited by E. U. Canning and CIA. Wright (London: Academic Press), pp. 109–122.
Nelson, G. S. 1975, in *Diseases Transmitted from Animals to Man* 6th edn, edited by W. T. Hubbert, W. F. McCulloch and P. R. Schnurrenberger (Springfield: Charles C. Thomas), pp. 620–640.
Nelson, G. S. 1977, *American Journal of Tropical Medicine and Hygiene*, **26**, 1093–1110.
Nelson G. S., 1982, *Symposium Zoological Society of London* **50**, 181–198.
Nelson, G. S. 1988, *Journal of Comparative Pathology*,**98**, 135–153.
Nelson, G. S., and Rausch, R. L., 1963, *Annals of Tropical Medicine and Parasitology*, **57**, 136–149.
Nelson, G. S., Teesdale, C. and Highton, R. B., 1962, *In CIBA Foundation Symposium on Bilharziasis* edited by G. E. W. Wolstenholime and M. O'Connor (London: J. A. Churchill Ltd).
Nelson, G. S., Guggisberg, C. W. T. and Mukundi, J., 1963, *Annals of Tropical Medicine and Parasitology* **57**, 332–346.
Nelson, G. S. Blackie, E. J., and Makundi, J., 1966, *Transactions of the Royal Society of Tropical Medicine and Hygiene*, **60**, 471–480.
Ongom, V. L. and Bradley, D. J., 1972, *Transactions of the Royal Society of Tropical Medicine and Hygiene* **66**, 835–851.
Ouma, J., 1988, *The transmission of Schistosoma mansoni in an endemic area of Kenya with special reference to the role of human defecation behaviour and sanitary practices*, PhD Thesis, University of Liverpool, UK.
Ozeretskovskaia, N. N. and Uspenski, S. M., 1957, *Meditsinskaya Parazitologiya i Parazitarnye Bolezni*, **26**, 152–159.
Pitchford, R. J. and Visser, P. S., 1962, *Transactions of the Royal Society of Tropical Medicine and Hygiene*, **56**, 126–135.
Polderman, A. M., 1974, *The Transmission of Intestinal Schistosomiasis in Begemder Province Ethiopia*, Thesis of Leiden University.
Pozio, E., La Rosa, G. and Rossi, P., 1989, *Parasitology Today* **5**, 169–170.
Prentice, M. A., 1984, *Annals of Tropical Medicine and Parasitology*, **78**, 112–127.
Rausch, R. L., 1970, In *Trichinosis in Man and Animals* edited by S. E. Gould (Springfield: Charles C. Thomas), pp. 348–373.
Rausch, R. L. and Nelson, G. S., 1963, *Annals of Tropical Medicine and Parasitology*, **57**, 127–135.
Rollinson, D. and Southgate, V. R., 1978, In *The Biology of Schistosomiasis* edited by D. Rollinson and A. J. G. Simpson (London: Academic Press), pp. 1–49.
Rowan, W. B., 1958, *American Journal of Tropical Medicine and Hygiene*, **7**, 374–381.
Schwabe, C. W., 1964, In *Veterinary Medicine and Human Health* (Baltimore: Williams and Wilkins).
Schwabe, C. W. and Abou-Daoud, K., 1961, *American Journal of Tropical Medicine and Hygiene*, **10**, 374–381.
Semple, A. B., Davies, J. B. M., Kershaw, W. E. and St. Hill, C. A., 1954, *British Medical Journal*, **1**, 1002–1006.

Shaikenov, B. and Boev, S. N. 1983. *Wiadomosci Parazytologiczne*, **29**, 595–608.
Simon, J. W., Riddel, N.J., Wong, W. T., An, A. C. S., Ko, R. C. and Templer, M.J., 1986, *Transactions of the Royal Society of Tropical Medicine and Hygiene*, **80**, 394–397.
Smyth, J. D., 1977, *Transactions of the Royal Society of Tropical Medicine and Hygiene*, **71**, 93–100.
Sundman, P. O., 1970, In *The Flight of the Eagle* (New York: Pantheon Books).
Theron, A., 1979, *Bulletin of the World Health Organization*, **57**, 971–975.
Thompson, R. C. A., 1986, In *The Biology of Echinococcus and Hydatid Disease* (London: George Allen and Unwin).
Thompson, R. C. A. and Lymbery, A.J., 1988, *Advances in Parasitology*, **27**, 40–258.
Thorborg, N. B., Tulinius, S., and Roth, H., 1948, *Acta Pathologica Microbiologica Scandinavia*, **25**, 788–794.
Wachira, T. M., 1989, *Studies on the epidemiology and control of Echinococcus granulosus in Kenya* PhD Thesis, University of Nairobi.
Warren, K. S., 1988, In *Parasitic Infections, Contemporary Issues of Infectious Diseases Vol. 7*, edited by J. H. Leech, M. A. Sande, and R. K. Root (New York: Churchill Livingstone), pp. 1–10.
Watson-Jones, D. L. and Macpherson, C. N. L., 1988, *Annals of Tropical Medicine and Parasitology*, **82**, 343–356.
Wilkins, H. A., 1977, *Annals of Tropical Medicine*, **71**, 51–58.
Wilkins, H. A., 1987, In *The Biology of Schistosomes* edited by D. Rollinson and A.J. G. Simpson (London: Academic Press), pp. 379–397.
Wilkins, H. A., Blumenthal, U.J., Hagan, P., Hayes, R.J. and Tulloch, S., 1987, *Transactions of the Royal Society of Tropical Medicine and Hygiene*, **81**, 29–35.
World Health Organization, 1981, *FAO/UNEP/WHO guidelines for Surveillance, Prevention and Control of Echinococcosis/Hydatidosis* (Geneva: WHO).
Wray, J. R., 1958, *East African Medical Journal* **35**, 37–39.
Wright, W. H., Jacobs, L. and Walton, A. C., 1944, *Public Health Reports*, **59**, 669–681.
Yokogawa, M., 1970, In *Recent Advances in Research on Filariasis and Schistosomiasis in Japan* edited by M. Sasa (Baltimore: Tokyo Park Press), pp. 231–255.
Zarlenga, D. S. and Murrell, K. D., *Proceedings of the 7th International Congress on Trichinellosis* edited by C. E. Tanner (in press).
Zenker, F. A., 1860, *Virchows Archiv Fur Pathologische Anatomie und Physiologie und Fur Klinische Medizin*, **18**, 561–572.

9. Human behaviour and the epidemiology of helminth infections: the role of behaviour in exposure to infection

D. A. P. Bundy and U. J. Blumenthal

INTRODUCTION

Parastic worms are among the most ubiquitous of human pathogens. In endemic areas infection prevalence may exceed 90%, while on a global scale some species (the gastrointestinal nematodes *Ascaris*, *Trichuris* and the hookworms) are estimated to infect more than a quarter of humankind (Warren and Mahmoud, 1984).

The two major routes of entry of infective stages, by ingestion or percutaneous penetration, are fundamentally dependent on the behaviour of the prospective human host. A third and less common route of infection, percutaneous injection by an arthropod vector, is more dependent on the behaviour of the intermediate host and will not be discussed here. Relevant human behaviours may be conveniently divided into those which are active, such as the ingestion of soil containing the infective stages of *Ascaris*, *Trichuris* or *Ancylostoma*, or permissive, such as exposing the skin to water containing the penetrative larvae of *Schistosoma* or soil containing the larvae of *Necator*. Whether these behaviours are active or permissive has important consequences for the approach to quantification, but appears to have little impact on the overall dynamics of transmission. Indeed the population dynamic patterns of the various helminth species exhibit remarkable similarities (Anderson and May, 1985).

A central determinant of the dynamics of *macro*parasites is that they do not multiply within the human host (*Strongyloides stercoralis* and *Capillaria philippinensis* are rare exceptions to this); each worm is the result of a separate infection event, and the number of worms (the intensity of infection) is a function of the frequency of infection. This can be contrasted with most *micro*parasites (protozoa, viruses and bacteria) where infection is an all-or-nothing event in which a single successful infection leads to colonization of the host. Thus for most helminths, but not microparasites, the intensity of human infection is related to the degree of exposure to infection.

This does not imply, however, that the intensity of worm infection is a simple function of the extent to which an individual's behaviours result in exposure. Humans have developed a battery of defence mechanisms specifically intended to prevent successful infection and which may, therefore, modify the effects of exposure. In the case of acquired immunity the consequences may be paradoxical, such that enhanced experience of exposure results in reduced infection success. In ecological terms, macroparasite populations are governed by an immigration–death process in which immigration is modified by host behaviour and parasite mortality is regulated by host resistance.

The immunology *versus* ecology debate has preoccupied helminthological research for more than two decades (Warren, 1973; Bradley and McCullogh, 1974), but has primarily focused on demonstrating a major regulatory (protective) role for the human immune response, perhaps anticipating the so far unfulfilled expectation of the development of protective vaccines. The behavioural component has received much less critical attention although its major practical importance is tacitly acknowledged: behavioural modification by health education and sanitation is considered in the planning of nearly all intervention programmes.

The aim of the present chapter is to examine the quantitative evidence for a role for infection exposure as a determinant of human helminthic infection. The role of 'permissive' exposure is examined by reference to human contact with water bodies containing the penetrative larvae of schistosomes, while the ingestion of soil-borne geohelminth eggs provides an illustration of 'active' exposure. We seek first to examine the problems of adequately quantifying such behaviours and then to demonstrate associations between the occurrence of a particular behaviour and the intensity of infection.

QUANTIFYING HUMAN BEHAVIOURS THAT RESULT IN INFECTION

Human behaviour is complex and is modified by a bewildering range of cultural, ethnic and gender-dependent factors. The relative importance of these factors has been the subject of considerable speculation (reviewed by Desowitz, 1981; Schad *et al.*, 1983; Bundy, 1988a), and is considered elsewhere in this volume (Chapter 8). Here we will consider only those behaviours which directly result in exposure to infective stages and which are amenable to quantification.

Quantifying human behaviour is facilitated in the case of schistosome transmission by the existence of distinct transmission sites, that is, points at water bodies where human water contact takes place and where infected snails are present (see Chapter 8). Observation can therefore be confined to such water contact sites, indeed in some studies water contact data are only included if they are from sites where infected snails are found (Butterworth *et al.*, 1985). Transmission of the geohelminths is more difficult to study since transmission sites cannot usually be so clearly defined. Studies have therefore concentrated on areas of unusually high transmission such as communal defaecation grounds (Schad *et al.*, 1983; Hominick *et al.*, 1987), or on behaviours likely to increase transmission, such as geophagia (Wong *et al.*, 1988). The quantification of behaviour leading to exposure has thus taken two paths: direct methods where transmission points can be identified, and indirect methods when it is difficult to accurately quantify the behaviour.

It is clear that for exposure to occur there are two preconditions; the presence of the infective agent and the presence of the behaviour which brings the person into contact with the infective agent. Both factors need to be measured to quantify exposure. In addition, there are qualitative and quantitative differences in behaviour which affect its exposure value. The terms 'exposure' and 'behaviour' are often used interchangeably but accurate quantification of exposure involves much more than the quantification of behaviour.

Direct assessment of exposure to schistosome infection

The dominant behaviours in the transmission of urinary and intestinal schistosomiasis are those which lead to direct contact with infected water, since infection occurs via percutaneous penetration by cercariae emitted from a freshwater snail intermediate host. Contact with infective water bodies may occur during domestic, occupational or recreational activity. Collecting water and washing clothes are important behaviours leading to female exposure in some areas (Kloos and Lemma, 1980; Kloos *et al.*, 1983), while contact while fishing or working in irrigated fields may lead to significant male exposure (Tayo *et al.*, 1980; Fenwick *et al.*, 1982). Recreational contact is most common in children as a result of swimming and playing in water, but may also occur when adults swim or bathe after work (Dalton and Pole, 1978; Blumenthal, 1985). In general, the most important behaviours leading to exposure to *S. haematobium* and *S. mansoni* infection are a consequence of domestic or recreational activity. This implies that children and adult females are more exposed to infection, except where specific occupational activities (such as irrigation or fishing) enhance the risk to adult males. The relative importance of these water contact activities may be modified by socioeconomic circumstances. Domestic water contact, in particular, is likely to be affected by the relative availability of protected water supplies. In this section we will consider how this complex of water contact patterns might be related to exposure to infection.

The first detailed investigations of water contact patterns involving direct observation were reported by Farooq and Mallah (1966) from the Nile delta. Since then, studies have been completed in several areas of the world and a variety of ecological situations: a citrus estate in Zimbabwe (Husting, 1970, 1983), a dam in Nigeria (Tayo *et al.*, 1980), river and stream sites in St. Lucia (Dalton, 1976), Ethiopia (Kloos and Lemma, 1980), Kenya (Butterworth *et al.*, 1984), the Philippines (Tiglao and Camacho, 1983) and South Africa (Kvalsvig and Schutte, 1986), the

Volta Lake in Ghana (Dalton and Pole, 1978), irrigation canals and the Nile in Upper Egypt (Kloos *et al.*, 1983) and seasonal pools in The Gambia (Blumenthal, 1985).

In most of the studies, calculation of exposure was based on the frequency and duration of water contact (Dalton, 1976; Dalton and Pole, 1978; Husting, 1983; Butterworth *et al.*, 1984), while others extended their estimation of exposure to include other variables in an 'index' of exposure. The importance of the proportion of body surface exposed during a particular water-contact activity has been examined by a number of authors. Kloos and Lemma (1980) incorporated surface area in an exposure index (the product of frequency, mean duration of water contact per activity and mean proportion of body surface exposed). However, although they recorded activity and frequency of contact for all individuals, mean duration of contact for each activity and the percentage body surface wetted were recorded only for a subsample. Kvalsvig and Schutte (1986) used a similar design, but observed duration of contact for each individual and used these data to calculate two comparative exposure indexes. In index A, exposure was calculated as the product of frequency, mean duration of contact and mean proportion of body surface exposed, whereas index B was simply the product of frequency and mean duration of contact. By comparing these indices the relative importance of different types of water contact behaviour could be examined (table 9.1). For example, swimming and washing the body became more important than washing clothes when body surface exposure was included in the index.

Table 9.1. Relative importance of different types of water contact using two different methods of calculation.

Activity	Index A[a]	Index B[b]	
Washing body/swimming	7·967	10·623	(major importance in summer)
Washing clothes	3·335	22·233	
Children playing/dangling legs	0·385	3·208	
Washing extremeties/crossing	0·029	0·145	
Washing objects	0·014	0·233	
Fetching water	0·006	0·100	

Data from Kvalsvig and Schutte (1986).
[a] Frequency × duration × body surface exposed.
[b] Frequency × duration.

Tiglao and Camacho (1983) extended this approach by using observations over a longer time period and recording both activity and duration of contact for each individual. The exposure index calculated took into account duration of contact and the mean proportion of body surface exposed for each contact episode, and summated this over the entire observation period. The adjustment of each record before summation to give the exposure of an individual represented a potentially significant advance in the methodology of exposure assessment. Unfortunately, their study did not separately report the exposure values and so it was not possible to assess accurately the value of their approach.

In the remainder of this section we will use a detailed data-set assembled in The Gambia (Blumenthal, 1985; Wilkins *et al.*, 1987) to examine the effects on exposure estimation of including different variables in the calculation of exposure indices.

To collect the data for the Gambia study the water contact of a cohort of individuals of all ages was monitored throughout one transmission season (6 months). Observers watched all sites at the seasonal pools where water contact took place; about 10% of all contact was observed (Wilkins *et al.*, 1987). Frequency and duration of contact, time of day and the activity performed were noted in all cases. Throughout the transmission season cercarial densities were regularly measured with a quantitative filtration technique and snail numbers and infection rates were monitored fortnightly at each site. These data made it possible to make quantitative estimates of the variation of cercarial densities by site, month and time of day. A correction factor was determined for each activity, based on the mean proportion of body surface exposed for each activity, with an allowance being made for the possible toxic effect on cercariae of the soap used in some activities. For each individual, an exposure value was calculated for each observed contact and the exposure value for all contacts summed to give one exposure value for the whole season for that individual.

This detailed data-set makes it possible to assess the effects of including different variables in the calculation of exposure indexes. In the first index of exposure which we consider, the duration of each contact is modified by a factor for the mean infectivity of the water over the season in the pool used (based on the data collected), and by a factor for the activity performed. Seasonal and diurnal variation in infectivity are deliberately excluded. Exposure to infection, E, is thus estimated for each individual using the relationship:

$$E = \Sigma(D_i \times P_i \times A_i) \qquad \text{index 9.1(a)}$$

where D_i = duration in minutes of each individual's i^{th} observed water contact; P_i = an estimate of the mean cercarial density over the season at the pool or site used, in cercariae/litre; and A_i = an activity factor whose value depends on an estimate of the proportion of body surface area exposed during each activity and modified for the probable effect of soap used in washing activity.

Two subsidiary indices are also calculated to separate the effects of the pool and activity factors on exposure:

$$E = \Sigma(D_i \times P_i) \qquad \text{index 9.1(b)}$$

$$E = \Sigma(D_i \times A_i) \qquad \text{index 9.1(c)}$$

Using index 9.1, the age-dependent pattern of exposure to infection is markedly different from the exposure pattern suggested by duration of water contact alone (figure 9.1). The exposure of females over 10 years old is substantially reduced by the index (more so for those over 15 years old than for those 10–14 years old) and the exposure of males 5–9 years old is increased. Therefore, the index results suggest that the exposure of adults of both sexes is substantially lower than the exposure of children 5–9 years old, whereas the duration of water contact of adult females is higher than that of children. The underlying reasons for these differences are indicated by comparing indices 9.1(b) and 9.1(c) (figures 9.1(c) and (d)). The pool factor alters the profile of mean exposure very little (only slightly decreasing the magnitude of the difference between adult females and female children) while the activity factor has a pronounced effect. The age dependency of the activity factor is mainly due to the common occurrence in the child age classes of activities involving a high proportion of body surface exposure (figure 9.2) (Blumenthal *et al.*, 1990).

In order to further refine the estimation of exposure a second index may be calculated which includes variation in water infectivity at each site due to diurnal and seasonal influences (based on the data collected), where exposure is given by:

$$E = \Sigma(D_i \times A_i \times PS_i \times T_i) \qquad \text{index 9.2}$$

where D_i = duration of contact (as in index 9.1(a)); A_i = an activity factor (as in index 9.1(a)); PS_i = an estimate of the mean cercarial density at bimonthly intervals at the pool or site used, in cercariae / litre; and T_i = a factor which varies according to the relative number of cercariae present at each hour of the day.

The age-dependent pattern of exposure to infection estimated by index 9.2 is not markedly different from the pattern described using index 9.1(a) (cf. figures 9.1 and 9.6(c)). The patterns of exposure of both males and females under 15 years old is modified, so that male exposure is apparently similar for all three age groups, whereas index 9.1(a) indicates a peak in the 5–9-year-old group. Index 9.2 increases the apparent exposure of females 5–9 years old relative to the exposure of females 10–14 years old. This may not, however, be a significant change since the variation between children within each age group is also increased such that there are no substantial age- or sex-related differences in exposure in those under 15 years old.

The high variance in exposure of children in each age group and in adult females (figure 9.3) suggests that caution should be exercised in interpreting patterns of exposure to infection when these are considered using mean values for different age groups, without taking into account the variation between individuals. Comparison of the mean exposure levels of different age groups gives an indication of age-related trends in exposure levels but obscures the wide variation between individuals. For example, some individual adults, particularly females, have exposure levels higher than those of many children, even though the mean exposure of adults is apparently much lower (figure 9.3(b)). Using index 9.2, one third of those in the upper exposure class are adult females whereas one third of those in the lowest exposure class are children.

In order to examine these individual age relationships the effect of using different indexes of exposure has also been explored for individuals (Blumenthal *et al.*, 1990). Cohort members were ranked according to their total duration of water contact and according to their exposure value calculated by each exposure index separately. They were then grouped to give three broad ranks of duration of exposure: lower (45 lowest rankings), middle (middle 45 rankings) and upper (46 highest rankings) where a high rank indicated high duration or exposure. The number of individuals in each exposure rank for each index was compared with that

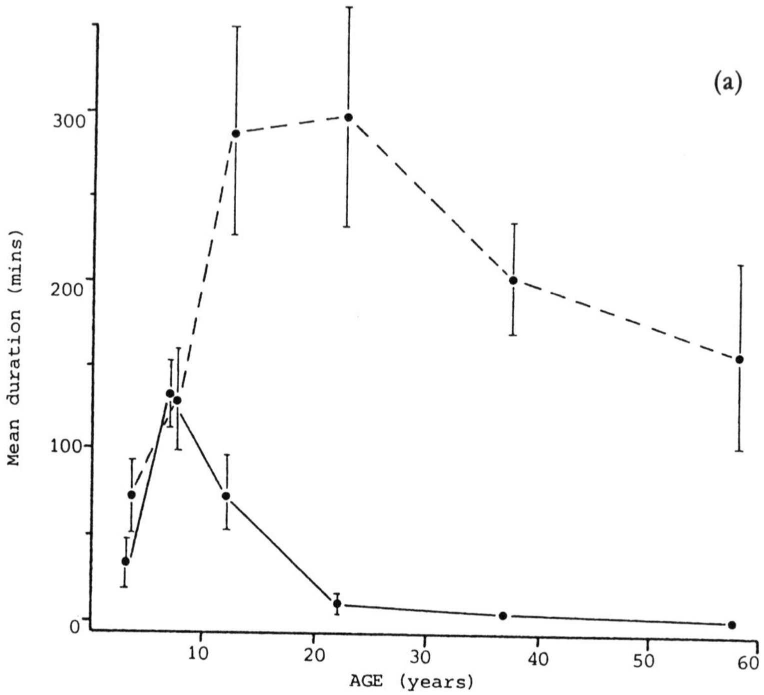
(a)
Mean duration (mins)
300
200
100
0
10
20
30
40
50
60
AGE (years)

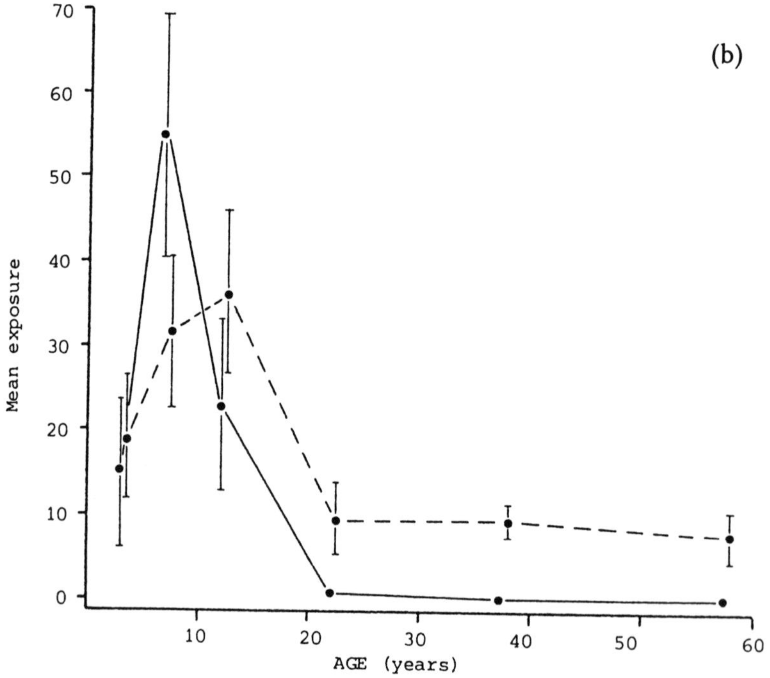
(b)
Mean exposure
70
60
50
40
30
20
10
0
10
20
30
40
50
60
AGE (years)

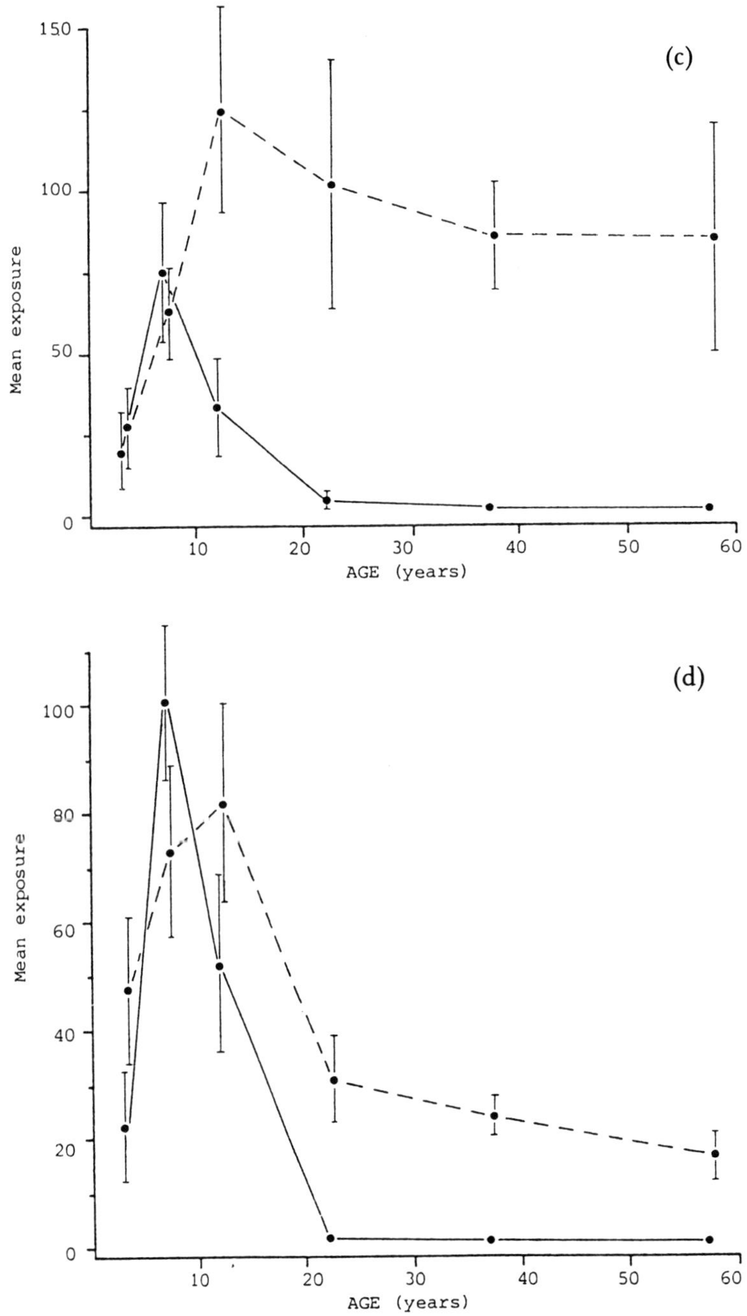

Figure 9.1. The different age profiles indicated by differing measures of exposure to *S. haematobium* infection in The Gambia. (a) Duration of water-contact. (b) Exposure index 9.1(a). (c) Exposure index 9.1(b). (d) Exposure index 9.1(c). Data from Blumenthal (1985). Data are shown as mean (SE) values for age classes of males (•——•) and females (•- - -•).

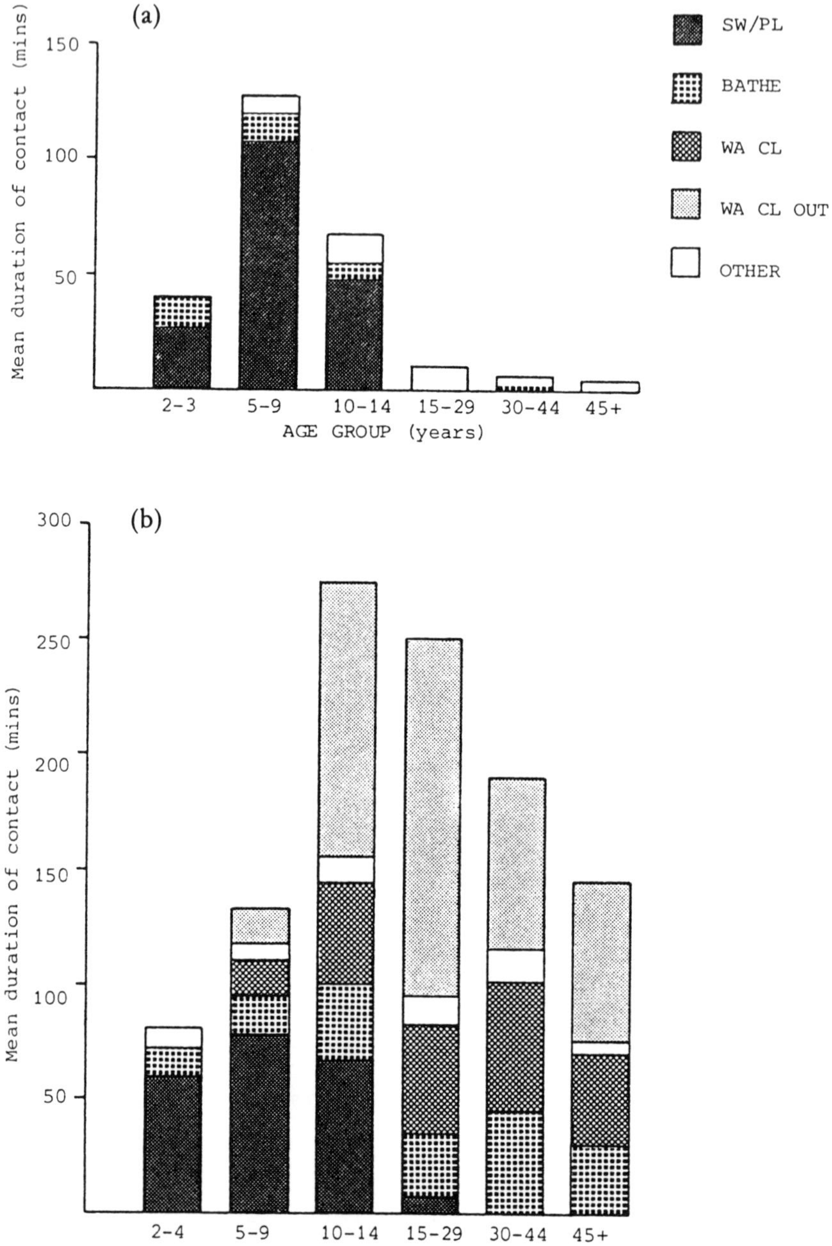

Figure 9.2. The proportion of time spent in different water contact activities by an age-stratified cohort in The Gambia. Males (a) tend to have lower durations of water contact than females (b). Data from Blumenthal (1985). SW/PL, swimming or other recreational activity; BATHE, bathing or washing limbs; WA CL, washing clothes in pool or collecting water; WA CL OUT, washing clothes in a basin; OTHER, any other form of water contact.

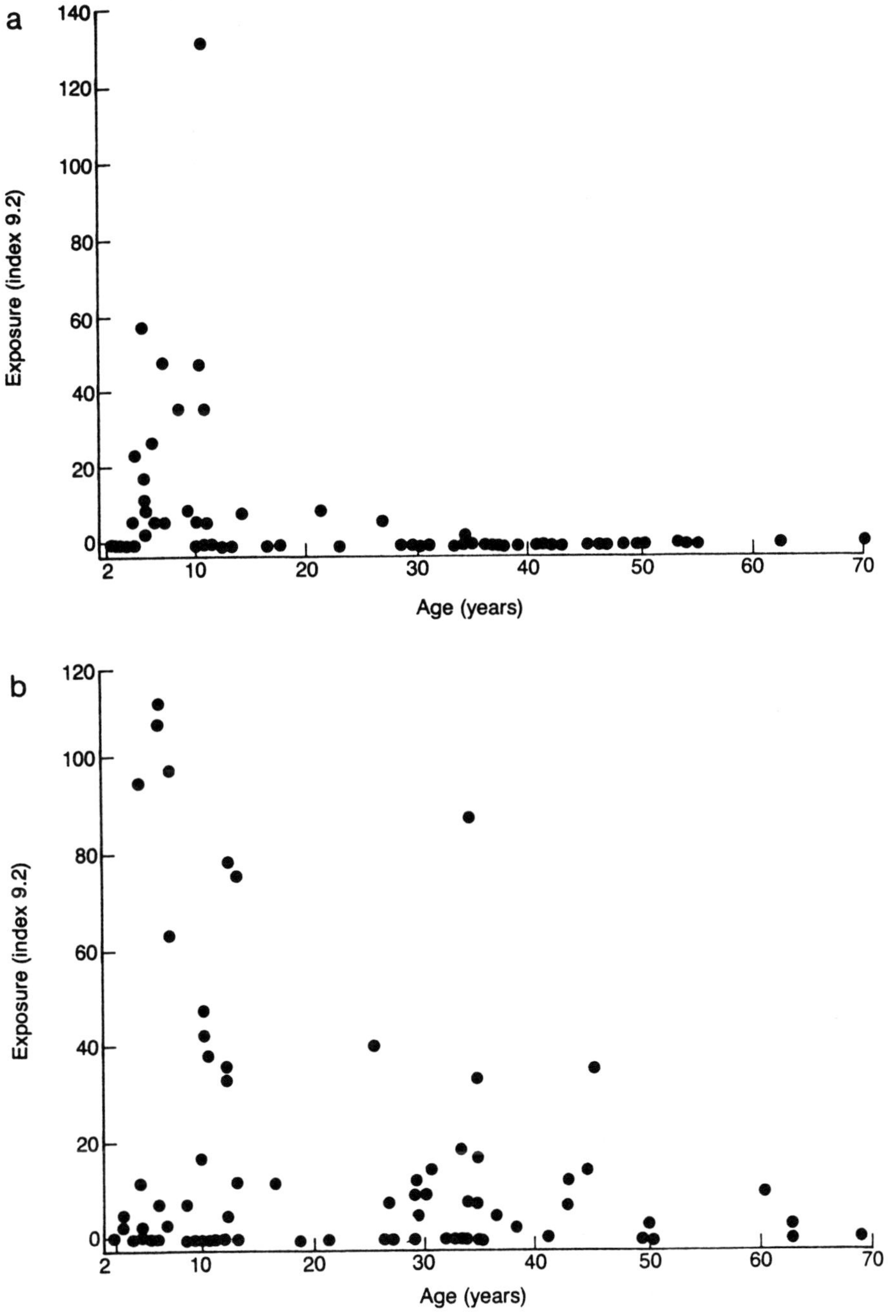

Figure 9.3 Degree of exposure calculated from index 9.2 for individuals in The Gambia cohort study (n = 136). Both male (a) and female (b) children exhibit considerable variation in exposure, while adult males tend to have uniformly low levels of exposure. Data from Blumenthal (1985).

obtained for duration of contact. The results of these analyses show, as was seen for the group means, that individuals could have their exposure ranking changed, depending on the index employed. These changes result in different individuals moving ranks to different degrees, and sometimes in different directions. For example, some individuals with a medium rank for duration could change to a high rank for exposure, while others change to a low rank. These pronounced changes in individual rankings are apparent for as much as 20% of the study group using some indexes, whereas changes in mean exposure of age groups have not been apparent. Furthermore, it appears that although the activity factor affects individuals in a particular age class together, the factors for site affect individuals of all ages.

The results of these studies clearly demonstrate that a measure of behaviour cannot be taken to be equivalent to a measure of exposure to infection. The central importance of including water infectivity variables in a measurement of exposure can also be demonstrated without reference to data on individuals by calculating a simple index of exposure potential where:

$$\text{Exposure potential} = \text{mean duration of water contact} \times \text{mean cercarial density}$$

Comparison of duration of water contact at each pool with 'exposure potential' at the same pool in the study area in The Gambia (figure 9.4) demonstrates how the majority of exposure (and therefore transmission) occurs at two pools although water contact occurs at all six pools.

The major conclusions about the quantification of exposure to infection from these analyses can be summarized as follows. First, an index of exposure is very sensitive to the variables included within it, so it is important to consider all factors known to modify exposure. Second, group data should be interpreted with care, since they obscure the, often marked, variation between individuals. Third, activity (as represented by proportion of body surface exposed) is the main factor that modifies age-group effects. Finally, the variable that has the greatest potential to change values of exposure, particularly of individuals, is site (here, water infectivity) and not activity.

Indirect assessment of exposure to geohelminth infection

Infection with the two most common geohelminths (*Ascaris lumbricoides* and *Trichuris trichiura*) occurs via ingestion of the parasite egg after it has undergone a period (weeks or months) of development to the infective stage. The uninfective eggs are passed out in host faeces which, in the absence of appropriate disposal procedures, are likely to be deposited in the peridomiciliary soil environment. During the developmental period the faeces are dispersed leaving the infective eggs as an intrinsic component of apparently uncontaminated soil (Beaver, 1975). Ingestion of this soil results in infection – hence the term geohelminth. Ingestion of recently deposited faeces, however hazardous in terms of other enteric infections, is not a route of infection with geohelminthiases.

Soil may be ingested accidentally as a contaminant of food, hands or utensils, or

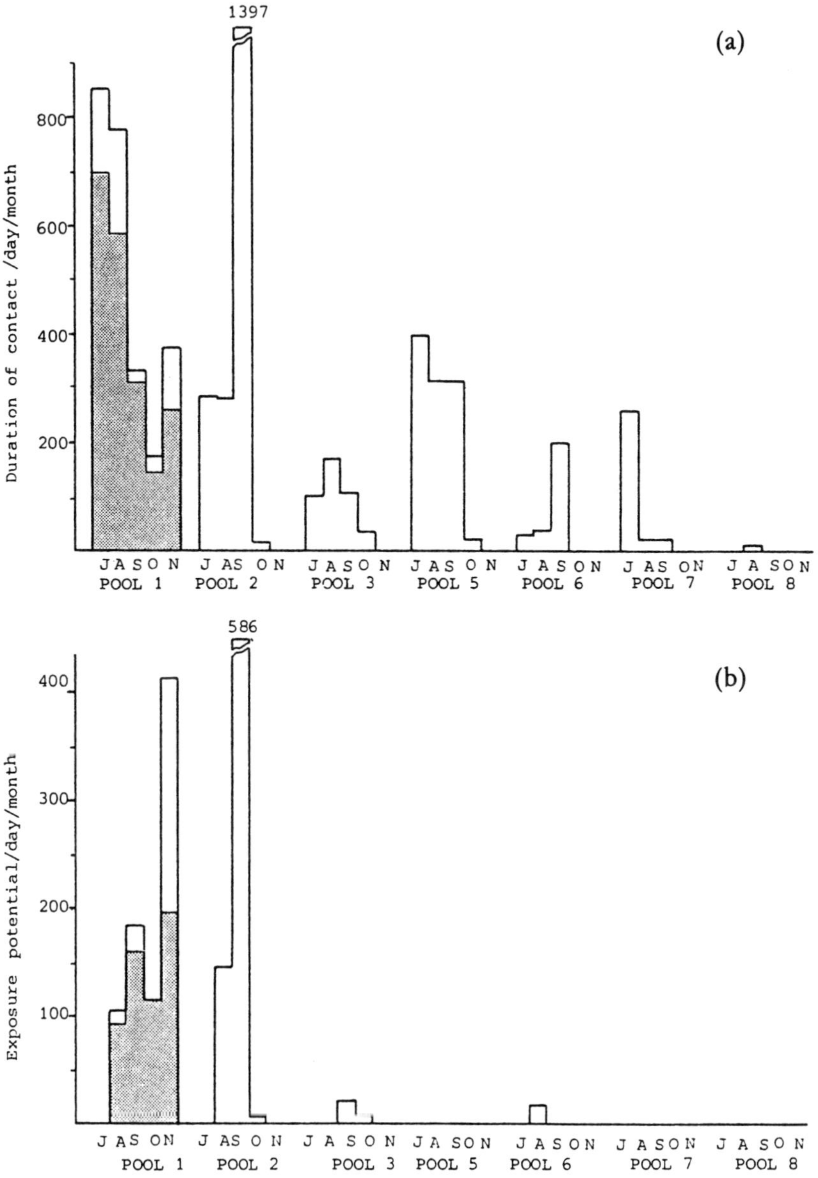

Figure 9.4. Water contact and exposure potential at seven pools in The Gambia during one transmission season (July–November). (a) Mean duration of water contact per day of observation per month. (b) Exposure potential (mean duration × cercarial density) per day per month. For Pool 1, shaded area represents side 1a, open area represents side 1b. Despite water contact at all pools, transmission mainly occurred at only two pools. Data from Blumenthal (1985).

purposefully as a consequence of the perversion of appetite known as geophagia. Geophagic behaviour is particularly common in the children of developing countries where geohelminthiases are endemic, perhaps because the conditions with which the practice is associated – deficiencies of trace elements, malnutrition, stress and psychological deprivation – are not uncommon in socioeconomically compromized populations (Lanzkowsky, 1959; Halstead, 1968; Singhi *et al.*, 1981). In some areas more than a third of the children may be geophagous. Thus in endemic regions the frequent practice of geophagia, the lack of adequate water for hygiene and food preparation, the use of human faeces as fertilizer and the absence of appropriate sanitation may conspire to facilitate parasite egg ingestion and geohelminth transmission.

Measurement of the rates of soil ingestion by these various routes has proven difficult to quantify directly. Early attempts to develop indirect methods, principally intended to confirm the diagnosis of geophagia, used radiological methods to demonstrate the presence of gross quantities of soil in the intestine (Vessal *et al.*, 1975). Such methods, while providing some degree of quantification, are clearly inappropriate for community applications.

More recently a new indirect procedure has been developed which uses the measurement of soil silica excreted in stool as an estimate of the quantity of soil ingested (Wong *et al.*, 1988). This indirect approach results in the measurement of the total amount of soil ingested, and thus encompasses both the purposeful ingestion of soil in geophagia as well as the accidental ingestion of soil from hands or food. The procedure is based on the assumptions that: (1) silica is a major component of soil, (2) silica passes quantitatively unchanged through the human gut, and (3) silica of soil origin in stool can be quantitatively differentiated from silica of dietary origin. Thus, provided the amount of silica in local soil and in the local diet are both known it is possible to estimate the rate of soil ingested from the amount of soil excreted over a given gut transit period.

The procedure was developed in Jamaica and has been validated for a West Indian population (Wong, 1988). Using a combination of 800 °C incineration to remove organic components from stool and 6 M HCl solution to remove soluble inorganic components, it is possible to retrieve 91% of soil-derived silica from stool specimens. This efficiency is achieved even with low concentrations of silica (<0·4%) in small (1 g) specimens of stool.

Volunteer experiments indicate that this efficiency is also achievable *in vivo* (figure 9.5). These experiments show that approximately 90% of ingested silica is excreted within 48 hours, a time period that corresponds closely with previous estimates of gut transit time for populations on a similar diet (Burkitt, 1972; Cummings, 1983).

Studies of children on a normal diet with no access to soil indicate that silica of dietary origin is a minor constituent of stool (mean 1·45 ± 0·72% dry weight). In contrast, children in two Jamaican communities had a mean stool silica concentration of 4·3% dry weight, with some individuals passing stools in which more than 20% of the dry weight was attributable to silica. Using these criteria it was estimated that up to 66% of the children studied had ingested soil within the

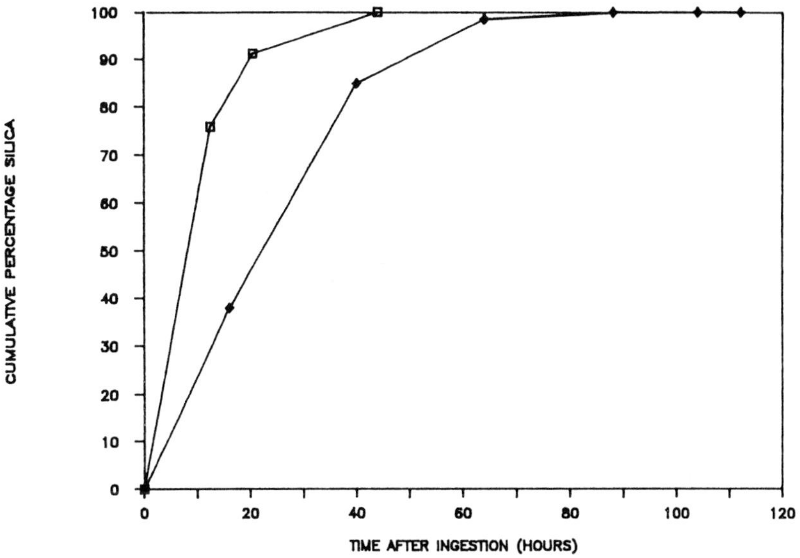

Figure 9.5. Relationship between soil ingested and silica recovered from stool. Proportional recovery of 4 gram of ingested silica from sequential stools passed by volunteers 1 (□) and 2 (◆). In both cases, approximately 90% of the silica was recovered from the first two stools passed, less than 48 hours after ingestion. Data from Wong *et al.* (1988).

previous 48-hour period. The average quantity of soil ingested was 0·25 gram per day, with a maximum of 10·3 gram per day. The higher value undoubtedly indicates overt geophagia, with the lower values probably reflecting accidental ingestion associated with inadequate hygiene.

Relating these rates of soil ingestion to infection exposure has proven difficult (Rukmono, 1980; Wong and Bundy, 1990). Although the mean number of eggs per gram of soil in the communities studied was relatively low (e.g. for *Trichuris* 4·0 eggs per gram of soil) the spatial distribution was highly overdispersed such that some areas contained dense aggregations of eggs (a maximum of 70 eggs per gram of soil for *Trichuris*). Without precise knowledge of the origin of soil ingested it is not possible to accurately estimate the degree of exposure. The mean values of soil ingestion rate and egg density would suggest an average exposure of one egg per day, while the maxima indicate that more than 700 eggs could be ingested in a 24-hour period.

This lack of precise understanding of the spatial pattern of contact with infective stages compromises the accuracy of the soil ingestion measure of exposure. It does not, however, invalidate the procedure. If it is assumed that soil selection for ingestion is an essentially random process, then individuals with high rates of ingestion have a higher probability of selecting dense aggregates of infective stages. Thus individuals with high rates of soil ingestion have a greater probability of exposure, although this will scale non-linearly. Empirical evidence tends to support this conclusion since intense infection with both toxocariasis and trichuriasis are statistically associated with a reported history of overt geophagia (Lariviere *et al.*, 1965; Glickman *et al.*, 1981; Gilman *et al.*, 1983).

CORRELATIONS BETWEEN EXPOSURE AND INFECTION

The purpose in developing the above quantitative procedures for estimating exposure is to examine the role of exposure as a determinant of human infection. In this section we seek associations between estimates of exposure, and observed levels of infection and reinfection in the same individuals or populations.

Schistosome water contact exposure and infection

The indices of exposure to schistosome infection involve no assumptions regarding host resistance or susceptibility to infection. They attempt to estimate the relative levels of exposure to infective cercariae, but not the rate of establishment of infection. Comparison between observed rates of exposure and infection might therefore be used to indicate whether the host mediates between exposure and establishment.

The age–intensity profiles of infection for both *S. mansoni* and *S. haematobium* are markedly convex (figure 9.6). Intensity of infection rises to a peak around 10 years old and declines to low levels in adulthood. The superficial similarity between the shape of the age–intensity curve and the age profile of the duration of contact curve has been suggested to indicate a direct relationship. Dalton and Pole (1978) compared two such curves (figure 9.7) and concluded from this, and a regression analysis, that age was less important than degree of exposure in explaining variation in infection intensity. This oft-quoted work was very influential in the development of exposure studies but the analysis and interpetation of the data has been called into question by Barbour (1985). The dangers of this simple comparative approach are illustrated by examining the data set on *S. haematobium* infection in The Gambia (figure 9.6). Here the pattern of duration of water contact is clearly very different from the pattern of infection, since females over 10 years old have very high levels of contact but relatively low levels of infection. Modifying the apparent exposure through the calculation of indices makes the age profiles superficially more similar, but major quantitative differences are still apparent. The mean egg output of adult females is lower than that of children (5–9 years old) by about a factor of 20, whereas the mean exposure of adult females is lower than that of children by only a factor of 3. A similar mismatch between the age profiles of exposure and infection has been described from analysis of other studies (Butterworth and Hagan, 1987; Bundy, 1988b), and is indeed apparent on closer inspection of the original graphs presented by Dalton and Pole (1978). This significant quantitative discrepancy between superficially similar curves has led to the development of more sophisticated approaches to examining the relationship between exposure and infection.

Another common defect in studies which compare established levels of human

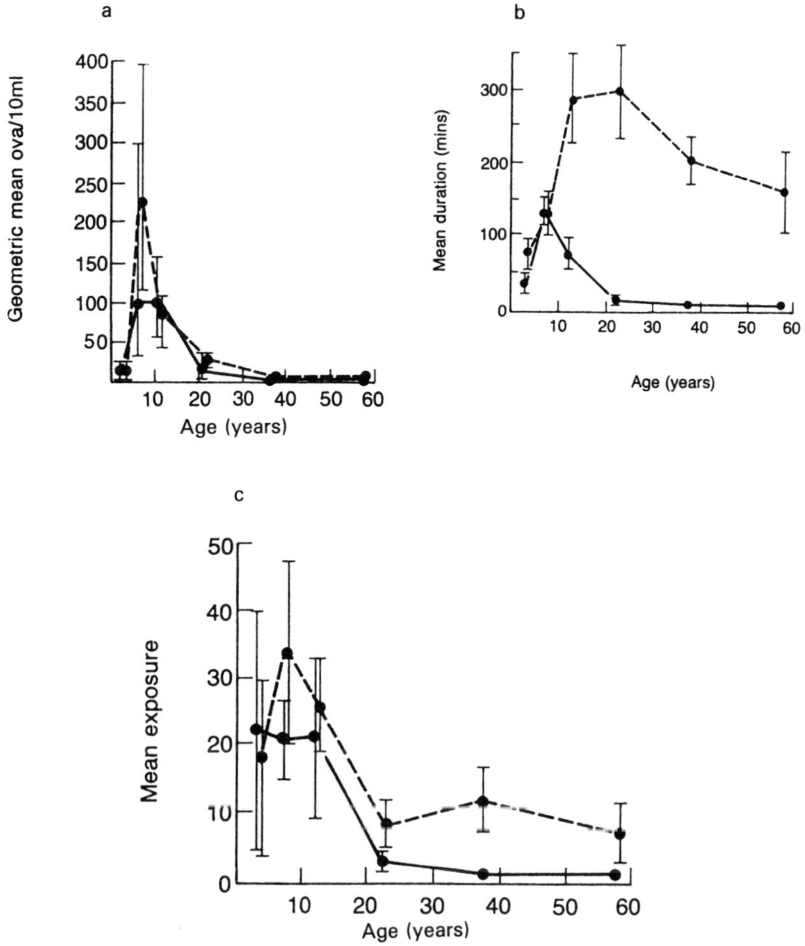

Figure 9.6. Comparison of (a) the age profile of *S. haematobium* infection intensity with (b) the age profile of mean duration of water contact and (c) the age distribution of exposure calculated by index 9.2. •——•, males; •– – –•, females. Data from Blumenthal (1985).

infection with observed water contact is the dissimilarity in time period. Such cross-sectional studies compare infection accumulated over many years with exposure measured over a period of weeks or months. This defect can be avoided by the use of longitudinal studies which compare exposure and acquisition of infection over the same time period, usually by simultaneously examining exposure and reinfection after chemotherapeutic cure. Major studies of this kind have been undertaken on *S. mansoni* by Butterworth and colleagues in Kenya and on *S. haematobium* by Wilkins and colleagues in The Gambia.

Butterworth *et al.* (1985) analyzed reinfection over 1 year in a group of 119 schoolchildren. Twenty-two children showed particularly high reinfection intens-

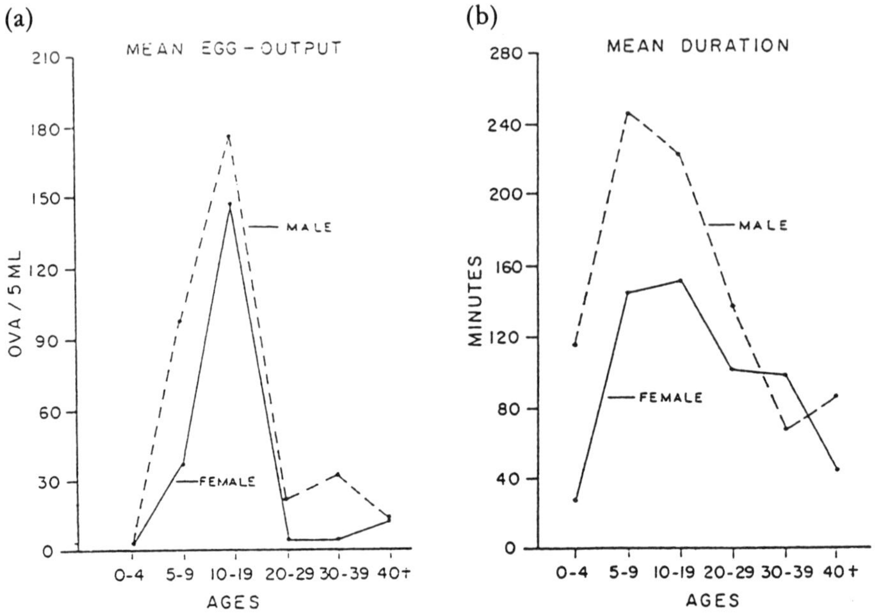

Figure 9.7. Comparison between a) the age profile of intensity of *S. haemotobium* infection and b) the age profile of the mean duration of water contact in Ghana. Data from Dalton and Pole (1978).

ities and were considered to be susceptible to infection while 20 children had particularly low reinfection intensities. There was no substantial difference in the 'relevant' water contact of the two groups, where relevance was defined as contact with sites found to contain infected snails during the 2 week period in which contact was observed. In particular, 35 of the 70 children showing low reinfection intensities had high levels of both total water contact and relevant water contact and hence were considered to be resistant to reinfection. Therefore, water contact patterns alone were not considered adequate to explain the different intensities of reinfection in the children. Since the mean age of the resistant group was 2 years greater than that of the susceptible group, it was suggested that resistance was an acquired and age-dependent phenomenon, and studies of immune responses showed various differences between the two groups. Thus in this small group of primary schoolchildren between 8 and 13 years old, there is no evidence for age-dependent changes in water contact.

The studies in The Gambia started with a cohort of children (Hagan *et al.*, 1985) and continued with a cohort of individuals across the age spectrum (Wilkins *et al.*, 1987). The profile of reinfection after treatment exhibits the expected convex pattern. However, separating the age–reinfection data on the basis of water contact or exposure reveals that the relationship between infection and exposure is complex (figure 9.8). While the mean intensity of reinfection clearly increases with increasing duration of contact in children 2–9 years old, in adolescents (10–14 years old) less increase in reinfection is apparent, and those with the highest levels of

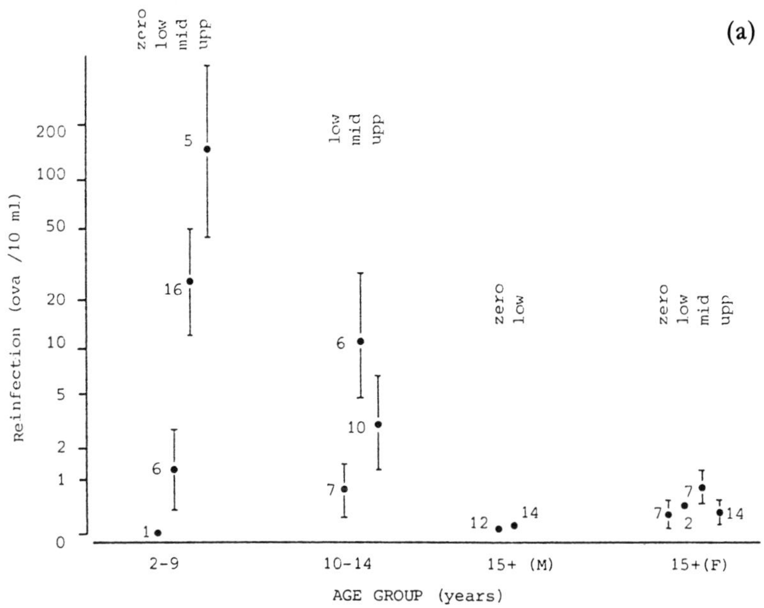

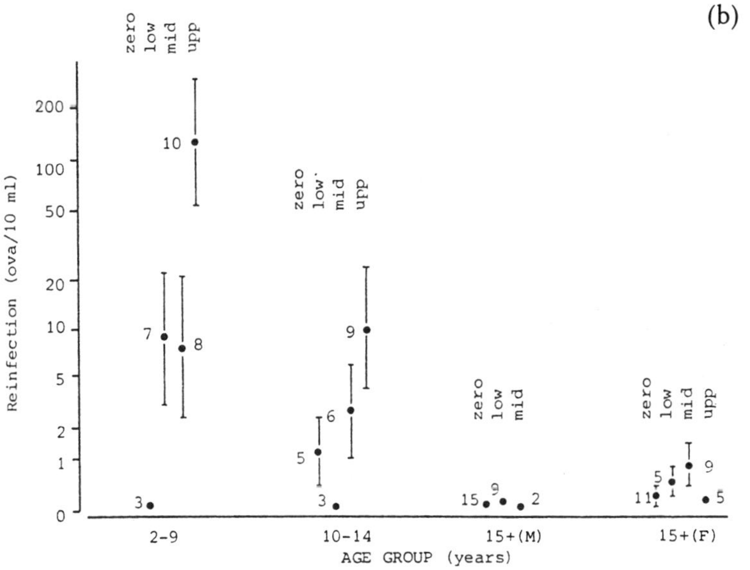

Figure 9.8. The relationship between mean intensity of reinfection and (a) mean duration of water contact or (b) exposure calculated using index 9.2. The individuals in each cohort were ranked, those with zero duration/exposure grouped in a 'zero' class, and the remainder grouped on the basis of the three equal-sized exposure bands (low, middle and upper). The data points represent the mean for each duration/exposure band. Data from The Gambia cohort study (Wilkins *et al.*, 1987).

contact do not have a greater intensity of reinfection than those with levels of contact in the middle rank (figure 9.8). In adult males, low levels of reinfection are associated with low levels of contact, but in adult females even the individuals who have high levels of exposure have a low mean intensity of reinfection.

When exposure, rather than water contact, is considered, more children are in the upper ranks of exposure than are in the upper rank of duration of contact (figure 9.8(b)). The mean intensity of infection of children increases with increasing exposure, though less markedly than with duration (cf. figure 9.8(a) and (b)). In adolescents, the mean intensity of reinfection also rises with increasing exposure but to a lesser extent than for 2–9-year-olds. Adult males have generally low exposure ranks and reinfection levels, and fewer adult women have high levels of exposure than have a high duration of water contact. Many adult women are in the middle rank for exposure to infection but have low levels of reinfection. Non-parametric analysis (Wilcoxon Rank test for all age groups and rank strata) of these results indicates that reinfection is significantly higher in children in the 2–9-year-old group than in the group over 10 years old ($z = 2 \cdot 12$; $P < 0 \cdot 01$ for duration, $z = 3 \cdot 48$; $P < 0 \cdot 001$ for exposure) (Wilkins *et al.*, 1987). It is also significantly higher in children (2–9 years old) than in adolescents (10–14 years old) when exposure to infection is taken into account ($z = 2 \cdot 41$; $P < 0 \cdot 05$), but not when duration of water contact is considered ($z = 1 \cdot 63$; $P > 0 \cdot 10$). Use of the exposure indices increases the significance of the difference compared with that shown when duration of water contact is taken into account. The importance of age in influencing levels of reinfection after treatment is therefore even more apparent when exposure to infection rather than duration of contact alone is considered.

Comparison of levels of exposure with the intensity of reinfection therefore indicates that reinfection levels significantly increase with increasing exposure in children, but not in adults. Exposure to infection is generally lower in adults than in children, but these lower levels of reinfection cannot be explained purely in terms of reduced exposure; factors related to age are clearly important. Indeed, the trend for reinfection to decrease with increasing age, even after allowance for variation in exposure, is highly significant ($P < 0 \cdot 001$).

These results have two important implications. First, they suggest that reinfection rates in adults are lower than would be predicted on the basis of exposure, and thus that hosts are increasingly able to modify their susceptibility with increasing age. This has been interpreted as evidence for an age-dependent increase in host resistance, perhaps due to the slow acquisition of an effective immune response due to repeated infection (Butterworth *et al.*, 1985; Wilkins *et al.*, 1987). A mammalian immune response to schistosome infection has been described in considerable detail (recently reviewed by Hagan, 1987) but the role of this response in limiting human infection remains unclear. Secondly, and equally important, the results of the study of Wilkins *et al.* (1987) and that of Butterworth *et al.* (1985) suggest that infection in the child age classes (2–9 years old) is not modified in this way. Thus the behaviours which result in exposure to infection may be a major determinant of the occurrence of schistosome infection in childhood.

Geohelminth soil-ingestion exposure and infection

It is consistently observed that the age–intensity profiles for geohelminth infection are markedly convex (figure 9.9). For both *A. lumbricoides* and *T. trichiura* the intensity rises rapidly in childhood to a maximum in the 5–10-year-old class, and then declines to attain a relatively stable and low level above 30 years old. The age–prevalence profiles, in contrast, exhibit little or no decline with age, indicating that although adults harbour fewer worms, similar proportions of adults and children are infected (Anderson and May, 1982; Bundy *et al.*, 1987a).

Studies of reinfection suggest that the age dependence of intensity is a function of the rate of acquisition of infection. When a population in a *T. trichiura* endemic area is successfully treated with an anthelmintic, the subsequent rate of reacquisition of infection exhibits marked age dependency (figure 9.10); the children in the 5–10-year-old class are reinfected much more rapidly than the adults (Bundy *et al.*, 1988). Age dependency in reinfection has also been shown for *A. lumbricoides* (Cabrera, 1981).

The intensity of infection is determined by the rate of acquisition of new infections and the rate of loss of existing infections. The observed age dependency of infection intensity and reinfection rate may, therefore, be due to enhanced exposure in children, reduced susceptibility in adults, or some combination of both processes (Anderson and May, 1982).

The relationship between the age profile of soil ingestion rate and *T. trichiura* infection intensity has been examined in an endemic community (figure 9.9). The results indicate that the mean proportion of silica in the stools of children exceeds the level attributable to dietary sources, whereas in adults average stool silica levels are largely attributable to diet. This suggests that children ingest more soil than adults and, as discussed above, have a potentially higher exposure to geohelminth infection. Comparison with the pattern of *T. trichiura* infection intensity in the same population (figure 9.9) shows that intensity rises with increasing rates of soil ingestion and declines as soil ingestion rates decrease. This suggests that the childhood peak in infection intensity may be related to enhanced infection exposure.

Preliminary studies of the age pattern of susceptibility to infection have also been undertaken for the same population using *T. trichiura* antigen ELISA (Bianco *et al.*, 1989). *Trichuris*-specific IgG, IgA and IgE serum levels exhibit similar and marked age dependency, rising in childhood and declining in adulthood in a pattern which mirrors that of *T. trichiuria* infection intensity (figure 9.9). Furthermore, during the rising phase in childhood all the antibody levels correlate positively with infection intensity, and it is only within the stable plateaux at 30 years old that an apparent effector association can be statistically demonstrated (potentially secretory IgA correlating negatively with intensity, and potentially blocking IgG4 correlating positively). This would appear to be unconvincing evidence of an effector role for the immune response in children, although it may be indicative of some form of resistance in the older age classes.

Taken together these studies suggest that the younger age classes have the highest rates of exposure and the least resistance to infection. These conclusions at the

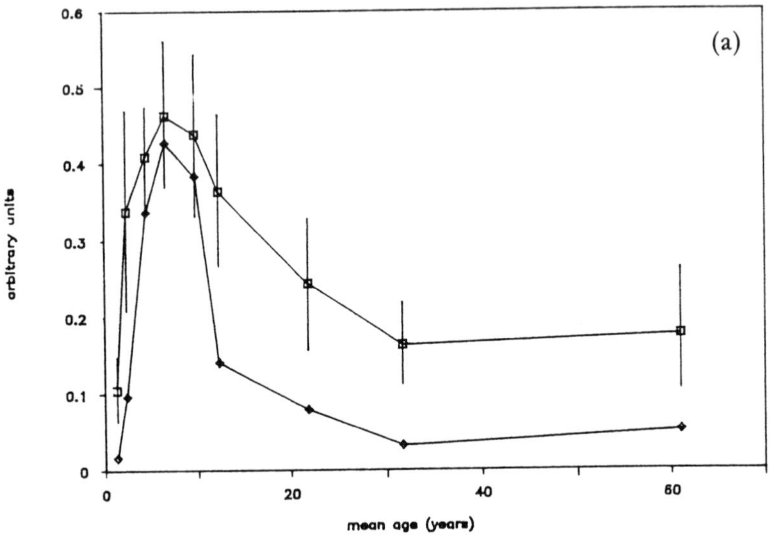

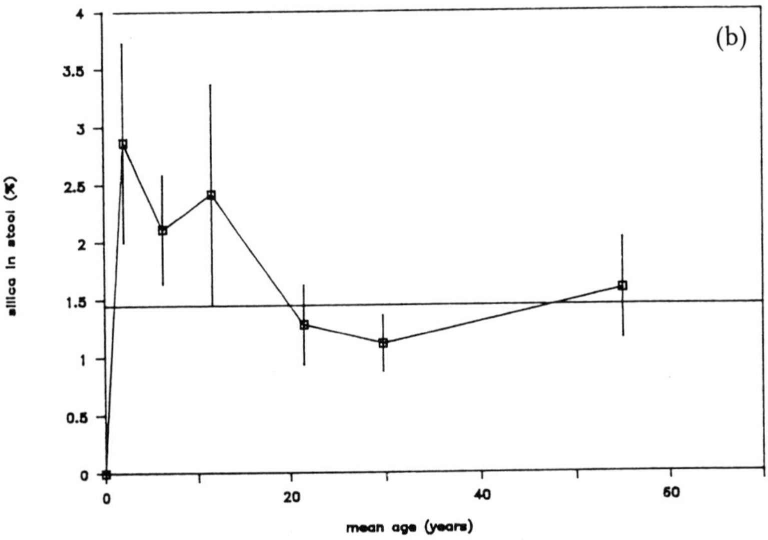

Figure 9.9. (a) Age dependency of *T. trichiura* infection intensity (□) and *T. trichiura* IgG antibody responses (◆). The age profiles of other isotypes and IgG subgroups (except IgG3) were similar. (b) Age dependency of geophagic behaviour as assessed by the proportion of silica in stool. The horizontal line represents the normative silica level of the diet (Wong *et al.*, 1988). The level of silica, and presumably soil-eating behaviour, corresponds with the variables shown in (a), which were measured in the same population. Data from Bundy (1988b).

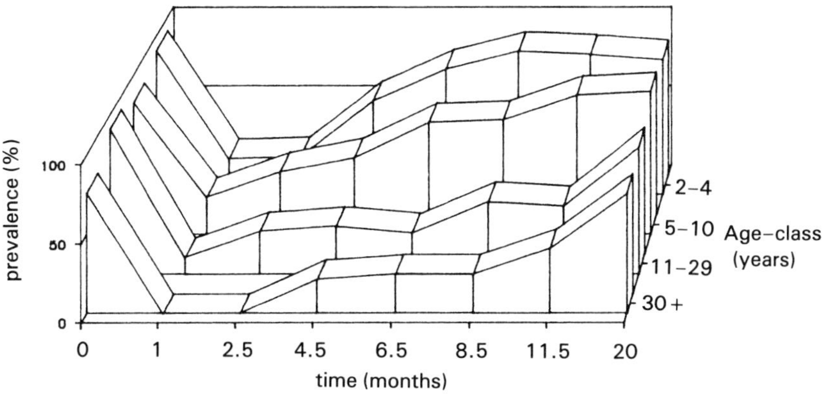

Figure 9.10. Age-dependent rate of reinfection with *T. trichiura*. Initial infection was assessed at time 0 and the population treated with an anthelmintic. The rate of reinfection was then monitored over a 20-month period following treatment. the child age classes reacquire infection more rapidly than do adults. Data from Bundy *et al.* (1988).

population level do not, however, take account of individual variation. While there is clearly variation in infection intensity between age classes, comparison within any age class reveals that there is further, and considerable, heterogeneity between individuals (Croll *et al.*, 1982; Bundy *et al.*, 1987b). For all helminth species examined the distribution of infection intensity is highly overdispersed; most individuals have few worms whereas a few hosts harbour disproportionately large worm burdens (figure 9.11(a)). Furthermore, community based studies of all the major geohelminths of humans have indicated that individuals are predisposed to a particular (high or low) intensity of infection; the size of the worm burden reacquired after successful treatment and the rate of its reacquisition are positively associated with the intensity of infection before treatment (Anderson, 1986; Bundy, 1986; Bundy and Cooper, 1988). The implication of these findings is that individuals consistently differ in their exposure or susceptibility to infection.

Differences in individual susceptibility to geohelminth infection have not been systematically sought. There is preliminary evidence that individual human infection sera have a differential ability to recognize *T. trichiura* antigen (Roach *et al.*, 1988), although the relevance of this finding to the child age classes in which an effector response appears not to be developed is unclear. In contrast, differences in individual exposure, as assessed by soil ingestion rates, have been clearly demonstrated (Wong *et al.*, 1988). It is apparent that the population distribution of exposure is highly overdispersed and closely mirrors observed frequency distributions of infection intensity (figure 9.11), although studies have not yet sought to correlate soil ingestion rates and infection intensity. It might be concluded, therefore, that individual differences in worm burden, particularly in the child age classes, are to a large extent determined by differences in exposure while the role of individual resistance is as yet undemonstrated.

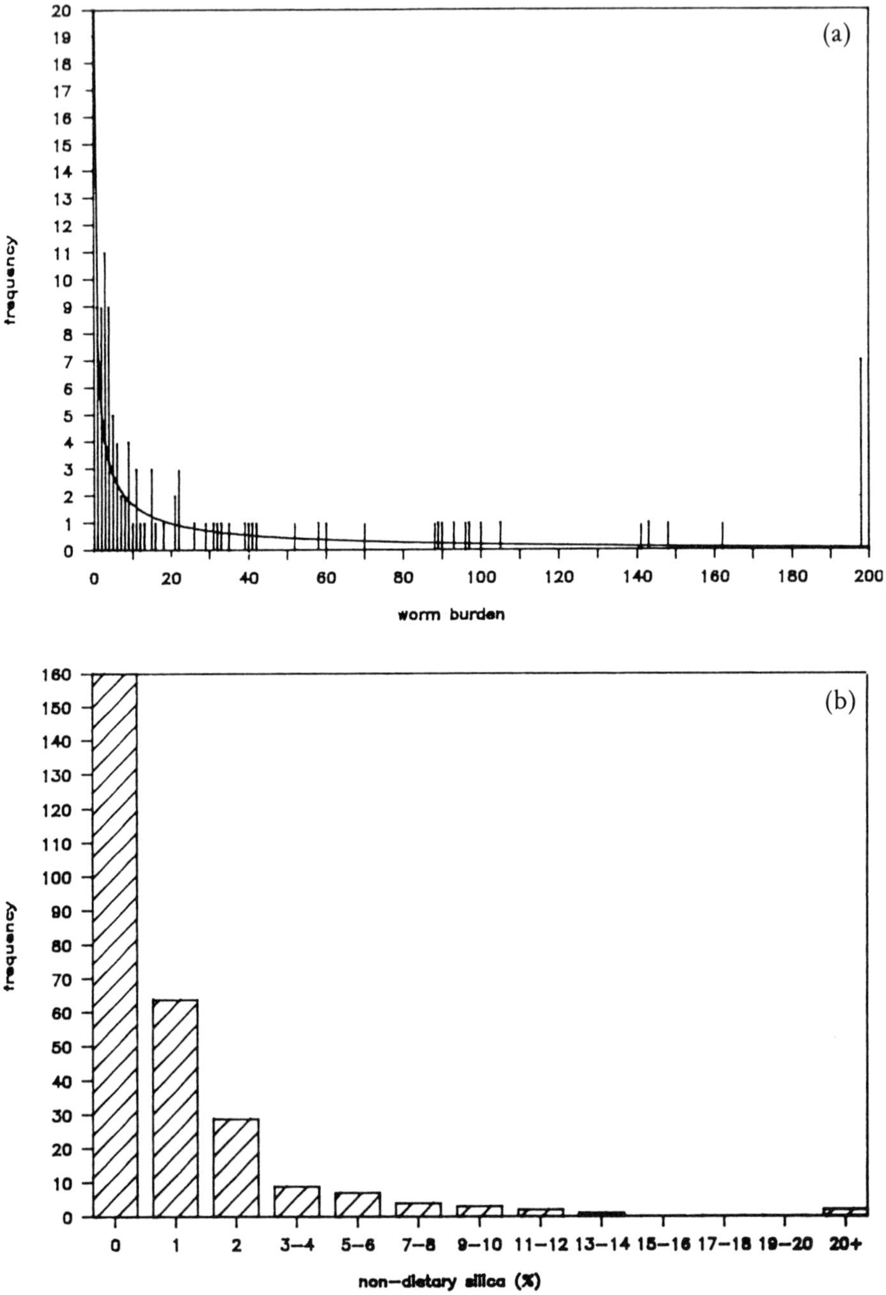

Figure 9.11. (a) Overdispersed frequency distribution of *T. trichiura* infection intensity. Data from Cooper and Bundy (1987). (b) Frequency distribution of soil-eating in children as assessed by levels of silica in stool which exceed normal dietary amounts. The distribution of this exposure is overdispersed and resembles that shown in (a).

IMPLICATIONS AND CONCLUSIONS

While the overall picture remains far from clear there is a trend which is discernible in all the human helminth infections which have been examined in detail: younger age classes have the highest intensities of infection, the highest exposure to infection and exhibit the least evidence of resistance to infection. This has the corollary that adults, who exhibit at least partial resistance to infection, are relatively unexposed. Since the morbid consequences of helminthic infection are largely related to infection intensity, this suggests that the major determinant of morbidity (which is also typically concentrated in the younger age classes) is exposure.

This conclusion has important practical consequences. It implies that naturally-developed immunity has little relevance for the prevention of disease in helminthic infection, and thus that the aim of vaccine development should be to stimulate effective immunity in the younger age classes in which natural immunity is least apparent; and that the major emphasis of parasite control should be to reduce infection exposure in the child age classes. For what is likely to be the majority of the population in most endemic areas it appears that exposure is the major determinant of infection intensity.

Given its potential importance it is perhaps surprising that infection exposure remains such an under-researched area. Yet the study of exposure remains an inexact and infant science: the first study on a role for water contact in exposure to schistosome infection was published only a decade ago, and the methodology for quantifying geohelminth exposure is less than 12 months old.

Two research areas are particularly deserving of further development. The first is the design of field studies which simultaneously assess exposure, infection and immune status in the same individuals. The studies of Butterworth and colleagues in Kenya have elegantly demonstrated how such an approach can provide insights into the human immune response to infection. Indeed it is only such combined studies which can provide convincing evidence that an immune response, however clearly demonstrated in the laboratory, actually has an influence on human infection. The second neglected area of research concerns the implementation of measures to reduce exposure. While there have been considerable advances in the development of the hardware necessary to enhance water supplies and sanitation, there has been relatively little attention paid to methods of reducing exposure through changes in behaviour. Health education has undoubtedly a major role to play in reducing exposure to parasitic infection, but effective education will have to await more precise understanding of how, when and why an individual becomes infected.

ACKNOWLEDGEMENTS

U.J.B. gratefully acknowledges the Medical Research Council for support of the studies described from The Gambia. The studies of geohelminth transmission were generously supported by grants (to D.A.P.B.) from the Wellcome Trust.

REFERENCES

Anderson, R. M., 1986, *Transactions of the Royal Society of Tropical Medicine and Hygiene*, **80**, 686–696.

Anderson, R. M. and May, R. M., 1982, *Nature* **297**, 557–563.

Anderson, R. M. and May, R. M., 1985, *Advances in Parasitology* **24**, 1–101.

Barbour, A. D., 1985, *Transactions of the Royal Society of Tropical Medicine and Hygiene*, **79**, 151–153.

Beaver, P. C., 1975, *Health Laboratory Service*, **12**, 116–125.

Bianco, A. E., Bundy, D. A. P., Lillywhite, J. E. and Didier, J. M., 1989 (in preparation).

Blumenthal, U. J., 1985, Transmission of *Schistosoma haematobium* in seasonal pools in The Gambia with particular reference to the role of human water contact, PhD Thesis, University of Liverpool, UK.

Blumenthal, U. J., Wilkins, H. A. and Tulloch, S., 1990, *Transactions of the Royal Society of Tropical Medicine and Hygiene* (in preparation).

Bradley, D. J. and McCullogh, F. S., 1974, *Transactions of the Royal Society of Tropical Medicine and Hygiene*, **80**, 706–718.

Bundy, D. A. P., 1986, *Transactions of the Royal Society of Tropical Medicine and Hygiene*, **80**, 706–718.

Bundy, D. A. P., 1988a, *Parasitology Today*, **4**, 186–189.

Bundy, D. A. P., 1988b, *Philosophical Transactions of the Royal Society*, **B 321**, 405–420.

Bundy, D. A. P. and Cooper, E. S., 1988, *Advances in Parasitology*, (in press).

Bundy, D. A. P., Cooper, E. S., Thompson, D. E., Anderson, R. M. and Didier, J. M., 1987a, *Transactions of the Royal Society of Tropical Medicine and Hygiene*, **81**, 85–94.

Bundy, D. A. P., Cooper, E. S., Thompson, D. E., Didier, J. M., Anderson, R. M. and Simmons, I., 1987b, *Epidemiology and Infection*, **98**, 65–71.

Bundy, D. A. P., Cooper, E. S., Thompson, D. E., Didier, M. M. and Simmons, I., 1988, *Parasitology*, **97**, 469–476.

Burkitt, D. P., 1972, *British Medical Journal*, **ii**, 556–561.

Butterworth, A. E., Capron, M., Cordingley, J. S., Dalton, P. R., Dunne, D. W., Kariuki, H. C., Kimani, G., Koech, D., Mugambi, M., Ouma, J. H., Prentice, M. A., Richardson, B. A., Arap Siongok, T. K., Sturrock, R. F. and Taylor, D. W., 1985, *Transactions of the Royal Society of Tropical Medicine and Hygiene*, **79**, 393–408.

Butterworth, A. E., Dalton, P. R., Dunne, D. W., Mugambi, M., Ouma, J. H., Richardson, B. A., Arap Siongok, T. K. and Sturrock, R. F., 1984, *Transactions of the Royal Society of Tropical Medicine and Hygiene*, **78**, 108–123.

Butterworth, A. E. and Hagan, P., 1987, *Parasitology Today*, **3**, 11–16.

Cabrera, B. D., 1981, *Collected Papers on the Control of Soil-Transmitted Helminthiasis. Vol. I.* (Tokyo: Asian Parasite Control Organisation) pp. 181–191.

Cooper, E. S. and Bundy, D. A. P., 1987, In *Intestinal Helminthic Infections* edited by Z. S. Pawlowski. *Bailliere's Clinical Tropical Medicine and Communicable Diseases*, **2**, 629–643.

Croll, N. A., Anderson, R. M., Gyorkos, T. W. and Ghardirian, E., 1982, *Transactions of the Royal Society of Tropical Medicine and Hygiene*, **76**, 187–197.

Cummings, J. H., 1983, *Gut*, **14**, 69–81.

Dalton, P. R., 1976, *Bulletin of the World Health Organization*, **54**, 587–595.

Dalton, P. R. and Pole, D., 1978, *Bulletin of the World Health Organization*, **56**, 417–426.

Desowitz, R. S., 1981, *New Guinea Tapeworms and Jewish Grandmothers: Tales of Parasites and People* (New York: W. W. Norton).

Farooq, M. and Mallah, M. B., 1966, *Bulletin of the World Health Organization*, **35**, 377–387.

Fenwick, A., Cheesmond, A. K., Kardaman, M., Amin, M. A. and Manjing, B. K., 1982, *Journal of Tropical Medicine and Hygiene*, **85**, 3–11.

Gilman, R. H., Chong, Y. H., Davis, C., Greenberg, B., Virik, H. K. and Dixon, H. B., 1983, *Transactions of the Royal Society of Tropical Medicine and Hygiene*, **77**, 432–438.

Glickman, L. T., Chaudry, I. U., Costantino, J., Clack, F. B., Cypress, R. H. and Winslow, L. 1981, *American Journal of Tropical Medicine and Hygiene*, **30**, 77–80.

Hagan, P., 1987, *The Biology of Schistosomes: From Genes to Latrines* edited by D. Rollinson and A. J. G. Simpson (London: Academic Press), pp. 295–320.

Hagan, P., Wilkins, H. A., Blumenthal, V. J., Hayes, R. R. and Greenwood, B. M., 1985, *Parasite Immunology*, **7**, 625–632.

Halstead, J. A., 1968, *American Journal of Clinical Nutrition*, **21**, 1384–1393.

Hominick, W. M., Dean, C. G. and Schad, G. A., 1987, *Transactions of the Royal Society of Tropical Medicine and Hygiene*, **81**, 978–986.

Husting, E. L., 1970, *Central African Medical Journal*, **16**, 5–10.

Husting, E. L., 1983, *Journal of Tropical Medicine and Hygiene*, **86**, 23–35.

Kloos, H., Higashi, G. I., Cattani, J. A., Schlinski, V. D., Mansour, N. S. and Murrell, K. D., 1983, *Social Science and Medicine*, **17**, 545–562.

Kloos, H. and Lemma, A., 1980, *Ethiopian Medical Journal*, **18**, 91–98.

Kvalsvig, J. D. and Schutte, C. H. J., 1986, *Annals of Tropical Medicine and Parasitology*, **80**, 13–26.

Lanzkowsky, P., 1959, *Archives of Disease in Childhood*, **34**, 140–148.

Lariviere, M., Satge, P. and Dun, V., 1965, *Afrique Medicale*, **32**, 441–446.

Roach, T. I. A,, Wakelin, D., Else, K. J. and Bundy, D. A. P., 1988, *Parasite Immunology*, **10**, 279–291.

Rukmono, B., 1980, In *Collected Papers on the Control of Soil-Transmitted Helminthiases*, Vol. I (Tokyo: Asian Parasite Control Organisation), pp. 125–128.

Schad, G. A., Nawalinski, T. A. and Kochar, V. K., 1983, In *Human Ecology and Infectious Disease*, edited by N. A. Croll and J. Cross (New York: Academic Press), pp. 187–223.

Singhi, S., Singhi, P. and Adwani, G. B., 1981, *Clinical Paediatrics* **20**, 783–785.

Tayo, M. A., Pugh, R. N. H. and Bradley, A. K., 1980, *Annals of Tropical Medicine and Parasitology*, **74**, 347–354.

Tiglao, T. V. and Camacho, A. C., 1983, *South-east Asian Journal of Tropical Medicine and Public Health*, **14**, 18–24.

Vessal, K., Ronaghy, H. A. and Zarabi, M., 1975, *American Journal of Clinical Nutrition*, **32**, 2129–2135.

Warren, K. S., 1973, *Journal of Infectious Diseases*, **127**, 595–609.

Warren, K. S. and Mahmoud, A. A. R., 1984, *Tropical and Geographical Medicine* (New York: McGraw-Hill).

Wilkins, H. A., Blumenthal, U. J., Hagan, P., Hayes, R. J. and Tulloch, S., 1987, *Transactions of the Royal Society of Tropical Medicine and Hygiene*, **81**, 29–35.

Wong, M. S., 1988, The role of environmental and host behavioural factors in determining exposure to infection with *Ascaris Lumbricoides* and *Trichuris trichiura*, PhD Thesis, Faculty of Natural Sciences, University of the West Indies.

Wong, M. S. and Bundy, D. A. P., 1990, *Transactions of the Royal Society of Tropical Medicine and Hygiene*, (in press).

Wong, M. S., Bundy, D. A. P. and Golden, M. H. N., 1988, *Transactions of the Royal Society of Tropical Medicine and Hygiene*, **82**, 621–625.

10. Influence of host behaviour on some ectoparasites of birds and mammals

M. D. Murray

INTRODUCTION

Arthropod ectoparasites have evolved a number of subtle and sensitive mechanisms for locating and facilitating their transmission to hosts. Hosts, in response, have adopted several defensive strategies to reduce the impact of the parasites. Attention has been focused chiefly on the regulatory effects of the host's immune system, but it is clear that the behaviour of hosts can have an effect on the establishment and proliferation of ectoparasites. This chapter reviews some of the relationships between host behaviour and ectoparasitic infestation among avian and mammalian host species.

Most of the ectoparasites of mammals and birds are either arachnids or insects, and the known physiological, ecological and behavioural adaptations are so varied that one's sense of wonder at future discoveries is assured. The response of the rabbit flea, *Spilopsyllus cuniculi*, to the hormones of a pregnant doe enables its breeding cycle to be synchronized to that of the rabbit (Rothschild and Ford, 1964). The tungid flea, *Tunga monositus*, burrows into the skin of its deermouse host, *Peromyscus maniculatus*, where the changes in feeding behaviour are timed to synchronize with the inflammatory reactions and tissue repair processes of the host (Lavoipierre *et al.*, 1979). The anopluran lice of the Weddell seals are adapted to survive prolonged periods at sea, the pressures when seals dive to 600 metres, and desiccation when their hosts are ashore in the arid Antarctic (Murray, 1976).

In order to assure survival, arthropod ectoparasites have evolved extremely sensitive mechanisms to detect vulnerable hosts. The sparse wide-ranging screw worm flies, *Chrysomya bezziana*, of the Old World and *Cochliomyia hominivorax* of the New World, successfully find hosts with open wounds around which to lay their eggs; odours produced by skin bacteria around the wound may play a role in stimulating oviposition (Eddy *et al.*, 1975). The tsetse flies of Africa find downwind plumes of odours from the breath of cattle and follow these to their hosts (Vale, 1977a, 1984). The black fly (*Simulium euryadminiculum*) of North America is attracted to odours from the tail region of the common loon (Fallis and Smith, 1964). The tick (*Ixodes matopi*) aggregates on twigs which have been marked by the klipspringer antelope, *Oreotragus oreotragus*, with secretions from the preorbital gland (Rechaw *et al*, 1978, Colborne *et al.*, 1981).

Many species of blood-sucking flies, mosquitoes and biting midges, live within restricted habitats into which a host comes to feed or drink. Adult stages of other species fly from the ground vegetation every night to disperse passively in the laminar winds above the ground to locate herds of hosts, where they obtain a bloodmeal or dung in which to oviposit; in this way they maintain daily contact with their mobile hosts and the biting midge, *Culicoides brevitarsis*, is a good example (Murray, 1987a). Stronger fliers such as the buffalo fly (*Haematobia irritans exigua*), seek their host and remain on the host's body even when not feeding, leaving only briefly to oviposit in fresh dung (Roberts, 1952).

Wherever a host pauses or rests regularly some ectoparasite may breed. Fleas and ticks, for example, wait in the tussock grass of Sub-Antarctic islands for the return of albatrosses and penguins which they infest each breeding season. Soft-bodied argasid ticks wait in the sand of coral cays for the return of seabirds, and in the sand of resting sites for kangaroos in inland Australia. Argasid ticks also wait in tree holes for nesting parrots, fleas wait in nests for the return of migrating house martins in Europe, and hemipteran bugs wait in nests for the return of cliff swallows in Northern America. Liponyssid mites await the return of starlings or pigeons to their old nests, and pupiperous flies await the return of bats to their breeding or roosting caves. Many more examples can be found in the various publications on ectoparasite/host associations (Rothschild and Clay, 1952; Nelson *et al.*, 1975; Waage, 1979; Marshall, 1981).

VISITING ECTOPARASITES

Where it is only the immature stages that are parasitic many ectoparasites visit a host solely to feed or to deposit eggs or larvae. Hosts react to their presence but there is little quantitative data available on the success of these reactions. One reason for this dearth of data is probably that the subject has been dismissed as obvious because we carry out many of these actions ourselves. The human hand and arm is a remarkably versatile structure with which we still pick, squeeze, rub, scratch (any part of the body), wave away, brush away, slap or swat ectoparasites, and after an initial experience it is not long before we commence to use other senses. We see attacks by midges, we hear the approach of mosquitoes, we feel the bite of a tabanid or a flea running over the skin, and we suspect the odours of unclean places, become wary, and move away. Animals and birds behave similarly.

The first 'line of defence' is the pelage or plumage. Mosquitoes feed on the bare feet and legs of herons, egrets and ibises once the plumage of chicks has developed (Webber and Edman, 1972). Midges can crawl through the haircoat of cattle but not the dense fleece of sheep which they bite around the head or feet (Muller and Murray, 1977). Man covers himself with clothes, some mammals with mud or water.

Vale (1977b) showed that more tsetse flies fed on sedated goats than on unsedated goats and attributed this to the differences in the movements of the head, legs and tails of the two groups. Fewer stable flies (*Stomoxys calcitrans*) feed on calves which vigorously switch their tails, stamp their feet or swing their heads (Warnes and Findlayson, 1987). Wild game alleviate attacks by blood-sucking flies by twitching the skin muscles, flicking ears and tail, head movements and lifting legs (Kangwangy, 1978).

Detailed studies have been carried out on the effects of behaviours of ciconiiform birds on the feeding success of mosquitoes (Edman and Kale, 1971; Edman *et al.*, 1972; Kale *et al.*, 1972; Webber and Edman, 1972). In cage experiments more mosquitoes engorged on the Green Heron (*Butorides striatus virescens*) and the Black-crowned Night Heron (*Nycticorax nycticorax*) than on the Little Blue (*Egretta caerulea*), the Louisiana Heron (*Egretta tricolor*), the Cattle Egret (*Bubulcus ibis*), the Snowy Egret (*Egretta thula*) or the White Ibis (*Eudocimus albus*). This was attributed to the intensity of the defense activities of each species for when birds were restrained, similar numbers of mosquitoes engorged on all species of birds. A variety of defence behaviours were seen: shaking, rubbing and scratching the head; rubbing, pecking, fluffing the body and flapping the wings: rubbing, pecking, stamping, slapping, shaking and lifting the legs and feet. Most of the behaviours were directed to protect the legs and feet. The increased frequency was directly related to the success in disturbing mosquitoes, and frequency increased with mosquito density. Green Herons and the Night Heron, which remain still for long periods when feeding, tolerated mosquito attack until the intensity became too great whereas the other species, which are more active in the wild, responded quickly and reduced the feeding success of mosquitoes more than threefold. Thus, there are considerable differences in the toleration of insect attack, in the behavioural responses used and in the intensity with which behavioural defense strategies are deployed. These varied responses are the next 'line of defence', and enable a host to pursue other important activities such as feeding. Animals clearly vary in their individual responses to attacks by visiting ectoparasites; such variation being expressed through differential sensitivity to pain and irritation and possibly through immunological reactivity (e.g. allergic) to parasite secretions during feeding.

When mechanical disturbance of arthropod activity fails to drive away potential invaders animals can resort to more radical evasive measures. Tsetse flies bite warthogs around the eyes and are disturbed by flicking of the ears, lifting the forelegs, or rubbing the head against objects; when their attack becomes too intense, however, the warthogs return to the bush to escape the annoyance (Vale, 1974). When their initial responses become inadequate, the horses of the Camargue gather into herds and seek respite from tabanids during the summer by moving to well-defined bare regions more exposed to the prevailing wind (Hughes *et al.*, 1981). Cattle under heavy attack from tabanids may commence to gallop around fields, and they may stampede into lakes to seek relief when under heavy attack by black flies. Many of these behaviours have also been described in reindeer, which may run against the wind, enter water or scrub, or climb to the snow in mountainous regions to escape attack by midges, mosquitoes, tabanids or botflies; where these avenues are not open to them they clump into dense herds (Breev, 1951). Breev showed that individual reindeer away from a herd were more heavily attacked than those within a herd and the reindeer on the edge of a herd were attacked more intensively than those within. Reducing the intensity of attack may be one selection pressure leading to a 'selfish herd' effect (Hamilton, 1971). Helle and Aspi (1983) created an artificial replica of individuals and herds using traps baited with carbon dioxide, and most ceratopogonids and simulids were captured in the isolated traps, fewer in the outer ring of traps of the 'herd' and fewest within the herd. It is doubtful, however, whether such reactions on the part of the host have much effect on the populations of a blood-sucking fly, unless the numbers of the species are few and feeding is disturbed by other insects or by the hosts' reactions to the insect complex attacking them (Waage and Davies, 1986). Nevertheless, escape behaviours probably do save the lives of hosts. It is not uncommon to find kangaroos blinded by swollen eyes and starving after the sudden emergence of the black fly, *Austosimulium pestilens*, following floods in inland Queensland (McCarthy, 1961).

Exploitation of host behaviour and learned evasive responses by hosts

The attack strategies of parasites sometimes make subtle use of host behaviour (see also Chapter 6) and the responses of hosts may depend on experience of attacks rather than being 'hard-wired'.

The responses of botflies are interesting in this context. *Cephenemyia jellisoni* and *C. apicata* squirt larvae into the noses of the Black-tailed Deer (*Odocoileus hemionus*), which respond by jerking back the head, sneezing, shaking the head, and lowering their noses to the ground. *C jellisoni* adopts a covert, inaudible 'stalking' flight and is undetected until larviposition whereas *C. apicata* land on or near to deer, arouse attention and curiosity, and lead the deer to expose the nose. Previous experience of larviposition results in deer recognizing the flies and taking evasive actions such as ignoring the flies and thus not exposing the nose, turning the head away, lowering the head, moving away and even biting the flies (Anderson, 1975). Reindeer in clumped herds place their heads between the bodies of the animals in front to escape *C. trompe* (Breev, 1951). Sheep attacked by the sheep nasal-fly, *Oestus ovis*, place their noses near to the ground, gather together, and may raise the dust with their feet. These behaviours combined with an immunological response usually result in older animals carrying fewer larvae. Horses react to the presence of botflies, *Gasterophilus* spp, in a similar manner, and cattle react dramatically to warble flies, *Hypoderma* spp. These latter responses are to oviposition as adult oestrid flies do not feed and the immature stages may not complete development within the host for several months. Perhaps some botflies simulate attack by species that cause immediate annoyance, and the behavioural responses of experienced hosts expose the naive and susceptible individuals for exploitation.

Many learned responses can be adopted quite quickly. Horses soon learn to stand in the smoke of a drover's camp-fire, and it only takes a few days for water buffaloes (*Bubalus bubalis*), used as bait animals, to rush across a paddock to seek the attention of an entomolgist with a sweep net which has been used to collect buffalo flies (A.L.Dyce, personal communication).

Recently Bucher (1988) discussed evidence for some species of birds using materials in their nests which may control the numbers of their ectoparasites. Many birds use the same nest each year, whether it be an individual or cooperative nest. Several species, 28 of 49 Falconiformes listed by Wimberger (1984), use green plant material in their nests, and he suggested that the aromatic constituents may control ectoparasites. Clark and Mason (1985) found that the European Starling (*Sturnus vulgaris*) selected the more odorous plants and demonstrated that the volatiles released from the leaves reduced the number of eggs of lice, *Menacanthus* spp., which hatched but did not affect survival of the blood-sucking fowl mite (*Ornithonyssus sylvarium*). Screech owls (*Otus asio*) normally feed on blind snakes (*Leptotyphlops dulcis*), killing them before bringing them to their nests, but Gelbach and Baldridge (1987) also reported that 18% of 77 nests examined contained live snakes and that the nestlings from these nests grew faster and survived better than snakeless broods. They postulated that the snakes may have reduced the number of ectoparasites and other fauna in the nests to the benefit of the chicks. The nests of birds and mammals are distinct habitats in which many organisms live. The effects of hosts' methods of remaking nests each year and of the resident predators on the ectoparasite fauna warrant more critical study.

RESIDENT ECTOPARASITES

The other major group of ectoparasites are those that complete their whole life cycle on their host. Ticks may spend much of their lives upon their hosts but leave to oviposit or moult. Lice and many mites, however, live permanently within the pelage of mammals or the plumage of birds. The majority of the critical studies on the ecology of these ectoparasites have been on mammalian lice. Their lives are intimately linked to the ecological adaptations of their hosts which may be small or large animals, solitary or gregarious, and sparsely or densely covered with hair or feathers.

Ectoparasites of small mammals

The house mouse (*Mus domesticus*) is infested by the anopluran louse (*Polyplax serrata*). The life cycle of the louse is typical in there being an egg, three nymphal stages and adult females and males. Eggs are attached to the hair close to the skin. Nymphs and adults move freely within the pelage, from the skin surface to hair tip.

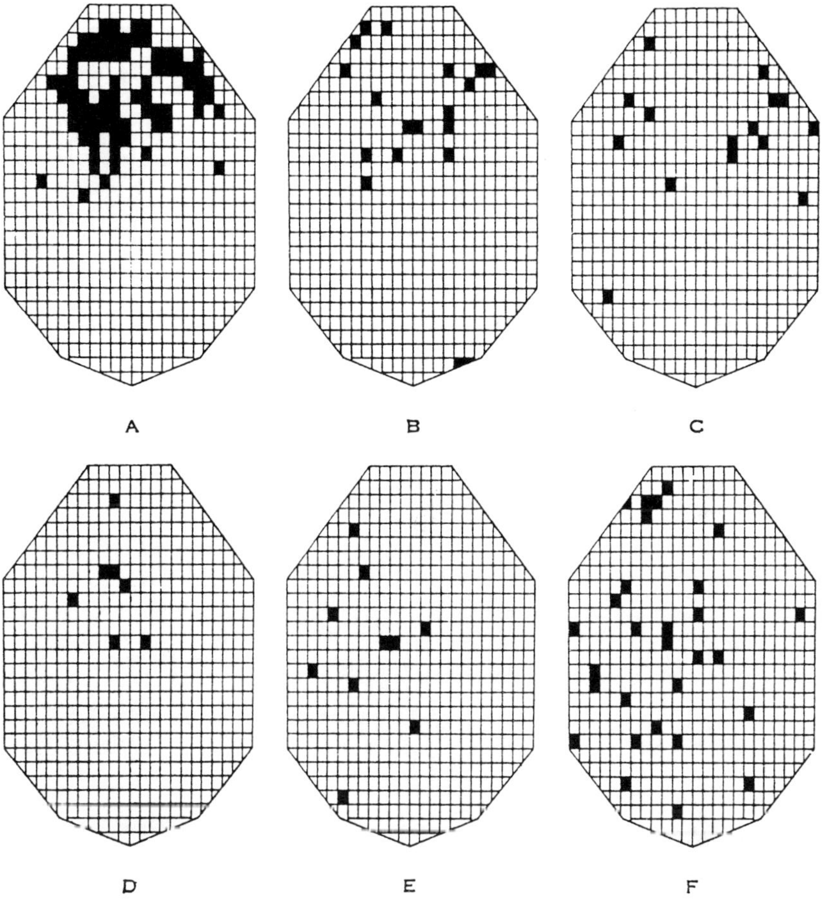

Figure 10.1. Distribution of *P. serrata* on the mouse (*Mus domesticus*), with normal grooming ability. The black areas are those in which one or more eggs or lice were found. (a) Eggs. (b) Females. (c) Males. (d) Stage 3 nymphs. (e) Stage 2 nymphs. (f) Stage 1 nymphs. Data from Murray (1961).

They feed at the skin surface where they obtain blood directly from venules. The various stages are not distributed evenly over the body of the mouse (Murray, 1961). Eggs, adults and Stage 3 nymphs are found on the fore-part of the body whereas some Stage 2 and many Stage 1 nymphs are also found on the hind-part (figure 10.1). If grooming is prevented by placing a collar around the neck of the mouse, lice and their eggs spread across the whole body (figure 10.2), and the number of lice may increase from 100 to more than 1000 within 4 weeks. Within 24 hours of being permitted to self-groom the louse numbers and distribution begin to return to normal. The same changes in distribution occur when the lower incisors are wedged apart (figure 10.3). The lower incisors, which are capable of lateral closure, are used to comb the hair and are sufficiently effective to control the distribution and abundance of lice on mice.

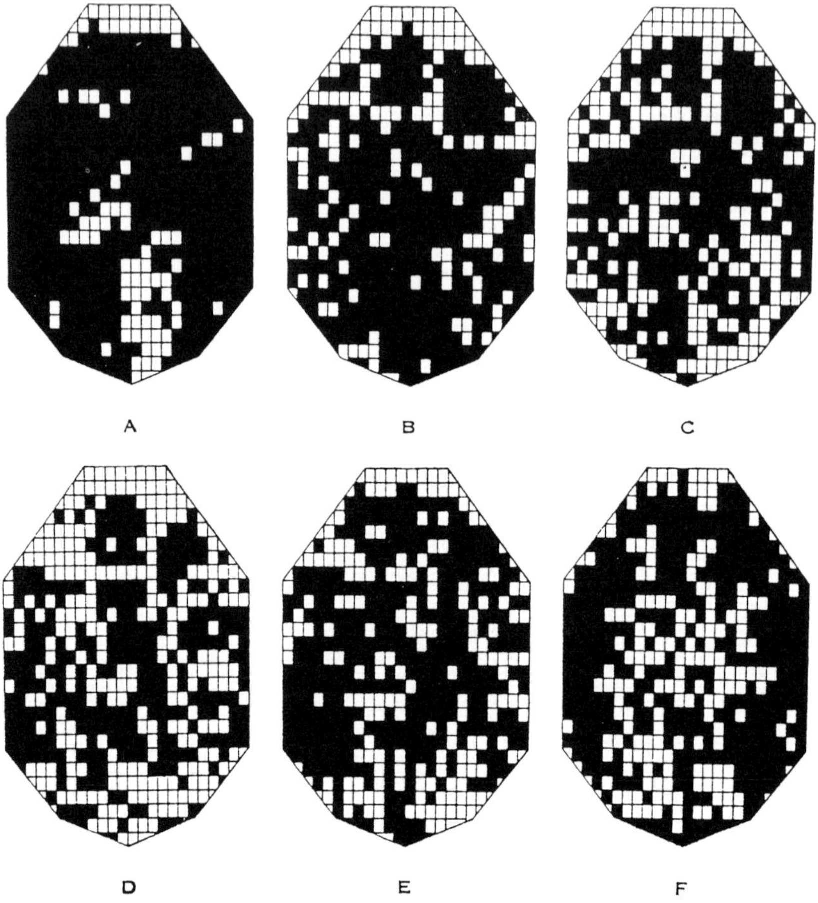

Figure 10.2. Distribution of *P. serrata* on a house mouse after grooming with the teeth has been prevented. Black areas as in figure 10.1. Data from Murray (1961).

Grooming with the fore- and hind-limb digits can also be effective. If the toes of the hind feet are absent, the number of lice on the fore-part of the body increases greatly (Bell *et al.*, 1962) but the overall population does not increase out of control. Mutual grooming further controls the number of lice on the fore-part of the body, and normal mice can control the number of lice on those with amputated hind feet (Bell and Clifford, 1964). The efficiency of mutual grooming declines when there is a disturbance in the social hierarchy of a group of mice (Lodmell *et al.*, 1970). Clearly then, mice groom themselves most efficiently by combing with their teeth while those regions that cannot be reached by the mouth are combed with their toes. Within a group of mice this is further supplemented by mutual grooming. The only stage of the life cycle which is not removed efficiently by grooming with the teeth on a mouse is the smallest (Stage 1) nymphal stage (figure 10.1). The most sensitive indicator of reduced efficiency of a grooming technique is

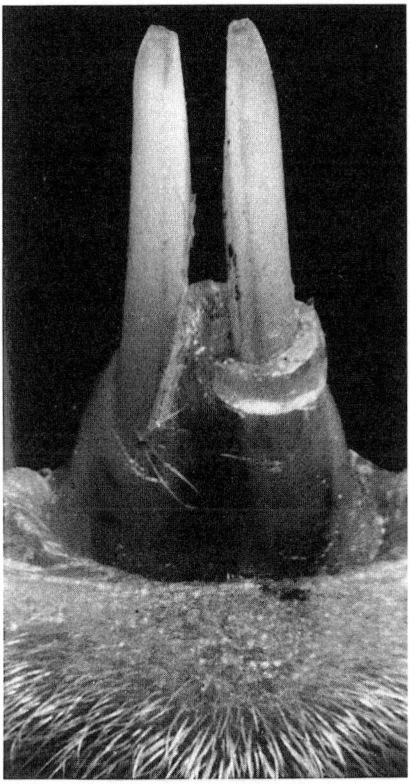

Figure 10.3 Lower incisors of *M. domesticus*. Small rodents comb through their fur with their lower incisors when grooming. If lateral closure of the incisors is prevented with a stout polythene ring, the 'comb' becomes inefficient and the number of lice increases rapidly. Data from Murray (1961).

an increase in the distribution of the parasite over the body. Thus, for grooming to control ectoparasites: (1) the technique should be efficient at removing vital stages of the life cycle, (2) the area groomed should be sufficient, (3) the time spent grooming should be adequate, and (4) the ectoparasite should not have predilection sites, and should move freely into the groomed regions.

Two other ectoparasites of mice, the mites *Myobia musculi* and *Myocoptes musculinus*, may be found all over the body but *M. musculi* is usually confined to the fore-part, where numbers are controlled by grooming with the hind toes (Weisbroth *et al.*, 1974), and the smaller *M. musculinus* is found on the hind-part. These distributions are quite distinct when both mites infect a mouse (figure 10.4). It would appear that where grooming is less efficient ectoparasites can become concentrated on specific parts of the body of their hosts, and thus associated with predilection sites. This may have an effect of reducing interspecific competition.

Lice also have restricted distributions on other small rodents (figure 10.5) which suggests that similar grooming techniques are employed here. Adults and nymphs of the biting louse, *Geomydoecus oregonus*, are found mostly on the head and

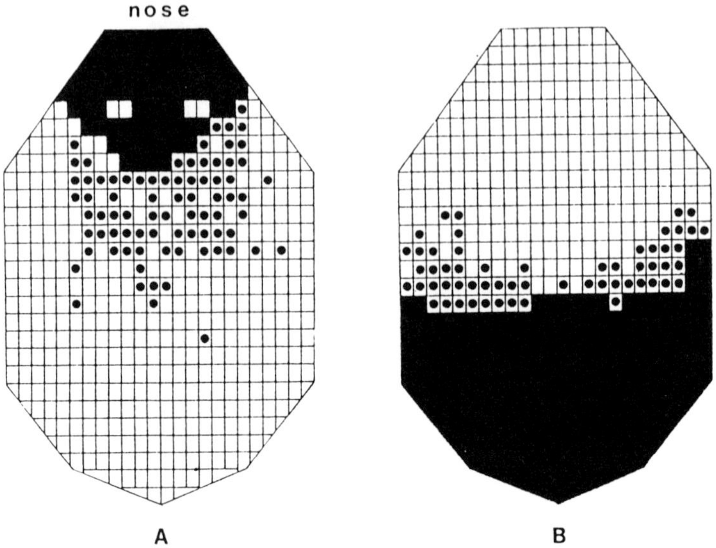

Figure 10.4. Distribution of two mites concurrently infesting a mouse, *(M. domesticus)*. *(a) Myobia musculi*. *(b) Myocoptes musculinus*. ■ = 4 + mites in each area; [●] = 1–3 mites. Data from Murray and MacIntyre (unpublished).

fore-part of the body of the gopher, *Thomomys bottae*, and eggs are mainly around the ears and eyes with some on the mid-dorsal back (Rust, 1974). Vysotskaya (1950) reported a seasonal change of *Hoplopleura acanthopus* over the body of the vole, *Microtus arvalis*; lice were found only on the fore-part of the body in autumn but were distributed over the whole body when they were abundant in early spring. She attributed these observations to behavioural changes by the vole in response to surrounding air temperatures (such as sleeping in a curled-up position in the winter enabling lice to proliferate unmolested), and to lice being lost with the pelage as the winter coat was shed. Slight changes in grooming efficiency could also explain these findings, grooming being less efficient with a thicker winter coat but returning to full efficiency with the thinner summer coat. There are many reports of seasonal changes in the abundance of lice on rodents which could be explained similarly.

One of the consequences of malnutrition is lethargy, and the increased severity of louse infestations on mammals given diets deficient in vitamins probably reflect the consequences of inadequate grooming (Buxton, 1947). Malnutrition may change behavioural priorities so that less time may be spent grooming when more time has to be spent feeding or caring for young.

The age and size of the host affects the amount of time spent grooming and may underlie changes in ectoparasite abundance and distribution. As a rodent becomes larger, for instance, grooming with two teeth must become less efficient purely because the increased area will require more time to be cleaned effectively, and the host may have other priorities in the allocation of its time. In keeping with this, the

Figure 10.5. Distribution of anopluran lice on wild small animals. (a) *P. affinis* on *Apodemus agrarius*. (b) *Polyplax serrata* on *Mus domesticus*. (c) *P. spinulosa* on *Rattus norvegicus*. (d) *Hoplopleura acanthopus* on *Microtus arvalis*. (e) *P. reclinata* on *Crocidura suaveolens*. Modified after Dubinin (1953).

distribution of lice on the larger rat, *Rattus norvegicus*, is more extensive than on smaller conspecifics (figure 10.5).

When *Polyplax serrata* are placed on the head of a mouse which has been prevented from grooming, they can be found all over the body within 24 hours (Murray, 1961). The lice are agile, undoubtedly an adaptation to escape grooming, but grooming is so efficient as to preclude association with predilection sites as apparently exist for smaller mites. However predilection sites have been identified in louse infestations on slightly larger rodents; *Hoplopleura acanthopus* is usually found on the heads and shoulder blades of the vole, *M. arvalis*, whereas *H. acanthoplus* is found around the base of the ears and root of the tail (Dubinin, 1953); *H. oenomydis* infest the neck and head of the rat, *R. norvegicus*, and *P. spinulosa* infest the back and haunch (Kaneko, 1959).

Other anti-parasite techniques employed by rodents are to bite and lick. When larval *Ixodes holocyclus* are placed on mice frenzied grooming with the teeth and feet commences once the ticks start to attach. This is followed by an immediate loss of ticks, approximately half of 20 larvae placed on each of five mice (Murray, unpublished data). The surviving larvae were attached on the fore- and hind-parts of the body in a ratio of 2:1 but after 60 hours a mean infestation of 2.5 ticks remained in a ratio of 4:1. When larger nymphal and female ticks were placed on mice to study the toxicity of the salivary secretions, it was necessary to place them on the heads of the mice to prevent removal by grooming because the mice bit the ticks and killed them. It was also necessary to keep mice in separate cages to prevent ticks being killed during mutual grooming (Murray and Koch, 1969). Larger ectoparasites may be killed by biting when they cannot be removed initially, and grooming can be directed to the specific site of irritation.

Hart *et al.* (1987) showed that genital-licking by rats mechanically washes the region, and furthermore the bacteriocidal effect of constituents of the saliva play a role in reducing genital infections. Licking could have wider effects if it influences the distribution and abundance of the bacterial flora of the skin as these bacteria are probably an important component of the diet of some ischnoceran lice of mammals (Murray and Edwards, 1987).

Comparative evidence that oral grooming/biting is important comes from the distribution of the louse, *P. reclinata*, on the shrew, *Crocidura suaveolens* (figure 10.5). The dentition of a shrew differs from that of a rodent and the grooming technique appears to permit lice to spread over the whole body of this smaller mammal.

Ectoparasites of large mammals

Large mammals, such as ungulates, are able to carry a sufficient depth of pelage to insulate them from the adverse climates of their habitats, and this, together with their smaller surface area:volume ratio, allows them to remain exposed to most weather conditions. Several species of lice may infest large animals concurrently, each differing in distribution over the body depending upon the season of the year. The body of a mammal may be divided into those regions that favour the breeding of lice and those that are unfavourable. The former are usually characterized by the presence of eggs. Consequently, the distribution of regions that favour breeding is intimately linked with conditions that favour oviposition and egg development.

A fibre must be present for the attachment of eggs. Some lice can only attach an egg to the fine hairs of the body whereas others can also use the coarse hairs (Murray, 1957c). Where different types of hairs grow on the same mammal, a corresponding difference in the distribution of lice may occur; the eggs of the small trichodectid lice of ungulates are found on the thin hairs of the body but those of the large haematopinid lice may also be found on the coarse hairs of the mane and tail. When an animal sheds its coat in the spring there may be a striking season-related reduction in louse populations living within body hair. Those in the mane and tail, however, show greater stability since these hairs are shed randomly but continuously throughout the year. Animals which appear rough and are in poor condition at the end of winter have frequently been reported to be lousier than those in good condition with sleek coats. Shedding the coat is delayed when animals are in poor condition (Yeats, 1955), and, as a consequence, lice are retained (Murray, 1957c) and their numbers continue to increase during this period. Any factor that leads to a delay in moulting can affect the abundance of lice; the position of an animal in the hierarchy of a herd, for instance, could influence its access to available food resulting in a loss of condition and retention of the coat and accompanying ectoparasites.

Temperature is the other primary factor influencing the suitability of an area for breeding. Temperatures usually need to be between 30 and 37°C for maximal oviposition and development and hatching of eggs. It is likely that the eggs of many ectoparasitic mites require the same temperature range. These temperatures are usually found close to the skin in the temperature gradient of the pelage, and the depth of suitable zone above the skin is determined by the temperature at the tip of the pelage which in turn is influenced by air temperature and solar radiation (Murray, 1957b). A Merino sheep has a dense compact fleece that collects dust at its tip which then acts like a black body in its ability to absorb solar radiation. On a summer's day in southern Australia, the tip temperature of a 5 cm fleece can reach 70°C within 20 minutes and the gradient within decline from 70°C at the tip to 45°C at the skin. The temperatures within much of the dorsal area of the fleece become lethal to lice and the temperature close to the skin can prevent egg-laying by female lice (Murray, 1968). Thus, much of the upper part of the body can become unfavourable for breeding. Given shade some animals seek shelter in the heat of the day but not all (figure 10.6). The individual behaviours of animals within a herd or flock are not the same, and these individual variations can be sufficient to influence the number of ectoparasites carried.

Some lice are found on the body of their host and others on the extremities. In temperate zones, three species of lice are found on sheep: *Bovicola ovis*, a biting

Figure 10.6 Herd of wildebeest *(Connochaetes taurinus)* sheltering in the shade of a tree. Photograph by J. Kirsch.

louse on the body, *Linognathus ovillus* on the hairy face of sheep from where dispersal is restricted by a wool barrier (Murray, 1963a), and *L. pedalis* on the feet. The temperatures required for breeding for all three species are over 32 °C (Murray, 1957a, 1960a,b, 1963a). To conserve body-core temperature in cold periods the blood supply to the extremities of mammals and birds is restricted, and the skin temperature drops to that of the atmosphere. This is a rapid response and usually occurs only when atmospheric temperature has dropped below 15 °C. The result is that an extremity such as a foot can change quickly from a favourable to an unfavourable habit for lice. The speed with which a population of foot lice can increase is thus determined by the length of time the skin temperature of a foot remains over 32 °C (Murray, 1960a). The sheep with the warmest feet can become the lousiest, a condition which is likely to be influenced by the activity of the sheep itself.

Other examples of host behaviour influencing the temperature of different parts of the body, and thus their suitability for ectoparasites come from seals. *Lepidophthirus macrorhini*, the louse of the southern elephant seal, *Mirounga leonina*, is found mostly on the hind flippers. Active breeding only takes place when the seals are ashore as the lice do not oviposit when immersed in water; breeding occurs on the hind flippers because seals dissipate heat through them, and as a consequence flippers tend to be warmer more often than elsewhere on the body. Futhermore, seals when at sea also dissipate heat through their flippers so opportunities are presented to the lice to feed and thus survive the prolonged periods the seal may be away from land. Elephant seals come ashore to moult and, on Macquarrie Island, mature females spend the moulting period in mud wallows, a wet site that does not favour oviposition by lice; it has been suggested that this change in behaviour of mature seals may account for their smaller louse populations when compared with immature seals (Murray and Nicholls, 1965). Differences in the periods spent ashore by different age classes of Weddell seals (*Leptonychotes weddelli*) could account for differences in the abundance of lice with host age (Murray *et al.*, 1965). Seals in poor condition come ashore for prolonged periods (J. Gerraci, personal communication) which probably accounts for their often being particularly lousy. White-sea seal pups (*Phoca groenlandica*) are kept ashore in enclosures before slaughter in the USSR, and during this time they become heavily infested with lice (Tsvetkova, 1986).

To maintain infestations of the cattle tick, *Boophilus microplus*, on cattle it is necessary to prevent licking the sites of attachment (Snowball, 1956). Subsequent heavy infestations can be rapidly reduced by allowing cattle to groom. Efficient grooming is probably one of the factors that explains why short-coated Zebu cattle of the tropics carry fewer ticks than do the introduced British breeds with their longer coats. In temperate zones, cattle are infested by *Linognathus vituli*, a slow-moving sucking louse, and infestations are usually confined to the ventral parts of the body. Cattle which are prevented from licking themselves become very lousy, and when permitted to groom the numbers of lice are reduced rapidly within a few days (Lewis *et al.*, 1967).

Cattle, and probably many ungulates, groom themselves with their tongue

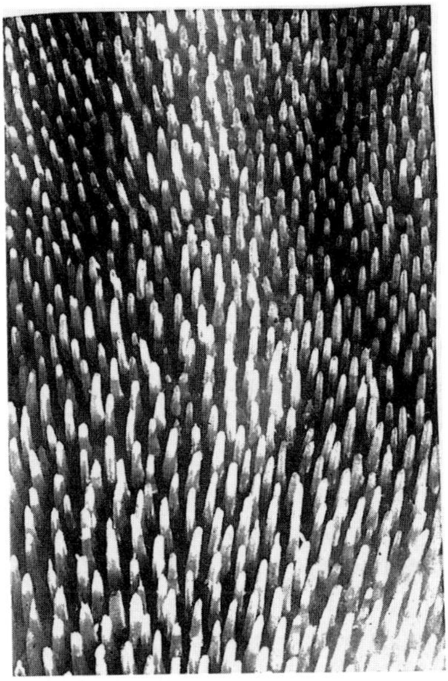

Figure 10.7 The papillae on the tongue of an ox. Many ungulates and carnivores groom themselves by licking with a rasping tongue. Photograph by I. Roper.

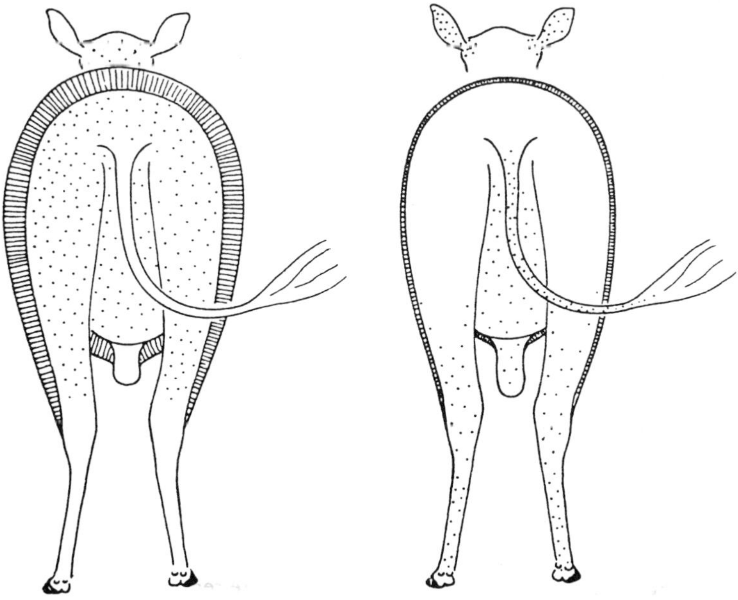

Figure 10.8. The seasonal changes in the environment of lice on ungulates. [symbol] = thickness of pelage; [symbol] = regions usually infested with lice.

which is covered with coarse papillae (figure 10.7) and acts like a broad comb. Efficiency is influenced by the type of hair coat, which is not uniform over the body. It also varies with the season depending upon whether the beast is carrying a long winter or short summer coat. Not all parts of the body are reached so some species of lice are found on the ventral parts of the body, under the neck or on the escutcheon while others escape from the body during the summer months to the ears and tail, where the temperatures become more favourable in the warmer weather. The efficiency of grooming varies seasonally and, combined with moulting and the resultant microclimatic changes in the pelage, plays an important role in reducing louse numbers (figure 10.8).

There are many examples of ungulates tolerating the presence of birds upon them, such as the Red-billed Oxpecker, *Buphagus erythrorhynchus*, of southern Africa which can spend 60% of their feeding time eating ticks on cattle or game animals. A mean of 408 ± 166.5 ticks were found in the stomachs of a sample of 53 birds, and it was also found that the ticks were fully digested in 24 hours (Bezuidenhout and Stutterheim, 1980). In controlled experiments in which 2–4 birds were caged with 2–4 tick-infested cattle for 6–10 days, over 90% of adult ticks were removed but fewer larval or nymphal ticks. The authors were unable to appraise whether these birds exerted any control over tick numbers under natural conditions but clearly their daily consumption of ticks can be great.

Ectoparasites of birds

The plumage of birds also harbours mites and lice, often several species on each individual. However, the extensive data on mammalian lice cannot be transposed to birds because the habitat within feathers is vastly different from that in hair. If the pelage of a mammal is likened to a grassland plain, the plumage of a bird resembles a forest. There are fewer feathers than hair and their fewer attachments to the skin are much stouter. Access to the skin is reduced but is is easier for lice to run over the skin. Feathers are ornate and provide many different habitats to which mites and lice have become adapted. Mite species with bizarre morphologies abound and new species are steadily described by our few devoted taxonomists. The lice also differ considerably in their morphology.

Much has been written on bird lice because of their potential value in appraising the phylogenetic relationships of their hosts but remarkably little has been written on their biology. Amblyceran and ischnoceran lice have been found on most species, and whereas the latter feed on feathers the former are more catholic in their choice of food, with some eating feathers, some imbibing blood and others eating the eggs and moulting stages of other lice and mites (Neslon, 1971). There have been a few studies on seasonal abundance (e.g. Baum, 1968).

Nelson and Murray (1971) studied the distribution and abundance of lice on the domestic pigeon, *Columba livia*. Four species of lice were investigated: *Columbicola columbae*, an elongated louse which lays eggs on the ventral surfaces of undercovert feathers of the wing and is found on the wings and fore-part of the body; *Colpocephalum turbinatum*, a small active amblyceran louse, many of which live in the shafts of primary, overcovert and secondary feathers of the wing, the eggs being laid on the ventral surfaces of these feathers; *Companulotes bidentatus compar*, a small flattened ischnoceran louse found in the fluffy part of body feathers and *Hohorstiella lata*, a large active amblyceran louse, which lays eggs in groups at the base of the shafts of head feathers of the gular area, and moves rapidly over the skin of the body. When the tips of the bills of pigeons were removed (which did not prevent the birds from feeding) the louse populations increased rapidly. Preening controlled the number of lice. There have been many reports of wild birds with damaged bills being lousy. *C. b. compar* and other species that live in the fluffy part of body feathers, are broad and flattened dorsoventrally whereas *C. columbae* and other wing lice, are thin, elongated and 'rest' between the barbs of wing feathers. On pigeons it was also found that all lice ate the lower portion of the body feathers, with the amblyceran lice supplementing this diet by predation (*C. turbinatum*) or by imbibing blood (*H. lata*). It would appear that some of the morphological adaptations of bird lice are not related to features of the site where they feed but are more related to areas safe from preening. Distributions such as those on the pigeon are common on different species of birds.

Birds spend much time in comfort behaviours which are associated with care of the plumage. Preening does control ectoparasites and is likely to vary in efficiency between species with different plumages and bill shapes. Birds whose eyesight is adapted to seeing small morsels of food (insect, seeds etc.) may be better at detecting ectoparasites. Small passerines may be more efficient than larger birds, and birds with small bills may remove ectoparasites better than those with elongated or specialized bills. Such morphological differences could lead to differences in the distribution and species of lice over the body. Birds oil or powder themselves from oil glands or powder-down patches but whether these affect or are an essential component of the diet of feather lice is unknown.

Specialized host behaviours

In addition to the many behaviours mentioned, animals indulge in several distinct and specialized behaviours, the functions of which are still far from clear.

Any person who has watched the excellent natural history films on televion will have seen the extensiveness of mutual grooming between adults of various species as distinct from the grooming of young by their parents. In addition to its important role in establishing and maintaining a particular relationship between the animals concerned, it can contribute to the control of ectoparasites particularly when applied to the head, a region animals have difficulty in grooming. It occurs extensively in rodents and primates, and studies on the mouse, *M. domesticus*, have shown that it can exert a considerable control on lice and tick infestations. Before the advent of efficient insecticides grooming was a common adjunct to human louse control measures (figure 10.9). It is widespread in certain groups of birds such as parrots and even albatrosses (figure 10.10) but the significance of mutual grooming in reducing ectoparasite numbers on birds has not been studied.

Figure 10.9 Family grooming. Photograph from Nuttall (1917).

Figure 10.10 Female Wandering Albatross (*Dionedea exulans*) preening the head of her mate. She was seen to remove a louse shortly after this photograph was taken. Photograph by M. D. Murray.

Several birds and mammals dust themselves, a behaviour by which 'dust' is driven into the plumage or pelage (Simmons, 1985). Birds usually use a hollow of dust and shower the dust onto their backs with their feet and wings. Some pick the dust up in their bills and throw it over their body. Ectoparasite control is one function that has been suggested for this activity. The habitat within the plumage or pelage is arid; Williams (1971) studying pheasant lice was the first to show that some ischnoceran lice need to obtain moisture from the air and Rudolph (1983) described the mechanism. Dust could place an extra stress on water conservation should it abrade the cuticle (this is the mechanism of some insecticides) or abrasion could occur as the bird subsequently preens to remove the dust. Critical experiments are required to demonstrate such effects.

Passerine birds may apply the body fluids of ants to the ventral surface of the tips of the primary feathers of the wings from where it may be applied to the head (Simmons, 1966, 1985). Both of these regions are 'inaccessible' for full preening. In the main, two non-stinging families (Formicinae and Dalichoderinae) of ants are used. Sometimes birds permit ants to swarm over their bodies. This behaviour is followed by bathing, oiling and preening, and thus appears to be part of overall feather maintenance. Dubinin (cited by Kelso and Nice, 1963) found that many of the feather-mites on the wings of Meadow Pipits (*Anthus pratensis*) died within 12 hours of anting whereas few died on the feathers from birds which had not indulged in this behaviour. Several chemicals with marked biological activites are known from ants: formic acid, various pheromones which mark trails, repellants to predators and bacteriocides to protect their cuticle. Again, more critical experiments are required to define the role this behaviour may have in parasite control.

Sunning is another well-documented behaviour of birds, in which the bird deliberately positions itself in the sun (Simmons, 1986). It is usually the wings that are exposed, and there could be an effect on ectoparasite numbers, should the wing temperatures become sufficiently high, probably by reducing the fertility of the parasite as has been shown for *B. ovis* on sheep (Murray, 1968).

Man

Man uses many techniques to groom, which he has improved by making tools. He can train the next generation and organize groups other than the immediate family, he can remove and sterilize his 'pelage' at any time of the day, he can develop insecticides and by much technical and organizational ability based on knowledge he can eradicate ectoparasites. Success in all these measures depends upon the hygienic practices of individuals and an awareness of the problem (Maunder, 1983). *Pediculus* spp. infest the head and body, the body lice laying eggs on and residing in clothing. Grooming of heads with fingers, combs and with insecticides has been practiced to control the head lice. Regular changes of clothes, with those that have been sterilized by boiling, dry cleaning or insecticides, are important in the control of body lice. All persons must indulge in this activity otherwise foci of infestations remain. Residual lice infestations are a problem of poverty where these basic requirements cannot be met or where persons indulge in practices which lead to a lack of motivation to live hygienically. The reproductive potential of the head and body lice of man are the highest known so the efforts of many can be rapidly undone by the neglect of a few. Occasionally, circumstances cause a decline in standards of populations as when natural disasters or war force many people closer together for prolonged periods, and in recorded history there are many instances of the overcrowding being followed by outbreaks of typhus, which is transmitted by lice.

The principles of control of many ectoparasites are well understood, and man can impose these principles upon his domesticated animals so effectively as to render them free of parasites. He has difficulty in imposing these standards upon himself because man is abundant with a world-wide distribution. It requires education followed by self-discipline now that many instinctive behaviours appear to have been lost, and because man considers, more vocally of recent years, that individuals have a right to behave in their own way this has assured the transmission of disease to his fellow beings. Even his better intentions have resulted in changing parasite relationships with himself and other host species. Species after species of ectoparasite have become resistant to chemicals used for their control, sometimes with unexpected speed. As a consequence, the depth of the gene pool, that has enabled the development of these adaptations to take place, has been revealed — and is a source of wonder to us.

Adaptive responses of parasites to host removal behaviour

The range of anti-parasite behavioural responses by hosts has led, as might be expected, to counter-adaptations among parasite species (see Chapter 1).

Agility enables the parasite to escape from the site being groomed and is particularly noticeable in fleas. Less agile species, such as lice, can prevent removal by conforming to the contours of the habitat (e.g. wing lice of birds), or by increasing the strength of the grasp of the tarsal claws (anopluran lice). Ischnoceran lice grasp hair or feathers with their mandibles, and may have a groove in the ventral surface of the head in which the hair being held lies (*Felicola subrostratus* of the cat whose host grooms itself with a rasping tongue, has such a groove). Lice can be difficult to remove with forceps once they have grasped a feather or hair firmly. Some amblyceran and anopluran lice have stout spines on the ventral surface of their heads which probably anchor the head to the skin when feeding but could also prevent removal (figure 10.11). Ticks which remain attached for several days have long well-barbed hypostomes or secrete substances to cement their mouth parts to the skin. Such ticks are difficult to remove. More subtle but of great evolutionary importance are the reproductive adaptations to compensate for these losses.

Where the number of eggs laid daily, the generation time and sex ratios are known it is possible to compare the reproductive potentials of lice (Murray, 1987b). Such comparisons are shown in figure 10.12, and clearly there are major differences in the reproductive adaptations of lice to survive losses. The lice of man, elephant seal and mouse have high reproductive potentials whereas those of linognathid and trichodectid lice are low. The lice on seals multiply rapidly whenever the seal is ashore to provide sufficient lice to survive the prolonged periods at sea, and this is accomplished by increasing the number of eggs laid daily. Human lice (*Pediculus* spp.) also lay many eggs daily and have the highest reproductive potentials known for lice; this would appear to be a response to the grooming efficiency of man. The louse on the house mouse lives in an environment where the option of increasing the size of its abdomen to accommodate more eggs does not exist because of the efficiency of the host's combing grooming technique that can be adjusted to the

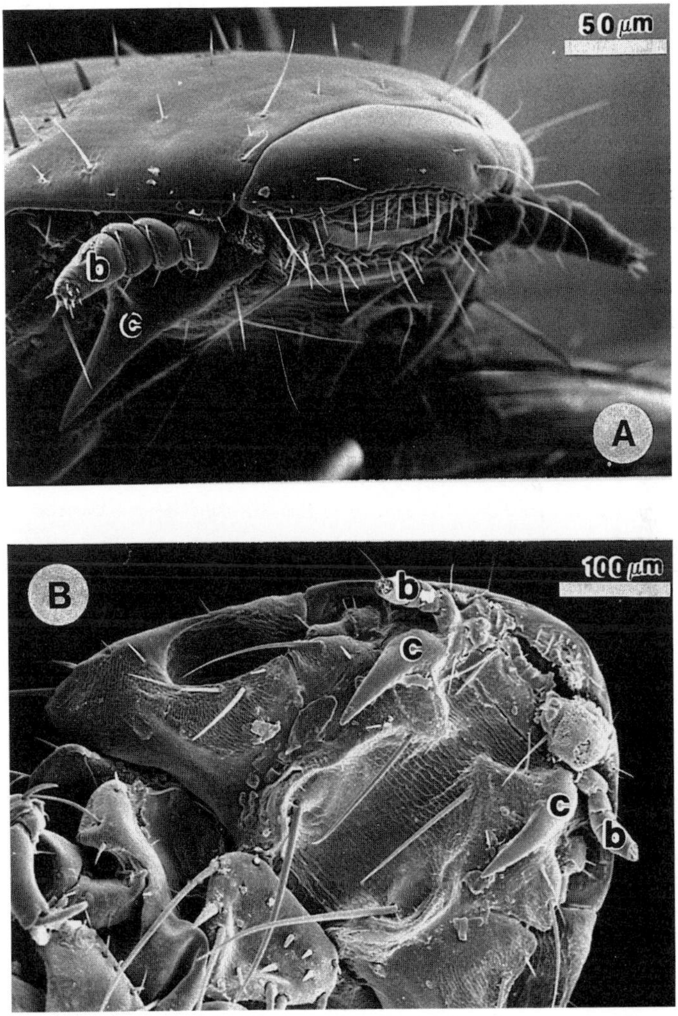

Figure 10.11 Head of *Heterodoxus maynesi*, a blood-imbibing louse of a rock wallaby: (A) anterolateral and (B) ventral aspects. (b) Maxillary palp. (c) Large stout spinous process. Data from Barker (1988).

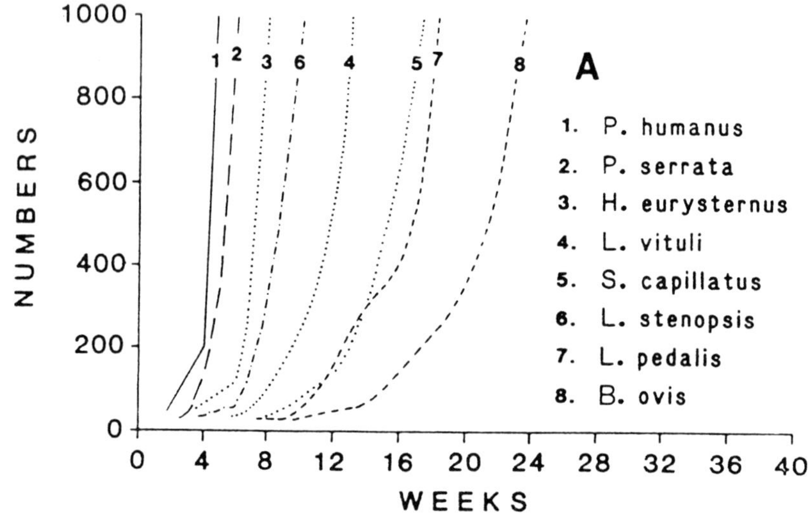

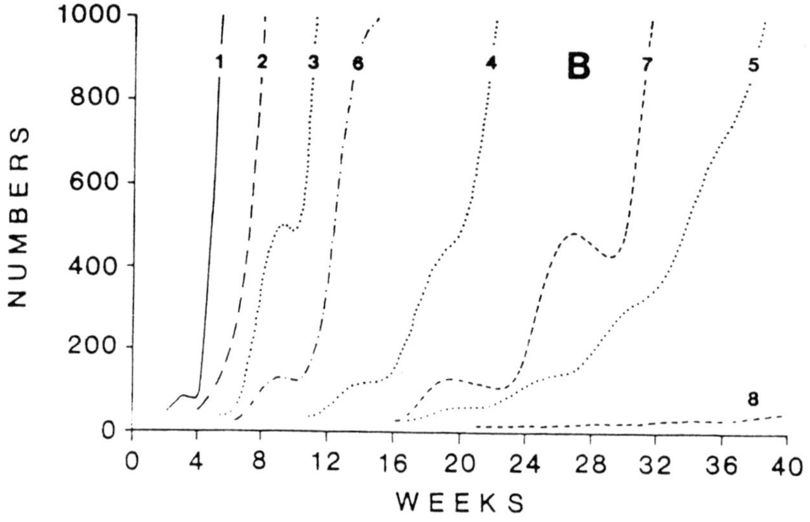

Figure 10.12. Calculated increase in the numbers of nymphs and adults from one fertilized female louse for (A) zero mortality when females live 28 days and (B) a mortality of 20% eggs and 30% nymphs when females live 14 days. The sex ratio is 1 : 1. In brackets are the host, generation time in days and number of eggs laid per day for: 1 = *Pediculus humanus* (man, 19, 10); 2 = *Polyplax serrata* (mouse, 13, 2); 3 = *Haematopinus eurysternus* (ox, 28, 4); 4 = *Linognathus vituli* (ox, 28, 1); 5 = *Solenopotes capillatus* (ox, 30, 0.6); 6 = *L. stenopsis* (goat, 29, 2); 7 = *L. pedalis* (sheep, 43, 1); 8 = *Bovicola ovis* (sheep, 35, 0.4). The curves for *Lepidophthirus macrorhini* (elephant seal, 21, 8) lie between 1 and 2 in (A) and (B), for *Polyplax spinulosa* (rat, 26, 2) are close to 3 in (A) and (B), for *Linognathus ovillus* (sheep, 31, 1) are close to 1 in (A) and between 4 and 7 in (B), and for *B. bovis* (ox, 30, 0.7) are close to 5 when parthenogenetic and 8 when not, in both (A) and (B). Data from Murray (1987b).

thickness of a single hair. Thus, its morphology appears to have been determined by the grooming technique of its host; it is a small louse with a long thin abdomen in which only two well-developed eggs can be accommodated at one time. The morphological adaptations of the abdomens of lice are governed by the limits set by the size of mature eggs (Murray, unpublished data). The reproductive potential of *P. serrata* has been increased by reduction of the generation time to 13 days, and this has been accomplished by reducing the developmental time of the nymphal stages. The more widespread distribution of *P. spinulosa* on the rat indicates that the grooming technique is less efficient, and this louse, which has a generation time of 26 days (Holmes, 1958) has a lower reproductive potential than *P. serrata* on the smaller mouse. The number of lice on a rat are under control because the reproductive potentials of lice are adapted to the level of losses they encounter on their hosts. On cattle there is a range of reproductive potentials among lice from the higher value of *H. eurysternus*, which can inhabit the harsher environment on cattle in the tropics where grooming of the short coat is efficient, to the low values of those on the more sheltered regions of the ventral body of cattle in temperate zones.

The low rate of multiplication of the trichodectid louse (*B. ovis*) of the domestic sheep is such that the increase in numbers during the favourable winter months is that expected from an unimpeded natural increase (Murray and Gordon, 1969). The number at the end of winter is directly related to the number at the beginning of winter which is determined by the losses suffered during the unfavourable summer. The greatest losses occur when a sheep is shorn, usually in the spring, after which factors need not be great to retard population growth because of the low reproductive potential of the louse. Factors such as rain and solar radiation, which cause changes of short duration to the microclimate, can suffice in the summer when numbers are low (Murray, 1963b, 1968). The length of the exposure to these conditions is greatly influenced by the host's behaviour in seeking shade or shelter and by its ability to obtain these resources in competition with others.

B. ovis feed on the statum corneum and lipids (Sinclair, 1983) and bacteria of the skin (Murray and Edwards, 1987), a diet which is probably common to most members of this group of lice. It is a diet not abundant in water, and these lice may not be able to increase greatly their daily egg production in contrast to anopluran lice which can get the additional water requirements from blood. Such nutritional restrictions may well have necessitated the development of parthenogenetic reproduction in the trichodectid lice on large mammals, where losses can be greater, to enable the survival of a small active louse whose abdomen can only accommodate a small stomach and a single mature egg which requires 24–26 hours to develop. *B. bovis* on cattle reproduces parthenogenetically in the summer months when numbers are low (Matthysee, 1946). Several factors have probably led to this method of reproduction, such as reduced opportunities to mate because of the large surface area in the habitat, particularly when densities have been reduced by mortalities due to microclimatic changes, and the increased efficiency of grooming in the summer. *G. oregonus*, the trichodectid louse of the gopher, has a mean generation time of 40 ± 6 days (Rust, 1974), and has a low reproductive potential. The absence of trichodectid lice from smaller mammals, where the predatory pressure of grooming is greater, could well be due to these lice being physiologically incapable of producing sufficient replacements.

It appears that many lice do not have high reproductive potentials and consequently, factors that cause small mortalities or declines in 'fitness' are important. It is in this context that the specialized behaviours of dusting, anting and sunbathing may eventually be shown to be significant in the regulation of natural populations of ectoparasites

Dispersal

Resident ectoparasites, which are flightless, spread from host to host when these hosts come into contact. Hosts that sleep together assure rapid transmission but often the opportunities are limited to when the young are being raised Domestication provides many more opportunities as man determines the hosts' behaviours and often brings them into prolonged contact in pens when they are drenched, shorn or overwintered. In the wild, such contacts may be limited to periods of reproduction and to times of stress, such as when animals gather around water-holes during a drought. Additional contacts can lead to different age distributions of infested hosts, and severities of infestation within populations. Thus, moderate to heavy infestations are not uncommon in domestic animals whereas they are the exception in wild populations. The reproductive potentials of lice have evolved to ensure survival at these low densities, and to ensure successful dispersal to new hosts, at intervals which can be as long as the generation time of the host. Dispersal is entirely dependent upon host behaviour.

Other flightless parasites, which remain on their hosts for short periods, are dispersed spatially by their hosts and their adaptations assure that some return to suitable habitats where they may survive several months. Mammalian argasid ticks feed rapidly, within 30 minutes, and thus many return to the substrate of a resting site before a host departs. Ixodid ticks, which may take 6 days to engorge, can survive in nests and burrows which are occupied by hosts regularly for several weeks each year, and specific host–parasite relationships may develop. Ticks may be dispersed widely, whereas their hosts range widely, but their ground habitat is not uniform. Some find suitable conditions and survive to create foci of infestation. These ticks often infest many different host species which undoubtedly enhances their ability to survive. Even long-distance dispersal may occur on migrating birds. Thus, on Macquarie Island in the sub-Antarctic, *Ixodes pterodromae* is found in nests in the burrows of prions whereas *I. uriae*, which is widely dispersed and survives under stones and in tussock, is found in association with other seabirds, albatrosses and especially penguins (Murray and Vestjens, 1967). In Australia, *I. ornithorhynchi* is found on the duck-billed platytpus (*Ornithorynchus anatinus*) which lives in burrows in river banks, and the widely distributed *I. tasmani* is found on a variety of large and small marsupials (Roberts, 1970). All are dependent upon host behaviour and transmission necessitates a host nesting or resting in or passing through an infested focus.

Only those ectoparasites which can fly are less dependent. Flies such as midges, which produce an abundance of offspring, initially drift in the wind to locate a host (Murray 1987a). Among the higher Diptera, however, are strong fliers which actively search for a host using well-developed eyes and sensors for detecting the odours that emanate from the breath of their hosts, as do tsetse flies (Vale, 1977a, 1984). Those that replenish water and energy reserves, as by feeding on nectar, can search for many days over a greater area; a marked screw-worm fly, *Cochliomyia hominovorax*, which causes myiasis of mammals, has been recaptured 180 miles from its point of release (Hightower *et al.*, 1965).

CONCLUSIONS

The selection of the ectoparasites discussed has been severely limited to some dipterous flies and lice yet many of the ways host behaviour can influence the ecology of ectoparasites have been demonstrated. There is a range of reactions to the visiting ectoparasites, from toleration and reducing the annoyance, to departure and escape. Even the most tolerant of hosts eventually resort to a behavioural defence. Some insects may exploit these reactions and the naive curiosity of the young to expose susceptible hosts. Many behaviours have become stereotyped but new responses can be learned quickly in this dynamic relationship between host and parasite.

Once an ectoparasite resides on the host its ecology is intimately linked with host behaviour, which may affect numbers indirectly as by altering the microclimate within the skin covering or directly by grooming. The most important of the predatory grooming techniques appear to be those associated with the mouth. These have undoubtedly led to additional morphological adaptations to prevent removal, but the more subtle ones compensate for losses by enhancement of the reproductive output. An apparent inability of some groups of lice to adapt to these sources of mortality may account for their absence from some hosts species. The opportunities for dispersal and, thus, ultimate survival are entirely dependent upon host behaviour.

One aspect that still awaits detailed examination is the effect of the application of saliva to the body when an animal licks itself. Factors that influence the distribution and abundance of the bacterial flora over the body could affect oviposition stimulants (Emmens and Murray, 1983) and the diets of ectoparasites (Murray and Edwards, 1987). Should licking have such effects, we could be underestimating the role of grooming in the evolution of ectoparasites.

ACKNOWLEDGEMENTS

I wish to thank the editors, Drs C.J. Barnard and J.M. Behnke, for several constructive suggestions. Dr S.C. Barker kindly gave permission to reproduce the scanning electron micrographs shown in figure 10.11 from his PhD Thesis

'Evolution in the *Heterodoxus octoseriatus* group (Boopidae:Phthriaptera) and the development of host associations with *Petrogale* (Marsupialia)'.

The following publishers have allowed figures to be reproduced: CSIRO Editorial and Publishing Unit, figures 10.1–10.3 from the *Australian Journal of Zoology*; Cambridge University Press, figure 10.9 from *Parasitology*; Permagon Press, figure 10.12 from *International Journal for Parasitology*.

REFERENCES

Anderson, J. R. 1975, *Canadian Journal of Zoology*, **53**, 977–992.
Barker, S. C., 1988, PhD Thesis, Macquarie University, Australia.
Baum, H., 1968, *Angewandte Parasitologie*, **9**, 129–175.
Bell, J. F. and Clifford, C. M., 1964, *Experimental Parasitology*, **15**, 340–349.
Bell, J. F., Jellison, W. L. and Owen, C. R., 1962, *Experimental Parasitology*, **12**, 176–183.
Bezuidenhout, J. D. and Stutterheim, C. J., 1980, *Onderstepoort Journal of Veterinary Research*, **47**, 51–75.
Breev, K. A., 1951, *Parasitologicheskii sbornik*, **13**, 343–354.
Bucher, E. H., 1988, *Parasitology Today*, **4**, 1–3.
Buxton, P. A., 1947, *The Louse* (London: Edward Arnold).
Clark, L. and Mason, J. R., 1985, *Oecologia*, **67**, 169–176.
Colborne, J., Norval, R. A. I. and Spickett, A. M., 1981, *Onderstepoort Journal of Veterinary Research*, **48**, 31–35.
Dubinin, V. B., 1953, *Parasitologicheskii sbornik*, **15**, 252–301.
Eddy, G. W., DeVaney, J. A. and Handke, B. D., 1975, *Journal of Medical Entomology*, **12**, 379–38.
Edman, J. D. and Kale, H. W., 1971, *Annals of the Entomological Society of America*, **64**, 513–516.
Edman, J. D., Webber, L. A. and Kale, H. W., 1972, *The American Journal of Tropical Medecine and Hygiene*, **21**, 487–498.
Emmens, R. L. and Murray, M. D., 1983, *Bulletin of Entomological Research*, **73**, 411–415.
Fallis, A. M. and Smith, S. M., 1964, *Canadian Journal of Zoology*, **42**, 723–730.
Gelbach, F. and Baldridge, R., 1987, *Oecologia*, **71**, 560–563.
Hamilton, W. D., 1971, *Journal of Theoretical Biology*, **31**, 295–311.
Hart, B. L., Korinek, E. and Brennan, E., 1987, *Physiology and Behaviour*, **41**, 321–325.
Helle, T. and Aspi, J., 1983, *Acta Zoologica Fennica* No. 175, 129–131.
Hightower, B. G., Adams, A. L. and Alley, D. A., 1965, *Journal of Economic Entomology*, **58**, 373–374.
Holmes, D. T., 1958, PhD Thesis, Ohio State University, USA.
Hughes, R. D., Duncan, P. and Dawson, J., 1981, *Bulletin of Entomological Research*, **71**, 227–242.
Kale, H. W., Edman, J. D. and Webber L. A., 1972, *Mosquito News*, **32**, 343–350.
Kaneko, K., 1959, *Japanese Journal of Experimental Medicine*, **15**, 252–301.
Kangwangy, T. N., 1978, *Proceedings of the 1st East African Conference of Entomology and Pest Control*, pp. 32–44.
Kelso, L. and Nice, M. M., 1963, *Wilson Bulletin*, **75**, 23–26.
Lavoipierre, M. M. J., Radovsky, F. J. and Budwiser, P. D., 1979, *Journal of Medical Entomology*, **15**, 187–217.
Lewis, L. F., Christenson, D. M. and Eddy, G. W., 1967, *Journal of Economic Entomology*, **60**, 755–757.
Lodmell, D. L., Bell, J. F., Clifford, C. M., Moore, G. J. and Raymond, G., 1970, *Experimental Parasitology*, **27**, 184–192.
McCarthy, P. H., 1961, *Australian Veterinary Journal*, **37**,334,
Marshall, A. G., 1981, *The Ecology of Ectoparasitic Insects* (London: Academic Press).
Matthysee, J. G., 1946, *Bulletin of the Cornell Agricultural Experimental Station*, No.832.
Maunder, J. W., 1983, *Proceedings of the Royal Institution of Great Britain*, **55**, 1–31.
Muller, M. J. and Murray. M. D., 1977. *Australian Journal of Zoology*, **25**, 75–85.
Murray, M. D., 1975a, *Australian Journal of Zoology*, **5**, 19–29.
Murray, M. D., 1957b, *Australian Journal of Zoology*, **5**, 173–182.
Murray, M. D., 1957c, *Australian Journal of Zoology*, **5**, 183–187.

Murray, M. D., 1960a, *Australian Journal of Zoology*, **8** 349–356.
Murray, M. D., 1960b, *Australian Journal of Zoology*, **8**, 357–382.
Murray, M. D., 1961, *Australian Journal of Zoology*, **9**, 1–13.
Murray, M. D., 1963a, *Australian Journal of Zoology*, **11**, 157–172.
Murray, M. D., 1963b, *Australian Journal of Zoology*, **11**, 173–182.
Murray, M. D., 1968, *Australian Journal of Zoology*, **16**, 157–172.
Murray, M. D., 1976, In *Marine Insects* edited by L. Cheng (Amsterdam: North-Holland Publishing Company), pp. 79–96.
Murray, M. D., 1987a, *Australian Journal of Zoology*, **35**, 559–573.
Murray, M. D., 1987b, *International Journal for Parasitology*, **17**, 191–197.
Murray, M. D. and Edwards, J. E., 1987, *Australian Veterinary Journal*, **64**, 277.
Murray, M. D. and Gordon, G., 1969, *Australian Journal of Zoology*, **17**, 179–186.
Murray, M. D. and Koch, J. H., 1969, *Journal of the Australian Entomological Society*, **8**, 187–188.
Murray, M. D. and Nicholls, D. G., 1965, *Australian Journal of Zoology*, **13**, 437–454.
Murray, M. D., Smith, M. S. R. and Soucek, Z., 1965, *Australian Journal of Zoology*, **13**, 761–771.
Murray, M. D. and Vestjens, W. J. M., 1967, *Australian Journal of Zoology*, **15**, 715–725.
Nelson, B. C., 1971, *Nature New Biology*, **232**, 255.
Nelson, W. A., Bell, J. F., Clifford, C. M. and Keirans, J. E., 1977, *Journal of Medical Entomology*, **13**, 389–428.
Nelson, W. A., Keirans, J. E., Bell, J. F. and Clifford, C. M., 1975 *Journal of Medical Entomology*, **12**, 143–166.
Nelson, B. C. and Murray M. D., 1971, *International Journal for Parasitology*, **1**, 21–29.
Nuttall, G. H. F., 1917, *Parasitology*, **10**, 383–405.
Rechav, Y., Norval, R. A. I., Tannock, J. and Colborne, J., 1978, *Nature London*, **275**, 310–311.
Roberts, F. H. S., 1952, *Insects Affecting Livestock* (Sydney: Angus and Robertson).
Roberts, F. H. S., 1970, *Australian Ticks* (Melbourne: CSIRO).
Rothschild, M. and Clay, T., 1952, *Fleas, Flukes and Cuckoos. A Study of Bird Parasites* (London: Collins).
Rothschild, M. and Ford, R., 1964, *Nature London*, **201**, 103–104.
Rudolph, D., 1983, *Journal of Insect Physiology*, **29**, 15–25.
Rust, R. W., 1974, *Oecologia*, **15**, 287–304.
Simmons, K. E. L., 1966, *Journal of Zoology, London*, **149**, 145–162.
Simmons, K. E. L., 1985, In *A Dictionary of Birds* edited by B. Campbell and E. Lack (Carlton: T. and A. D. Poyser), pp. 19; 101–104; 161; 162; 569–570.
Simmons, K. E. L., 1986, *The Sunning Behaviour of Birds* (Bristol: The Bristol Ornithological Club)
Sinclair, A. N., 1983, PhD Thesis, University of New South Wales.
Snowball, G. F., 1956, *Australian Journal of Agricultural Research*, **7**, 227–232.
Tsvetkova, A. V., 1986, *Veterinariya Moscow* (cited in *Review of Applied Entomology*, 1987, **53**, No.414).
Vale, G. A., 1974, *Bulletin of Entomological Research*, **64**, 589–594.
Vale, G. A., 1977a, *Bulletin of Entomological Research*, **67**, 297–303.
Vale, G. A., 1977b, *Bulletin of Entomological Research*, **67**, 635–649.
Vale, G. A., 1984, *Bulletin of Entomological Research*, **74**, 143–152.
Vysotskaya, S. O., 1950, *Parasitologicheskii sbornik*, **12**, 73–79.
Waage, J. K., 1979, *Biological Journal of the Linnaen Society*, **12**, 187–224.
Waage, J. K. and Davies, C. R., 1986, *Journal of Animal Ecology*, **55**, 171–180.
Warnes, M. L. and Findlayson, L. H., 1987, *Medical and Veterinary Entomology*, **1**, 53–57.
Webber, L. A. and Edman, J. D., 1972, *Animal Behaviour*, **20**, 228–232.
Weisbroth, S., Friedman, S., Powell, M. and Scher, S., 1974, *Laboratory Animal Science*, **24**, 510–516.
Williams, R. T., 1971, *Journal of Experimental Biology*, **55**, 553–568.
Wimberger, P. H., 1984, *Auk*, **101**, 615–618.
Yeates, N. T. M., 1955, *Australian Journal of Agricultural Research*, **6**, 891–902.

Index

Index